Computational Intelligence Applications in Cyber Security

The book provides a comprehensive overview of cyber security in Industry 5.0, data security in emerging technologies, block chain technology, cloud computing security, evolving IoT and OT threats, and considerable data integrity in healthcare. The impact of security risks on various sectors is explored including artificial intelligence in national security, quantum computing for security, and AI-driven cyber security techniques. It explores how cyber security is applied across different areas of human life through computational modeling. The book concludes by presenting a roadmap for securing computing environments, addressing the complex interplay between advanced technologies and emerging security challenges, and offering insights into future trends and innovations for sustainable development.

This book:

- Analyzes the use of AI, support vector machines, and deep learning for data classification, vulnerability prediction, and defense.
- Provides insights into data protection for Industry 4.0/5.0, cloud computing, and IoT/OT, focusing on risk mitigation.
- Explores block chain's role in smart nations, financial risk management, and the potential of quantum computing for security.
- Examines AI's applications in national security, including India's AI strategy and securing smart cities.
- Evaluate strategies for data integrity in healthcare, secure IoT platforms, and supply chain cyber security.

The text is primarily written for senior undergraduate, graduate students, and academic researchers in the fields of electrical engineering, electronics and communication engineering, computer engineering, and information technology.

Computational Intelligence Applications in Cyber Security

Edited by
Suhel Ahmad Khan, Mohammad Faisal,
Nawaf Alharbe, Rajeev Kumar and
Raees Ahmad Khan

CRC Press
Taylor & Francis Group
Boca Raton London New York

CRC Press is an imprint of the
Taylor & Francis Group, an **informa** business

Designed cover image: shutterstock

First edition published 2025
by CRC Press
2385 NW Executive Center Drive, Suite 320, Boca Raton FL 33431

and by CRC Press
4 Park Square, Milton Park, Abingdon, Oxon, OX14 4RN

CRC Press is an imprint of Taylor & Francis Group, LLC

© 2025 selection and editorial matter, Suhel Ahmad Khan, Mohammad Faisal, Nawaf Alharbe, Rajeev Kumar and Raees Ahmad Khan; individual chapters, the contributors

Library of Congress Cataloging-in-Publication Data
Names: Khan, Suhel Ahmad, editor.
Title: Computational intelligence applications in cyber security /
edited by Suhel Ahmad Khan, Mohammad Faisal, Nawaf Alharbe,
Rajeev Kumar, Raees Ahmad Khan.
Description: First edition. | Boca Raton, FL : CRC Press, 2025. |
Includes bibliographical references and index.
Identifiers: LCCN 2024027981 (print) | LCCN 2024027982 (ebook) |
ISBN 9781032470597 (hardback) | ISBN 9781032865751 (paperback) |
ISBN 9781003514312 (ebook)
Subjects: LCSH: Computer security–Technological innovations.
Classification: LCC QA76.9.A25 C619 2025 (print) | LCC QA76.9.A25 (ebook) |
DDC 005.8–dc23/eng20240809
LC record available at https://lccn.loc.gov/2024027981
LC ebook record available at https://lccn.loc.gov/2024027982

ISBN: 9781032470597 (hbk)
ISBN: 9781032865751 (pbk)
ISBN: 9781003514312 (ebk)

DOI: 10.1201/9781003514312

Typeset in Sabon
by Newgen Publishing UK

Contents

1   A transition from Industry 4.0 to Industry 5.0: securing
    Industry 5.0 for sustainability   1

    SANDEEP KUMAR VERMA, VISHAL VERMA, AND
    MD TARIQUE JAMAL ANSARI

2   Security estimation of software by support vector
    machines: defence perspective   16

    AMITABHA YADAV, SYED ANAS ANSAR, MANU GAUTAM,
    SAKSHI PANDEY, TUSHAR CHANDRA, AND KHUSHI NAIK

SARITA SHUKLA, BINEET KUMAR GUPTA, PALLAVI SOMVANSHI, AND MUKESH MISHRA

ERAM FATIMA SIDDIQUI AND SANDEEP KUMAR NAYAK

KAVITA SAHU, R. K. SRIVASTAVA, ABHISHEK KUMAR PANDEY, AND KOTAIAH BONTHU

Preface

In our fast-paced, technology-driven world, the convergence of computational intelligence and cybersecurity has emerged as a pivotal force in shaping the digital landscape. This book, aptly titled *Computational Intelligence Applications in Cyber Security* embarks on a fascinating journey through the intricate interplay of these two realms. Each chapter within these pages offers a nuanced exploration of the multifaceted challenges that computational intelligence techniques encounter in the face of evolving cyber threats.

The odyssey commences with Chapter 1 setting the stage by emphasizing the need for sustainable and human-centric approaches in advanced industrialization. This opening chapter lays the groundwork for a broader understanding of how computational intelligence becomes a linchpin in securing critical infrastructures and systems.

Subsequent chapters delve into specific domains of cybersecurity, casting a spotlight on unique aspects of the field. Chapter 2 navigates the application of support vector machines in software security, particularly in defense contexts, highlighting the critical role of data security in this sector.

The exploration expands into the challenges and solutions presented by Industry 5.0 in Chapter 3. This chapter investigates cutting-edge technologies such as digital twins, edge computing, collaborative robots, and blockchain as security enablers for Industry 5.0 ecosystems, showcasing the fusion of creativity and cognitive abilities in enhancing industrial productivity and safety.

Blockchain technology, a transformative force over the last three decades, takes center stage in Chapter 4. This chapter elucidates the fundamentals of blockchain and envisions its potential role as a collective platform for national and international development goals, examining the progress made by various countries in leveraging blockchain for citizen welfare services.

In Chapter 5 the focus narrows to financial risk management in the banking sector. This chapter investigates the correlation between blockchain implementation and risk metrics, providing valuable insights into the benefits of adopting blockchain in managing financial risks.

The exploration continues with an in-depth analysis of cloud computing security in Chapter 6. This section critically examines security concerns associated with cloud computing, offering a comprehensive overview of the factors influencing cloud computing security, especially in defense applications.

The book widens its scope to address security challenges in emerging technologies such as the Internet of Things (IoT) and Operational Technology (OT) in Chapter 7. This chapter provides a thorough examination of security plans, prevalent threats, and major security attacks and protections for IoT and OT systems.

Healthcare information systems, a critical component of modern healthcare, are discussed in "Chapter 8. The emphasis is on data integrity concerns in the healthcare industry and proposes the integration of blockchain for prioritizing data integrity in healthcare.

The convergence of IoT, cloud computing, and fog computing takes center stage in Chapter 9. This chapter proposes a secure and low-latency IoT environment through fog computing, addressing security risks and enhancing real-time collaboration and safety.

Chapter 10 delves into the evolving landscape of healthcare web applications. Against the backdrop of the COVID-19 pandemic, it provides insights into prevalent threats, recent data breaches, and guidance for securing healthcare information systems.

As supply chain management undergoes digital transformation, Chapter 11 discusses the pivotal role of blockchain technology. It envisions blockchain as a common platform for governments to achieve national and international development goals, emphasizing its potential for creating smart countries.

The critical issue of software vulnerabilities is addressed in Chapter 12. This chapter explores the use of deep learning algorithms for predicting software vulnerabilities, crucial for preemptive actions to secure systems before potential exploitation.

The book takes a closer look at the challenges faced by artificial intelligence in national security in Chapter 13. It underscores the importance of transparency, accountability, privacy protection, and human oversight for the responsible deployment of AI technologies in safeguarding national interests.

Quantum computing, an emerging frontier, is explored in "Chapter 14. This section introduces the basics of quantum computing, its applications in network security, and its potential to redefine modern cryptography.

Chapter 15 delves into the growing dependency on the internet and the consequent increase in cyber-attacks. It discusses techniques for gathering and analyzing information, potential threats, and advanced prevention strategies based on AI and other advanced technologies.

Efficient cloud forensics is a critical need in the digital age, as discussed in "Chapter 16. The book provides insights into the challenges faced by digital forensic investigators in the context of cloud computing, discussing organizational, technological, and legal issues.

Smart cities, reliant on IoT, are addressed in Chapter 17. The focus is on the imperative of securing IoT-based applications in smart cities, given the increasing threats to these interconnected systems.

The book concludes with Chapter 18. This chapter delves into the quantum age and its impact on software security, with a particular emphasis on Grover's algorithm and its potential to enhance security measures in the era of quantum computing.

Computational Intelligence Applications in Cyber Security offers a holistic exploration of the dynamic landscape where computational intelligence confronts the intricate challenges of cybersecurity. The diverse array of chapters, each addressing a unique aspect of this intersection, provides readers with a comprehensive understanding of the current state, challenges, and future directions in the field. As we navigate through the pages of this book, we invite you to immerse yourself in the unfolding narrative of how computational intelligence becomes a stalwart guardian in our digital realms, securing the foundations of our interconnected world.

About the editors

Suhel Ahmad Khan is currently working as Assistant Professor in the Department of Computer Science, Indira Gandhi National Tribal University (A Central University), Amarkantak, Madhya Pradesh. He has ten years of teaching and research experience. His research interests include software engineering, software security, security testing, cyber security, and network security. He has completed one major research project with PI funded by UGC, New Delhi. He has published numerous papers in international journals and conferences including IEEE, Elsevier, IGI Global, Springer etc. He is the author of the books entitled *Software Security: Concept & Practices* and *Software Durability: Concept & Practices*, published by CRC Press in 2023. Suhel Ahmad Khan is an active member of various professional bodies IAENG, ISOC-USA, IACSIT, and UACEE.

Mohammad Faisal is currently working as Associate Professor and Head of the Department in the Department of Computer Application, Integral University, Lucknow, India. He has more than 15 years of teaching and research experience. His areas of interest are software engineering, requirement volatility, distributed operating system, cyber security, and mobile computing. He has published a book *Requirement Risk Management: A Practioner's Approach* published by Lambert Academic Publication, Germany, ISBN: 978-3-659-15494-2. He has published quality research papers in journals, national and international conferences of repute. He is contributing his knowledge and experience as member of Editorial Board/Advisory committee and TPC in various international journals/conferences of repute. Mohammad Faisal is an active member of various professional bodies such as IAENG, CSTA, ISOC-USA, EASST, HPC, ISTE, IAENG, and UACEE.

Nawaf Alharbe is currently working as Vice Dean of the Applied College, Badr, Taibah University, Kingdom of Saudi Arabia. He was the Head of the Department of Computing at Taibah University from 2016 to 2019. He completed his PhD degree in Computer Science from the Faculty of Computing, Engineering and Sciences, Staffordshire University, United

Kingdom, where he also earned five excellence certificates (during the PhD course 2012–2015), Saudi Arabian Cultural Bureau, United Kingdom. He has completed one project on Smart Hospital Management Information System for Healthcare Transformation in Saudi Arabia from 2013 to 2016 and one major project is currently running on Health Informatics. Overall, he has authored/co-authored more than 25 international publications including journal articles, conference proceedings, book chapters, and books. Nawaf Alharbe's research interest focuses on knowledge management systems in smart healthcare operations using emerging technology such as RFID, ZigBee, Internet of Things (IoT), cloud computing, artificial intelligence and machine learning, big data and data processing, knowledge engineering, knowledge harvesting, and healthcare applications.

Rajeev Kumar is currently working as Associate Professor in the Department of Computer Science & Engineering, Shri Ramswaroop Memorial University, Barabanki, Uttar Pradesh, India. Kumar has more than nine years of research and teaching experience. He has published numerous papers in international journals and conferences including IEEE, Elsevier, IGI Global, Springer, etc. His research interests include different areas of security engineering and computational techniques.

Raees Ahmad Khan (Member, IEEE, ACM, CSI, etc.) is currently working as Professor in the Department of Information Technology, Dean of School for Information Science & Technology, Babasaheb Bhimrao Ambedkar University (A Central University), Lucknow, India. Khan has more than 20 years of teaching and research experience. He has published more than 300 research publications with good impact factors in reputed international journals and conferences including IEEE, Springer, Elsevier, Inderscience, Hindawi, IGI Global, etc. He has published a number of national and international books (authored and edited) (including Chinese language). His research interests include different areas of security engineering and computational techniques.

Contributors

Rohit Aggarwal
Department of Computer Science & Engineering
Meerut Institute of Engineering & Technology
Meerut, Uttar Pradesh, India

Kalamuddin Ahmad
Department of Computer Application
Integral University
Lucknow, Uttar Pradesh, India

Masood Ahmad
Department of Computer Application
Integral University
Lucknow, Uttar Pradesh, India

Shish Ahmad
Department of Computer Science and Engineering
Integral University
Lucknow, Uttar Pradesh, India

Tasneem Ahmed
Department of Computer Application
Integral University
Lucknow, Uttar Pradesh, India

Jamal Akhtar Khan
Department of Computer Application
Lovely Professional University
Jalandhar, Punjab, India

Alka
Babasaheb Bhimrao Ambedkar University
Lucknow, Uttar Pradesh, India

Ishrat Amaan
Department of Computer Application
Shri Ramswaroop Memorial University
Lucknow, Uttar Pradesh, India

Syed Anas Ansar
School of Computer Applications
Babu Banarasi Das University
Lucknow, Uttar Pradesh, India

Md Tarique Jamal Ansari
Babasaheb Bhimrao Ambedkar University
Lucknow, Uttar Pradesh, India

Sandeep Kumar Verma
Babasaheb Bhimrao Ambedkar University
Lucknow, Uttar Pradesh, India

Narayan P. Bhosale
Indira Gandhi National Tribal University
Amarkantak, Madhya Pradesh, India

Kotaiah Bonthu
Department of Computer Science
Central Tribal University of Andhra Pradesh
Vizianagaram, Andhra Pradesh, India

Tushar Chandra
Student, Babu Banarasi Das University
Lucknow, Uttar Pradesh, India

Gaurav Chaudhari
Indira Gandhi National Tribal University
Amarkantak, Madhya Pradesh, India

Siddharth Dabade
National Forensic Sciences University
Gandhinagar, Gujarat, India

Dhananjay Deshpandey
MBA-ESG Business School
Pune, Maharashtra, India

Nafees Akhter Farooqui
Department of Computer Science and Engineering
Koneru Lakshmaiah Education Foundation
Guntur, Andhra Pradesh, India

Manu Gautam
Faculty, UIET
Babasaheb Bhimrao Ambedkar University
Lucknow, Uttar Pradesh, India

Bineet Kumar Gupta
Shri Ramswaroop Memorial University
Barabanki, Uttar Pradesh, India

Mohd Haroon
Department of Computer Science and Engineering
Integral University
Lucknow, Uttar Pradesh, India

Mohammed Ishrat
Department of Computer Science and Engineering
Koneru Lakshmaiah Education Foundation
Guntur, Andhra Pradesh, India

Mohit Joshi
Banaras Hindu University
Varanasi, Uttar Pradesh, India

Afsaruddin Khan
Department of Computer Science and Engineering
Dr A.P.J. Abdul Kalam Technical University
Lucknow, Uttar Pradesh, India

Jalaluddin Khan
Department of Computer Science and Engineering
Koneru Lakshimaiah Education Foundation
Guntur, Andhra Pradesh, India

Manoj Kumar Mishra
Rajiv Gandhi South Campus
Banaras Hindu University
Mirzapur, Uttar Pradesh, India

Mukesh Mishra
Shri Ramswaroop Memorial University
Barabanki, Uttar Pradesh, India

Mohd Nadeem
Department of Computer Science and Engineering,
Shri Ramswaroop Memorial University,
Lucknow, Uttar Pradesh, India

Khushi Naik
Student, National PG College
Lucknow, Uttar Pradesh, India

Sandeep Kumar Nayak
Babasaheb Bhimrao Ambedkar University
Lucknow, Uttar Pradesh, India

Abhishek Kumar Pandey
Center for Security, Theory and Algorithmic Research,
International Institute of Information Technology
Hyderabad, Telangana, India

Sakshi Pandey
Student, Babu Banarasi Das University
Lucknow, Uttar Pradesh, India

Prabhash Chandra Pathak
School of Computer Application
Babu Banarasi Das University
Lucknow, Uttar Pradesh, India

Wasiur Rhmann
Department of Computer Application
Lovely Professional University
Jalandhar, Punjab, India

Basudeo Singh Roohani
Department of Computer Science & Engineering
IMS Engineering College
Ghaziabad, Uttar Pradesh, India

Kavita Sahu
Department of Computer Science and Information Systems
Shri Ramswaroop Memorial University
Lucknow, Uttar Pradesh, India

Amal Krishna Sarkar
Sanjay Gandhi Postgraduate Institute of Medical Sciences
Lucknow, Uttar Pradesh, India

Nitin Sharma
Department of Information Technology
Ajay Kumar Garg Engineering College Ghaziabad,
Uttar Pradesh, India

Eram Fatima Siddiqui
Department of Computer Application
Integral University
Lucknow, Uttar Pradesh, India

Aditya Pratap Singh
Department of Computer Science and Engineering
Shri Ramswaroop Memorial University
Barabanki, Uttar Pradesh, India

Archana Singh
Department of Computer Science and Engineering
Shri Ramswaroop Memorial University
Lucknow, Uttar Pradesh, India

Sarita Shukla
Shri Ramswaroop Memorial University
Barabanki, Uttar Pradesh, India

Pallavi Somvanshi
School of Computational and Integrative Sciences
Jawaharlal Nehru University
New Delhi, India

Prabhat Kumar Srivastava
Department of Computer Science & Engineering
IMSEC
Ghaziabad, Uttar Pradesh, India

R. K. Srivastava
Department of Computer Science
Dr. Shakuntala Misra National Rehabilitation University
Lucknow, Uttar Pradesh, India

Saurabh Srivastava
Department of Computer Application
Integral University
Lucknow, Uttar Pradesh, India

Manish Madhava Tripathi
Department of Computer Science and Engineering
Integral University
Lucknow, Uttar Pradesh, India

Nayyar Ali Usmani
Puma Se
Puma Way 1, Herzogenaurach, Germany

Satya Bhushan Verma
Department of Computer Science and Engineering
Shri Ramswaroop Memorial University
Barabanki, Uttar Pradesh, India

Vishal Verma
Babasaheb Bhimrao Ambedkar University
Lucknow, Uttar Pradesh, India

Vijay Kumar Vishwakarma
Indira Gandhi National Tribal University
Amarkantak, Madhya Pradesh, India

Virendra P. Vishwakarma
University School of Information, Communication & Technology
Guru Gobind Singh Indraprastha University
New Delhi, India

Abhay Kumar Yadav
University School of Information, Communication & Technology
Guru Gobind Singh Indraprastha University
New Delhi, India

Amitabha Yadav
National PG College
Lucknow, Uttar Pradesh, India

About the book

Computational Intelligence Applications in Cyber Security is a comprehensive exploration of the evolving landscape where computational intelligence meets the critical domain of cybersecurity. The book encompasses a diverse range of chapters, each addressing specific facets of the intersection between computational intelligence techniques and the challenges posed by cyber threats.

The journey begins with a profound examination of the transition from Industry 4.0 to Industry 5.0, emphasizing the need for sustainable and human-centric approaches in the age of advanced industrialization. This sets the stage for understanding the broader implications of computational intelligence in securing critical infrastructures and systems.

One of the focal points of the book is Chapter 2. This chapter delves into the application of Support Vector Machines (SVM) in categorizing and securing software, particularly in defense contexts. The exploration of SVMs as a tool for fast and effective training of regression and classification models is vital for understanding the defense sector's emphasis on data security.

Moving forward, the book shifts its focus to the challenges and solutions presented by Industry 5.0. Chapter 3 investigates technologies such as digital twins, edge computing, collaborative robots, and blockchain as security enablers for Industry 5.0 ecosystems. This exploration emphasizes the fusion of creativity and cognitive abilities to enhance productivity and safety in industrial settings.

Blockchain technology, a transformative force in the last three decades, takes center stage in Chapter 4. This section not only elucidates the fundamentals of blockchain but also envisions its potential role as a collective platform for national and international development goals. The progress made by various countries in leveraging blockchain for citizen welfare services is thoroughly examined.

Financial risk management in the banking sector receives specialized attention in Chapter 5. Here, the book investigates the correlation between blockchain implementation and risk metrics, providing valuable insights into the benefits of adopting blockchain in managing financial risks.

The exploration of cloud computing security is brought to the forefront in Chapter 6. This section critically examines security concerns associated with cloud computing and provides a comprehensive overview of the factors influencing cloud computing security, especially in defense applications.

The book expands its scope to address security challenges in emerging technologies such as the Internet of Things (IoT) and Operational Technology (OT). Chapter 7 provides a thorough examination of security plans, prevalent threats, and major security attacks and protections for IoT and OT systems.

Healthcare information systems, a critical component of modern healthcare, are discussed in Chapter 8. The emphasis is on data integrity concerns in the healthcare industry and proposes the integration of blockchain for prioritizing data integrity in healthcare.

The convergence of IoT, cloud computing, and fog computing takes center stage in Chapter 9. This chapter proposes a secure and low-latency IoT environment through fog computing, addressing security risks and enhancing real-time collaboration and safety.

Chapter 10 explores the evolving landscape of healthcare web applications. Against the backdrop of the COVID-19 pandemic, it provides insights into prevalent threats, recent data breaches, and guidance for securing healthcare information systems.

As supply chain management undergoes digital transformation, Chapter 11 discusses the pivotal role of blockchain technology. It envisions blockchain as a common platform for governments to achieve national and international development goals, emphasizing its potential for creating smart countries.

The critical issue of software vulnerabilities is addressed in Chapter 12. This chapter explores the use of deep learning algorithms for predicting software vulnerabilities, crucial for preemptive actions to secure systems before potential exploitation.

The book takes a closer look at the challenges faced by artificial intelligence in national security in Chapter 13. It underscores the importance of transparency, accountability, privacy protection, and human oversight for the responsible deployment of AI technologies in safeguarding national interests.

Quantum computing, an emerging frontier, is explored in Chapter 14. This section introduces the basics of quantum computing, its applications in network security, and its potential to redefine modern cryptography.

Chapter 15 dives into the growing dependency on the internet and the consequent increase in cyber-attacks. It discusses techniques for gathering and analyzing information, potential threats, and advanced prevention strategies based on AI and other advanced technologies.

Efficient cloud forensics is a critical need in the digital age, as discussed in the chapter 16. The book provides insights into the challenges faced by digital forensic investigators in the context of cloud computing, discussing organizational, technological, and legal issues.

Smart cities, reliant on IoT, are addressed in Chapter 17. The focus is on the imperative of securing IoT-based applications in smart cities, given the increasing threats to these interconnected systems.

The book concludes with Chapter 18. This chapter delves into the quantum age and its impact on software security, with a particular emphasis on Grover's algorithm and its potential to enhance security measures in the era of quantum computing.

Computational Intelligence Applications in Cyber Security offers a holistic exploration of the dynamic landscape where computational intelligence confronts the intricate challenges of cybersecurity. The diverse array of chapters, each addressing a unique aspect of this intersection, provides readers with a comprehensive understanding of the current state, challenges, and future directions in the field.

Chapter 1

A transition from Industry 4.0 to Industry 5.0

Securing Industry 5.0 for sustainability

Sandeep Kumar Verma, Vishal Verma, and Md Tarique Jamal Ansari

1.1 INTRODUCTION

The use of steam and water power as the main sources of fuel during the First Industrial Revolution was the first step in the transition from hand production to machine production. Because new technology could take a long time to implement, it refers to the time period from 1760 to 1820, or perhaps as far back as 1840. This had social consequences, including the textile sector, iron industry, agriculture, mining, and the growing middle class. In the Second Industrial Revolution, electricity was born, which is also known as a technical revolution between 1871 and 1914. It was completely disruptive when computers were first introduced in the Industry 3.0 revolution. People of that time worked the manual pen–paper-based system, making an entry in a record register as opposed to working on a system. Instead of a good opportunity to work on computers, people took it as a threat. For the first time in 2011, the German government promoted computerization in the manufacturing process and designing process with the help of Industry 4.0, which was part of a good effort [1]. In the same year, the title and idea of this innovation, which is based on the high-tech strategy to allow intelligent decision-making in real time, were published during the Hannover Exhibition. Industry 4.0 is based on cyber–physical connectivity systems to initiate, share, process, analyze, and monitor intelligent action for different processes in the industry as well as to make a decision and behave in accordance with them. These functions can be carried out by machines that make it intelligent. These intelligent devices can regularly monitor activities, identify, and anticipate issues, and also recommend preventative measures and corrective action. This perspective is required for all Industry 4.0 technologies and services, with integration serving as the key consideration, Internet of Things (IoT), cloud computing (CC), cognitive computing, cyber–physical systems (CPS) are the main key concept of the research area since its introduction in 2011 in the Hannover fair [2]. The industrial revolution with respect to time, technologies and respective properties is depicted in Table 1.1.

DOI: 10.1201/9781003514312-1

1

Table 1.1 Evolution from Industry 1.0 to Industry 5.0 [3]

IR	Duration	Characterized	Details
IR 1.0	1780	Mechanization	Water and steam-related production system
IR 2.0	1870	Mass production	Electrification/assembly lines
IR 3.0	1970	Automated production	Electronic devices and computer used for automation
IR 3.5	1980	Globalization	Offshoring manufacturing to low-cost economies
IR 4.0	2011	Digitization	CPS-based connectivity to automate the further process [4]
IR 5.0	2022	Personalization	HCPS based system with man–machine collaboration, cognitive computing is used [5].

Industry 4.0 aims to integrate computer systems and operational innovation across enterprises and supply chains to build cyber–physical production systems (CPPS). The progress of migration from Industry 4.0 to Industry 5.0 is basically due to human involvement in CPS. After the Fourth Industrial Revolution, an extended version of the industrial paradigm known as Industry 5.0 emerged, which led to a discussion over the new paradigm's implementation process and motivations. The smart factory is a key component of Industry 4.0 where machines, devices, and systems are interconnected and communicate with each other to build CPPS [6]. The usage of Industry 4.0 has technically enhanced human–machine collaboration; however, taking into account the social, environmental, and sustainable elements, the fundamental role of humans in the technological advancements of Industry 4.0 must be taken into consideration [7]. Particular focus was paid to the value and function of employees during the coronavirus disease 2019 (COVID-19) epidemic, and this rethinking—now known as the new Industry 4.0 paradigm—was born at that time [8]. As a result, Industry 5.0 was conceptualized, expanding Industry 4.0 in a constructive manner with social and environmental elements [9]. In terms of skills, knowledge, and the capacity to properly collaborate with robots and machines, Industry 5.0 emphasizes both the adaptability and environmental effects of manufacturing methods.

It is necessary to keep in mind that Industry 5.0 is a creative and developing idea with many different definitions in the literature. The chapter uses the Industry 5.0 pillars established by the European Commission (EC) as of January 2021 as a point of reference [10], which, taking into account the future of European industry, strikes a balance between economic progress and stability. In any circumstance, given recent occurrences, the protracted COVID-19 epidemic that contributed to weather variation, and the recent confrontation between Russia and Ukraine, Industry 5.0 is becoming increasingly significant globally. Therefore, Industry 5.0's present goal is to remedy the shortcomings of Industry 4.0 by reconsidering social trends, which were not addressed by paradigms like the social justice approach or several of the Sustainable Development Goals (SDGs) of the United Nations [11].

Industry 5.0 is mostly the combination of similar technologies used by Industry 4.0 like blockchain [12], edge and fog computing, artificial intelligence (AI), augmented reality (AR), IoT [13], and digital twins [14] or some advancements in an area like energy efficiency or smart materials. In Industry 5.0, the application and design of such an innovation extend beyond their simple industrial exploration with three prominent principles. Let us focus on promoting the principles of human-centeredness, sustainability, and resilience.

Although some Industry 4.0 applications take into consideration some human-centric aspects (for example, operator safety), recent literature exhibits reference models and their approaches explaining how Industry 5.0 developments should be managed [15].

According to the EC [16], we should concentrate on determining what benefits these new technologies can provide for workers rather than considering how they can be employed. As a result, Industry 5.0 can be seen as an addition to Industry 4.0's key characteristics, which also include social equity and pertinent social considerations like environmental effects. Therefore, such improvements enable a refocus on Industry 4.0 concepts to build a future that is both human-centered and environmentally sensitive. The foundation of IoT is to establish the interconnection of physical devices and other objects that are embedded with sensors and connected via network connectivity. The combination of the digital and physical worlds plays an important role in creating intelligent systems that can effectively communicate with the environment as well as with the other systems and make better decisions They will also gather data in real time, which will be processed and saved on the cloud.

1.2 THE BASIC DRIVING TECHNOLOGIES INVOLVED IN INDUSTRY 4.0 AND INDUSTRY 5.0

Some technologies that play important roles in both the industrial revolutions are as follows:

- CC refers to the supply of a variety of facilities via the internet, such as storage, software, high processing power, quick data access, intelligence analytics, network, and other services on servers [17]. Cloud technology is critical to maximizing security strength. CC will be used to govern IoT networks as their cyber–physical and cognitive capabilities increase [18].
- The IoT stands for a new phase of interconnected network devices of physical objects—"things"—embedded with sensors, machine learning (ML), software, and other technologies in order to create a

holistic environment for exchanging data via the cyberspace with the help of these interrelated devices [19].

- The CPS is based on the most recent generation of digital systems, which may integrate computational power and physical capabilities to give a real-time interface with humans via innovative modularity [20], [21].
- AI integrated with CPS aids decision-making by handling the volume of data generated by various sensing nodes, analyzing it, and generating reports. It supports nonlinear and multistage production, as well as predictive analysis and complex decision-making [22].
- AR is the most important recent technological advancement that offers powerful apparatuses to assist operators with assembly-level tasks, context-aware support, interaction [acting as human–machine interface (HMI)] and data visualization, interior localization, quality control, material management, or maintenance applications [23].
- In any other application area, simulation is the most commonly used approach to design, optimize processes, improve safety, and analyze building systems. The usage of simulation and execution of other similar technologies in manufacturing technology is also part of the modern industrial revolution. Simulation is critical in the business for advancing and deploying technologies such as CPS, virtual reality (VR), AR, smart factories, digital twins, and the IoT [24]. Furthermore, in the context of managing such technologies, simulation aids in the operation, design, and optimization of processes in factories.
- Robotic process automation (RPA) must have massive benefits, including the use of AI techniques and algorithms to enhance the precision and performance of RPA procedures in data retrieval, recognition, categorization, predicting, and process optimization. RPA tools improve their functionality by utilizing text mining techniques, natural language processing methods techniques, and artificial neural network algorithms, for information retrieval and the resulting optimization process and forecasting scenarios in refining the organizations' business and operational processes [25].

Industries must develop, adapt, and accept greener and technological shifts to be competitive and prosperous. Based on this definition, Europe defines three fundamental core pillars of Industry 5.0, depicted in Figure 1.1: human-centricity, sustainability, and resilience [26].

In human-centricity, a person has to always be the prime player in all the decision-making structures, keeping in mind the structure of policy-making in a harmonious society. Machines or robots cannot replace humans in the industry due to enormous personalization demands from prosumers. There would not be any digitization or automation without human involvement.

Industry 4.0	Industry 5.0
Cyber Physical System(CPS) based Connectivity	Focus on Customer experience and Satisfaction
Mass Customization	Mass Personalization
Intelligent Supply chain management	Distributed and Responsive Supply chain Management
Smart service based Product	Experienced (User Interactive) activated Products
Dehumanization from factories	Return of manpower
Virtual environment	Real environment

Figure 1.1 Core Values of Industry 5.0.

Human presence is the main factor due to it having the better capability of fault tolerance [27]. Workers are positioned like pillars in the middle of the production chain in Industry 5.0. To always prioritize humans in Industry 5.0, an inclusive work-based environment should be built, focusing primarily on physical health, people's well-being, mental health, and eventually the fundamental rights of workers, such as autonomy, privacy, and human dignity. Industrial workers should continue to focus on upskilling and reskilling themselves to improve their career options and work–life balance [28]. Human-centered design should not be one-sided. Otherwise, the technologies will be unable to demonstrate their full potential. Analyze crucial socio-environmental data using a human-centered AI thinking approach for AI-enhanced AI-based strategic planning with the goal of guiding worldwide sustainable development outcomes.

Sustainability refers to the development of circular processes that evaluate, reuse, acknowledge, and reutilize natural resources, hence reducing surplus and environmental impact and eventually leading to a circular economy with optimized resource effectiveness and efficiency. Industry 5.0 is a way to use resources more efficiently so that manufacturers can meet the demands of the present. The business model is also supported by mutual cooperation between humans and machines; we must reduce waste while increasing production. Combined with innovative initiatives, local production strengthens and sustains the economy. Sustainability takes into account three core elements: the economy, the environment, and the people. Compared to the last industrial revolution, Industry 5.0 aspirations prioritize human-centricity to meet societal requirements, resulting in an imbalanced approach to sustainability. The major focus of Industry 5.0 is to enable sustainable development, integration of human creativity, problem-solving skills, resilience, and human-centricity with the goal of sustainable development in mind, environmental conservation through zero waste and sustainable products is critical [29].

Figure 1.2 Human–Cyber Physical System Design Factor for Industry 5.0.

Resilience highlights the need to build a sustainable and high level of agility in industrial production, to improve and insulate it from hindrances, and to ensure that key infrastructure is supplied and assisted in times of crisis. The future industry must be adaptable enough to respond to (geopolitical shifts and handle natural disasters quickly. Resilience highlights the agility and flexibility that a production plant must maintain in response to changes in the market [30]. There is increasing demand among customers to acquire high-tech breakthroughs and related resources that meet their needs, and the market itself is adapting to the ever-changing, specific demands of the manufacturing industry, which is one of the most difficult tasks. Manufacturing processes are gradually shifting from customized products to personalized products and services on a broad scale. It is important to note that as humans become less dependent on robots for primary jobs. Human collaboration promotes issue resolution linked to task and process flow and improves intellect and inventiveness [31].

Industry 5.0 is based on the human–cyber–physical system (HCPS), a next-generation industrial development used for the smooth integration of information technology (IT) and its application is mentioned in Figure 1.2. Efficient functionality should be developed to meet the needs of those application areas and customized to their specific requirements. For instance, a smart system for smart grids must be developed to collect and analyze data from the electrical network to increase its efficiency and effectiveness. Cyber security also plays a crucial role in Industry 5.0 in the age of advanced technology adoption. It detects and prevents unauthorized access that could pose major harm to the system. An appropriate design methodology with the help of an improved design tool ensures that the specific requirement is according to the user's need.

1.3 NEED FOR INDUSTRY MIGRATION

According to the EC, Industry 5.0 is the next phase of Industry 4.0; however, there are some issues. Industry 5.0 is an extended version of Industry 4.0 and outlines a new approach to how it might be used to improve our lives, going beyond productivity and efficiency as the primary goal. Figure 1.3 is responsible for exploring the simplified solutions of Industry 5.0. Industry 4.0 is a digital networking revolution. It is a vision of human–robot collaboration, coworking, and sustainability that provides a concept of joyful living through the balanced use of smart manufacturing, automation, and robotics [25].

1.4 THE TRANSITION FROM INDUSTRY 4.0 TO INDUSTRY 5.0

Technology advancements that significantly boost earnings, regardless of environmental and social indicators, are what define the fourth industrial revolution. Industry 4.0 denotes the incorporation of IT into the structure of

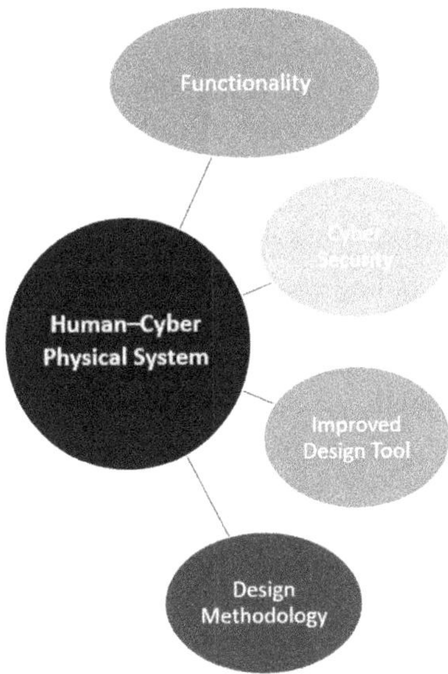

Figure 1.3 Need for Industry 5.0.

the industrial value chain and is governed by intelligent resources (machines and equipment) that positively impact the manufacturing process. Industry 4.0 is a new dimension and inventive style of working that is continually opening up new opportunities and responsibilities for industry professionals. Customers and business partners engage in IT-based intelligent business processes to create new value utilizing high-quality services; therefore quick and effective communication is critical in business. Companies may control their activities and act on them in a timely and effective manner because of intelligently controlled monitoring of the gadget in realtime. While businesses and other sectors are still working to deal with the problems posed by Industry 4.0, rapid digitization and a growing field of technological breakthroughs in IT have led to Industry 5.0, which is already on the doorstep.

The evolution of the concept of Industry 5.0 is also impacted by European political agendas. EC published an article in January 2021 entitled "Industry 5.0—towards a sustainable, human-centric and resilient European industry". To remain highly competitive, the European industry must continue to innovate. It must look forward, overcome the social and economic obstacles provided by the COVID-19 crisis, and create a "new normal model" with a more sustainable and environmentally sensitive business in which to invest. Industry 4.0 is primarily concerned with bringing together faster data networks with the assistance of robots and interconnected devices within the workplace environment to increase production in the factory and carry out everyday tasks, more smoothly, which robots can do far more efficiently than humans[3]. As a result, it provides numerous benefits including cost savings, zero mistakes with AI-dependent robots, and fast delivery of more tailored items.

The Industry 5.0 concept is still being directed by academics and industry today, but the European industry has already established the basis for it as a result of the EC's impact on such a notion in the coming years. The proposed concept has been viewed as an attempt to correct some shortcomings of Industry 4.0, but not all of them have been thoroughly considered or have proven contentious for failing to address crucial issues such as sustainability and social fairness. According to the EC, the foundation of Industry 5.0 must be connected with societal goals, while also considering social progress and economic growth. As a result, Industry 5.0 focuses on sustainable manufacturing and the well-being of industrial operators.

Industry 4.0 is commonly referred to as a technology-driven approach to manufacturing and production, having focused on the digital transformation of industrial processes using advanced technologies such as IoT, whereas Industry 5.0 is a value-driven or socially driven industry. With Industry 4.0, we observe a movement from the concept of CPS to the concept of cyber–physical–human systems (CPHS). The successful coexistence of Industry 4.0 and Industry 5.0 would result in increased efficiency and output while avoiding the displacement of human labor from the industrial sector. A vision of Industry 5.0 is in which robotic technologies collaborate with human minds as allies rather than rivals. According to this view, Industry 5.0 will create more job chances rather than diminish them [10]. According to a recent publication, both frameworks can coexist, with Industry 5.0 supplementing the existing Industry 4.0 paradigm with a focus on humans, a role that has been highlighted during the COVID-19 pandemic.

The shift should meet important societal expectations and concentrate on three areas of development: human-centered, resilient, and sustainable, focusing on the foundations of sustainable development and enhancing the standard of living. The rapid economic transition toward a human-centric, digitally empowered society and knowledge economy necessitates adjustments in investment methods as well as suitable government regulations. People must consider whether industry transformation is truly the best approach to "raise income", "increase output", and improve the average person's quality of life. The effective transition allows the social and economic impact of the Industry 4.0 to Industry 5.0 transition process to be reflected through the adoption of the 4Cs: creativity, collaboration, communication, and critical thinking. These 4Cs are essential skills for individuals and organizations looking to navigate this transition effectively. It is hoped that as a result of this development, we will be able to better appreciate the value of human capital.

While such transformation is thought to boost the efficiency of human capital in society, communications technology plays a unique role in the transition process, which prioritizes sustainability over productivity. Industry 4.0 must prioritize sustainable growth in economic, environmental, and social dimensions by rethinking the technology model to incorporate human-centeredness, resilience, and sustainability, in smart processes in order to meet the demands of the rapidly rising Industry 5.0. Despite the fact that Industry 5.0 has been human-centered, some people have expressed opposition. This is because the rate of technology development is far faster than the rate of employee response.

Industry 4.0 represents the present state of industrial automation and digitization, which involves the use of advanced technologies such as real-time data processing, AI, Big Data, ML, CC, and the IoT to create smart factories that can optimize production processes, improve quality, and increase efficiency. Industry 5.0, on the other hand, is still an emerging concept that is not yet fully defined. However, it is generally seen as an evolution of Industry 4.0 that seeks to blend human intelligence with advanced automation technologies to create a more collaborative, flexible, and sustainable manufacturing environment. Industry 5.0 will incorporate the human element and promote more social, ethical, and ecological concerns. While there is no one-size-fits-all roadmap for migrating from Industry 4.0 to

Table 1.2 Industry 4.0 vs. Industry 5.0 Comparison

Industry 4.0	Industry 5.0
CPS-based connectivity	Focus on customer experience and satisfaction
Mass customization	Mass personalization
Intelligent supply chain management	Distributed and responsive supply chain management
Smart service-based product	Experienced (user interactive) activated products
Dehumanization from factories	Return of manpower
Virtual environment	Real environment

Industry 5.0, some key considerations for companies looking to make this transition might include the following (Table 1.2):

- Reevaluating their business model and value proposition to better align with Industry 5.0's focus on human-centered manufacturing
- Investing in technologies such as VR, AR, and collaborative robots (cobots) that can help create a more human-centric work environment
- Developing new organizational structures and management practices that can support more decentralized decision-making and a more collaborative work culture
- Prioritizing sustainability and environmental concerns in their production processes and supply chains; reskilling and upskilling their workforce to ensure they have the necessary skills and knowledge to thrive in the new, more collaborative manufacturing environment.

The dynamic development of the Industry 4.0, which is the main concern that the adoption of its technologies connected with cyber–physical changes in the manufacturing sector, has raised concerns among the government and society about the future dehumanization of the industry. As more businesses rely on AI, they will be confronted with more data that is generated at a quick speed and presented in a variety of formats. Industry 4.0 saw the birth of industrial IoT, with automation, blockchain, ML, CPS connection, and real-time data becoming inextricably linked. From technological, real-time interoperability, organizational, robotics and automation, and managerial standpoints, the Fourth Industrial Revolution poses a range of issues for manufacturing organizations.

With the revolutionary changes brought about by new IT technologies, changes in job areas and future production system growth necessitate the development of new skills by personnel. The most significant barrier in established and emerging countries is the lag in technical advances and digitization, socioeconomic disparity, a lack of resources, and the adoption of Industry 4.0.

1.5 PARADIGM IN SOCIETY 5.0

An evolution of society 1.0 from society 5.0 is mentioned in Figure 1.4. Society 5.0 originated in the age of disruptive technology in Industry 4.0. This shift in societal paradigms presents issues for businesses in maintaining their operations. The concept of Society 5.0 envisions the development of a human-centric society where all industries and other sectors of the society utilize AI, IoT, robotics, Big Data, and other innovative technologies to address the major issues of Industry 4.0. It helps to balance economic growth. As a result, this shift will touch all elements of human life, ushering in Industry 4.0 and Society 5.0. The current infrastructure encourages pragmatism and energy efficiency to facilitate rapid growth within the lifestyle framework. IoT connects physical and cyberspace, serving as the foundation for Society 5.0. Through a high level of cyber–physical convergence, the Japanese vision of Society 5.0 envisions the construction of a human-centered society, one that provides people with a pleasant living and a high level of pleasure.

Society 5.0 is a society in which all industries and other sectors of society use AI, ML, CC, IoT, robots, Big Data, and other technologies and innovations to overcome critical challenges—not only for progress, but also to aid in the integration of virtual and physical spaces, so that the benefit of the facility to solve social problems can be provided to all who demand it. Society 5.0 fosters such values in society where one can find opportunities anywhere and at any time by utilizing one's various abilities, and where people can live more resilient, secure, and challenging lives, and where human life sustains in harmony with nature, and the problems posed by Industry 4.0 are easily solved.

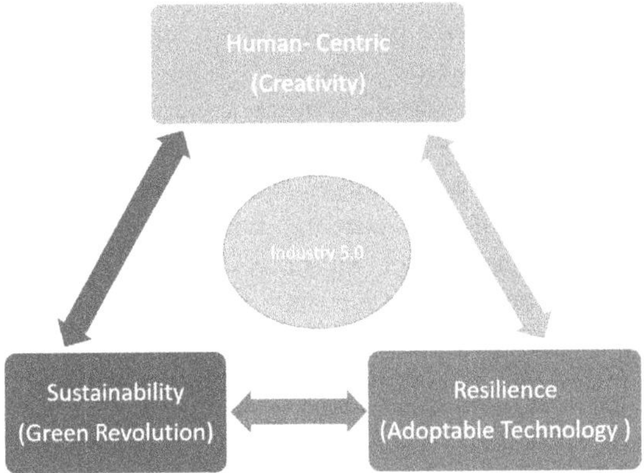

Figure 1.4 A Journey from Society 1.0 to Society 5.0.

In the middle of the Society 5.0 period, it is critical to demonstrate the concept of the nation as a highly desired future society capable of controlling and balancing the potential of AI-based social intelligence to solve societal challenges. Industry 4.0 is primarily concerned with the digital transformation of industrial processes and the integration of digital technologies into manufacturing and industrial processes. Society 5.0 is concerned with the broader social and economic implications of digital transformation and the potential for technology to create a more equitable and sustainable society. Nonetheless, this industry is proficient in 4.0 technologies, which are the foundation of the paradigm and enable society to 5.0. Society 5.0 allows the advancement of technology into social life and promotes the creation of sustainable technology with appropriate respect and prosperity toward humanization. Health, urban planning, transportation, agriculture, industry, ability, innovative culture, technology, and education will all be impacted by Society 5.0. Industry 5.0 demonstrates the opportunity to rearrange the quality of its global leadership via global collaboration, freedom, revolutionary technology, security, ethics, and digital economic growth to establish norms and standards.

1.6 CONCLUSION AND FUTURE SCOPE

The research we conducted provides insight into Industry 4.0 in companies and related technologies, focusing on the main issue, the absence of human factor engagement, and adopting HCPS in the Industry 5.0 era. The transformation to Industry 5.0 focuses on preparing the workforce and employees with the capabilities they need to adjust to the changing demands of the current digital age. Furthermore, unlike prior to Industry 4.0, successful technological implementations showed a lack of attitude to adopt new technology by employees as a human aspect in the processes, which is the primary reason for industrial migration. The major goal is to evaluate all potential user problems when using current technologies. It is important to note that the migration from Industry 4.0 to Industry 5.0 is likely to be gradual and will require a significant investment of time, money, and resources. However, companies that are willing to embrace this shift and make the necessary changes will be well-positioned to thrive in the future of industrial manufacturing. Based on the published scientific research paper, governments and organizations both work according to EC policies to adopt such efforts; or we may conclude that Industry 5.0 is not just a trend but rather a long-term vision for the future of manufacturing and industrial processes. It is a commitment to innovation.

REFERENCES

[1] M. C. Zizic, M. Mladineo, N. Gjeldum, and L. Celent, "From Industry 4.0 towards Industry 5.0: a review and analysis of paradigm shift for the people, organization and technology," *Energies*, vol. 15, no. 14, p. 5221, Jul. 2022, doi: 10.3390/EN15145221.

[2] N. Jafari, M. Azarian, and H. Yu, "Moving from Industry 4.0 to Industry 5.0: what are the implications for smart logistics?," *Logistics*, vol. 6, no. 2, p. 26, Apr. 2022, doi: 10.3390/LOGISTICS6020026.

[3] B. Ozkeser, "Lean innovation approach in Industry 5.0," *The Eurasia Proceedings of Science Technology Engineering and Mathematics*, vol. 2, no. 2, pp. 422–428, Aug. 2018, [Online]. www.epstem.net/en/pub/epstem/issue/38904/455975 (accessed: Feb. 04, 2023.

[4] L. Da Xu and L. Duan, "Big data for cyber physical systems in Industry 4.0: a survey," *Enterprise Information Systems*, vol. 13, no. 2, pp. 148–169, Feb. 2018, doi: 10.1080/17517575.2018.1442934.

[5] P. K. R. Maddikunta *et al.*, "Industry 5.0: a survey on enabling technologies and potential applications," *Journal of Industrial Information Integration*, vol. 26, p. 100257, Mar. 2022, doi: 10.1016/J.JII.2021.100257.

[6] L. Wang and G. Wang, "Big Data in cyber–physical systems, digital manufacturing and Industry 4.0," *International Journal of Engineering and Manufacturing*, vol. 6, no. 4, pp. 1–8, Jul. 2016, doi: 10.5815/IJEM.2016.04.01.

[7] D. Sharma, G. Singh Aujla, and R. Bajaj, "Evolution from ancient medication to human-centered Healthcare 4.0: a review on health care recommender systems," *International Journal of Communication Systems*, 2019, doi: 10.1002/dac.4058.

[8] X. T. R. Kong, H. Luo, G. Q. Huang, and X. Yang, "Industrial wearable system: the human-centric empowering technology in Industry 4.0," *Journal of Intelligent Manufacturing*, vol. 30, no. 8, pp. 2853–2869, Dec. 2019, doi: 10.1007/S10845-018-1416-9/TABLES/2.

[9] "Enabling technologies for Industry 5.0." https://research-and-innovation.ec.europa.eu/knowledge-publications-tools-and-data/publications/all-publications/enabling-technologies-industry-50_en (accessed Jan. 25, 2023).

[10] S. Nahavandi, "Industry 5.0 – a human-centric solution," *Sustainability*, vol. 11, no. 16, p. 4371, Aug. 2019, doi: 10.3390/SU11164371.

[11] P. Fraga-Lamas and T. M. Fernández-Caramés, "Leveraging blockchain for sustainability and open innovation: a cyber-resilient approach toward EU green deal and UN sustainable development goals," *Computer Security Threats*, Sep. 2020, doi: 10.5772/INTECHOPEN.92371.

[12] A. Verma *et al.*, "Blockchain for Industry 5.0: vision, opportunities, key enablers, and future directions," *IEEE Access*, vol. 10, pp. 69160–69199, 2022, doi: 10.1109/ACCESS.2022.3186892.

[13] S. Zeb *et al.*, "Industry 5.0 is coming: a survey on intelligent nextG wireless networks as technological enablers," 2022.[Online].https://arxiv.org/abs/2205.09084(accessed: Jan. 27, 2023).

[14] P. Stavropoulos and D. Mourtzis, "Digital twins in Industry 4.0," *Design and Operation of Production Networks for Mass Personalization in*

the Era of Cloud Technology, pp. 277–316, Jan. 2022, doi: 10.1016/B978-0-12-823657-4.00010-5.

[15] M. Ghobakhloo, M. Iranmanesh, M. F. Mubarak, M. Mubarik, A. Rejeb, and M. Nilashi, "Identifying Industry 5.0 contributions to sustainable development: a strategy roadmap for delivering sustainability values," *Sustainable Production and Consumption*, vol. 33, pp. 716–737, Sep. 2022, doi: 10.1016/J.SPC.2022.08.003.

[16] "Industry 5.0 – publications office of the EU." https://op.europa.eu/en/publication-detail/-/publication/468a892a-5097-11eb-b59f-01aa75ed71a1/ (accessed Jan. 31, 2023).

[17] A. Adel, "Future of industry 5.0 in society: human-centric solutions, challenges and prospective research areas," *J Cloud Comp,* vol. 11, no. 40, 2022. https://doi.org/10.1186/s13677-022-00314-5

[18] M. Javaid, A. Haleem, R. P. Singh, M. I. Ul Haq, A. Raina, and R. Suman, "Industry 5.0: potential applications in COVID-19," *Journal of Industrial Integration and Management*, vol. 5, no. 4, pp. 507–530, Nov. 2020, doi: 10.1142/S2424862220500220.

[19] S. Madakam, R. Ramaswamy, and S. Tripathi, "Internet of things (IoT): a literature review," *Journal of Computer and Communications*, vol. 03, no. 05, pp. 164–173, 2015, doi: 10.4236/JCC.2015.35021.

[20] Y. Liu, Y. Peng, B. Wang, S. Yao, and Z. Liu, "Review on cyber-physical systems," *IEEE/CAA Journal of Automatica Sinica*, vol. 4, no. 1, pp. 27–40, Jan. 2017, doi: 10.1109/JAS.2017.7510349.

[21] L. Monostori, "Cyber-physical production systems: roots, expectations and R&D challenges," *Procedia CIRP*, vol. 17, pp. 9–13, Jan. 2014, doi: 10.1016/J.PROCIR.2014.03.115.

[22] M. Javaid, A. Haleem, R. P. Singh, and R. Suman, "Artificial intelligence applications for Industry 4.0: a literature-based study," *Journal of Industrial Integration and Management*, vol. 7, no. 1, pp. 83–111, Mar. 2022, doi: 10.1142/S2424862221300040.

[23] P. Fraga-Lamas, T. M. Fernández-Caramés, Ó. Blanco-Novoa, and M. A. Vilar-Montesinos, "A review on industrial augmented reality systems for the Industry 4.0 shipyard," *IEEE Access*, vol. 6, pp. 13358–13375, Feb. 2018, doi: 10.1109/ACCESS.2018.2808326.

[24] M. M. Gunal, and M. Karatas, "Industry 4.0, digitisation in manufacturing, and simulation: a review of the literature," *Simulation for Industry 4.0*, pp. 19–37, 2019, doi: 10.1007/978-3-030-04137-3_2.

[25] J. Ribeiro, R. Lima, T. Eckhardt, and S. Paiva, "Robotic process automation and artificial intelligence in Industry 4.0 – a literature review," *Procedia Computer Science*, vol. 181, pp. 51–58, Jan. 2021, doi: 10.1016/J.PROCS.2021.01.104.

[26] X. Xu, Y. Lu, B. Vogel-Heuser, and L. Wang, "Industry 4.0 and Industry 5.0 – inception, conception and perception," *Journal of Manufacturing Systems*, vol. 61, pp. 530–535, Oct. 2021, doi: 10.1016/J.JMSY.2021.10.006.

[27] Y. Lu *et al.*, "Outlook on human-centric manufacturing towards Industry 5.0," *Journal of Manufacturing Systems*, vol. 62, pp. 612–627, Jan. 2022, doi: 10.1016/J.JMSY.2022.02.001.

[28] "Industry 5.0: towards more sustainable, resilient and human-centric industry." https://research-and-innovation.ec.europa.eu/news/all-research-and-innovation-news/industry-50-towards-more-sustainable-resilient-and-human-centric-industry-2021-01-07_en (accessed Feb. 05, 2023).

[29] A. Nayyar and A. Kumar, Eds., *A Roadmap to Industry 4.0: Smart Production, Sharp Business and Sustainable Development*, 2020, doi: 10.1007/978-3-030-14544-6.

[30] L. W. W. Mihardjo, Sasmoko, F. Alamsyah, and Elidjen, "Boosting the firm transformation in Industry 5.0: experience-agility innovation model," *International Journal of Recent Technology and Engineering*, vol. 8, no. 2, Special Issue 9, pp. 735–742, Sep. 2019, doi: 10.35940/IJRTE.B1154.0982S919.

[31] "Fifth revolution: applied AI & human intelligence with cyber-physical systems – PDF free download." https://docplayer.net/145940723-Fifth-revolution-applied-ai-human-intelligence-with-cyber-physical-systems.html (accessed Feb. 18, 2023).

Chapter 2

Security estimation of software by support vector machines

Defence perspective

Amitabha Yadav, Syed Anas Ansar, Manu Gautam, Sakshi Pandey, Tushar Chandra, and Khushi Naik

2.1 INTRODUCTION

The technique of categorizing data resources depending on how sensitive they are to information is known as data classification. Organizations can determine two important things by classifying data – permitted access and security measures to use when transferring and storing it [1]. Classification can also help create regulatory requirements for data protection. In general, data classification assists businesses in better managing their data for confidentiality, regulatory, and cyber security in defence. A company should categorize the data that it generates, maintains, and stores. Large-business environments, on the other hand, necessitate it even more. This is because big enterprises have a lot of data assets that are dispersed over many locations, including the cloud [2]. Administrators must keep an eye on this data and audit it to make sure the proper access and authentication measures are in place. Using data classification, administrators may determine where sensitive data are kept and how they should be accessed and shared. The first stage in almost any data compliance process is classification, which is crucial. Under the Health Insurance Portability and Accountability Act (HIPAA), General Data Protection Regulation (GDPR), Family Educational Rights and Privacy Act (FERPA), and other legal requirements, data must be labelled in order for security and authentication measures to limit access [3]. When data is tagged, it is safeguarded and organized. Additionally, the method reduces the likelihood of data duplication, reduces storage costs, enhances performance, and maintains data tracking as the data gets exchanged [4]. Accurate data classification is the first step in developing effective Data Loss Prevention (DLP) rules and procedures [5]. You must first classify your data to identify the data in each file to establish effective DLP rules. Any data that is stored can be categorized. As you discover and examine your data, you must ask a number of questions to categorize it. As you examine each section of your data, make use of the examples provided in Figure 2.1.

DOI: 10.1201/9781003514312-2

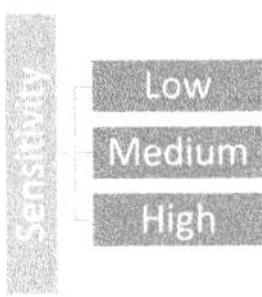

Figure 2.1 Data Classification According to Sensitivity.

Sensitivity high: The data must be secured and monitored to be shielded from threat actors. It generally falls under regulatory laws as it requires strong security controls and restricts the number of individuals that may access the data.

Sensitivity medium: The term "medium risk" refers to documents and information which cannot be publicly disclosed but do not constitute a considerable risk in the event of a data breach. This then requires access controls, but it is accessible to a broader range of individuals.

Sensitivity is low: Most of the time, this data is publicly available that does not require extensive security measures to avoid a data breach. To better protect and manage data, data categorization techniques work hand in hand with other technologies. Data classification helps administrators retrieve lost data and, presumably, the cyber-criminal in the event of a data breach.

2.2 DATA CLASSIFICATION-BASED TECHNOLOGIES

Different data classification-based technologies are available and depicted in Figure 2.2. The details of these technologies are as follows: Amazon Web Services (AWS) Identity and Access Management (IAM): With the use of IAM solutions, administrators may control both who and what has access to data. It is possible to create groups of users with similar permissions. Groups are given varying degrees of authority, and they are controlled collectively. When a user is removed from a group and leaves it, they lose all of their permissions. This method of grouping and organizing simplifies network permission management [6].

Data encryption: Any sort of crucial data resources must be secured using encryption both in transit and at rest. At-rest data is any information that is retained on a storage device, most commonly a hard drive. Data that is moving through a network is referred to as "data in motion." When data is encrypted, it renders intercepted information illegible [7].

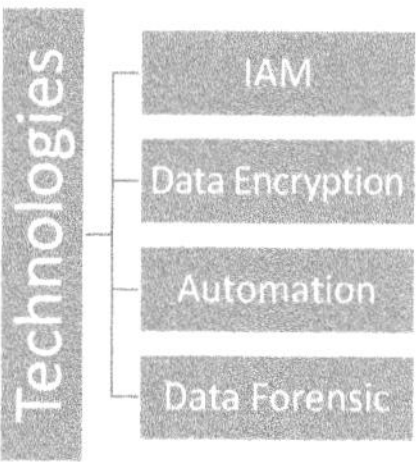

Figure 2.2 Data Classification Based on Technologies.

Automation: Automation locates, classifies, and labels data for operational analysis in conjunction with monitoring technologies. Some artificial intelligence (AI)- and machine learning (ML)-enabled technologies can recognize, label, and categorize data automatically. The technology can also assist in detecting potential theft threats. Administrators can use a classification model for data to provide rights and prevent specific threats from accessing stored data using IAM [8].

Data forensics: Data forensics is the procedure of determining what actually happened and who compromised the network. Following a security breach, data forensic work collects and retains material for subsequent investigation. Data forensics frequently consists of two parts. With the collection of data by automated systems, a human analyst detects abnormalities and initiates an investigation [9].

Data classification levels: You can more effectively categorize your data with these levels in mind. When classifying data, there are commonly four categories used:

- Public Records: The general public has access to this data locally or online.
- Public data require little protection because disclosing it would not be against compliance.
- Email messages, memos, and intellectual property are just a few examples of the kind of information that should only be available to internal staff [10].
- Data that is secret comparable to internal-only data, confidential data must have permission to access it. A specific staff person or authorized outside vendor may be given clearance [11].
- Constrained data often entails government information that is exclusively accessible to authorized individuals.
- If sensitive information becomes public, a company's revenue and reputation may suffer irreparable damage.

2.3 DATA CLASSIFICATION PROCESS

As you determine it is indeed appropriate to classify data to meet compliance requirements, the initial step you should undertake is to follow protocols to assist with data integrity and confidentiality, categorization, and selection [12]. Every approach is put in place in accordance with the architecture that best protects data and your organization's compliance requirements. The following are the general steps for classifying data and are mentioned in Figure 2.3.

Conduct a risk analysis: A threat assessment focuses on the sensitive nature of the information and the manner in which an adversary could get beyond network safeguards [13].

Create classification standards and guidelines: If more data is generated in the long term, a categorization policy allows for the simplification of a recurring process. This streamlines the procedure for staff employees and reduces the potential for errors [14].

Data sorting: After you have policies and a risk assessment in place, sort your data according to how sensitive it is, who should have access to it, and any compliance penalties if it is made public [3].

Implement controls: The restrictions you apply should require that every person and resource that wants access to data submit an authentication and authorization login request. Users should only be given permission if they require the data to execute a job function, and access should be granted on a "need to know" approach [11].

Monitor access and data: Data monitoring is essential for compliance and the security of your data. An attacker may have months to exfiltrate network data without being detected. With the proper monitoring controls, danger can be identified, reduced, and removed from the network more quickly.

The location of your data's storage must be known to employ the proper cyber security measures. By identifying data storage sites, it is possible to determine the type of cyber security that is necessary to protect data. Sort and name the data you have: once the data has been recognized, it may need

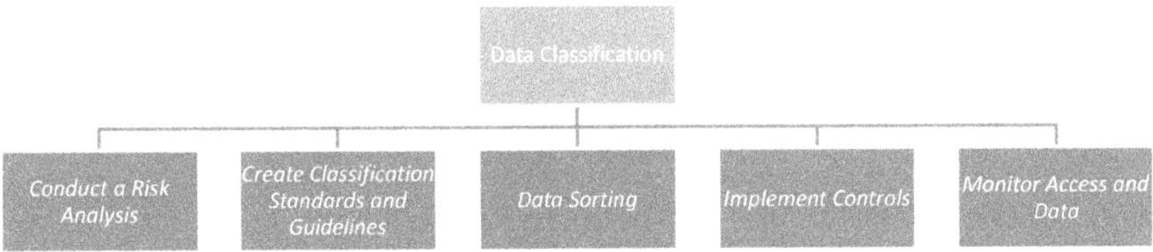

Figure 2.3 Data Classification Process.

to be classified. Software from a third party, which aids in data classification and tracking, makes this stage simpler.

2.4 ARTIFICIAL INTELLIGENCE FOR DATA CLASSIFICATION

Though AI data classification involves human intervention, most operations may be automated. Proof point created a data categorization engine that makes predictions with 99 per cent accuracy rate to integrate automation with decision-making. The engine regularly scans and reviews new documents that are introduced to the environment. AI automation allows an organization to continuously recognize, categorize, and secure its documents [15]. Machine learning algorithms can forecast document labels and evaluate their precision. The reviewer is presented with a "confidence level" so that they can reevaluate the application model before doing another set of information classifications. If the model displays low accuracy, human reviewers can improve accuracy by updating the model with more diverse sets of files. Using the new data, the engine will retrain itself to produce fresh, optimal outcomes [16]. Proof point built its engine to make sure that users have only access to the documents that require performing their duties.

Proof point's AI-powered data classification software drastically reduces the overhead for a task that can take months. All of your files are automatically scanned, their contents are determined, the proper categories and classification levels are assigned, and you are then given the option to select the right level of security [17].

2.4.1 Methodology

Support vector machines (SVMs) are a remarkable group of algorithms for supervised learning that can train regression and classification models fast and successfully in practice. A lot of beautiful mathematics is used in the derivation of different parts of the training algorithm, which we will examine in later postings. Convex optimization and Hilbert space theory are also the foundations of SVMs. For the time being, we will merely briefly go over the underlying theory of soft-margin kernel SVMs [2]. This presentation is a little faster than others because the usual method starts with hard-margin linear SVMs before moving on to the kernel trick and soft-margin formulation. Recent advancements in computer hardware, as well as the increased availability and affordable of computer components such as RAM as well as GPU, have had an impact on business and technology [18]. There is a lot of labelled data that is easily available and being produced every day. Focus on data analysis and predictive modelling to make the company less susceptible to unforeseen events [19]. -The growth of companies that use data science alone to produce tools as a significant component of their operations and the growing significance of having adequately assessed data support business choices. Huge amounts of data generated by the Internet of Things (IoT) must constantly be evaluated in order for various gadgets to work properly and receive updates.

Predictive modelling is increasingly being used to solve common problems. These and other considerations have contributed to the wider acceptance of using machine learning algorithms to address common corporate challenges. Although numerous algorithms can be used in the framework of machine learning, data scientists must first understand their distinct characteristics, benefits, and best practices before employing them in crucial business challenges. SVM is an advanced, well-known, and old algorithm. The SVM classifier is among the best linear and non-linear binary classifiers shown in Figure 2.4. SVM regressions are also thought to be a viable alternative to more standard regression approaches such as linear regression analysis [20].

The top data and analytics institution are focusing on teaching both individuals and organizations about AI and data science as they pertain to the enterprise. It is run by faculty members with extensive experience from McKinsey, IIT, IIM, and FMS.

In theory, the vector support machine algorithm, also known as the SVM algorithm, is basically linear. The SVM algorithm differs from others, in that it can tackle both problems with classification and regression by using an SVM classifier as well as an SVM regression. Always keep in mind, nevertheless, that the SVM classifier is the root of the supporting vector machine concept and, overall, the best method for dealing with classification issues. It can be comparable to linear or logistic regression because it is primarily a linear technique. To identify data points that belong to two distinct classes, an SVM classifier, for example, produces a line (plane or hyper-plane,

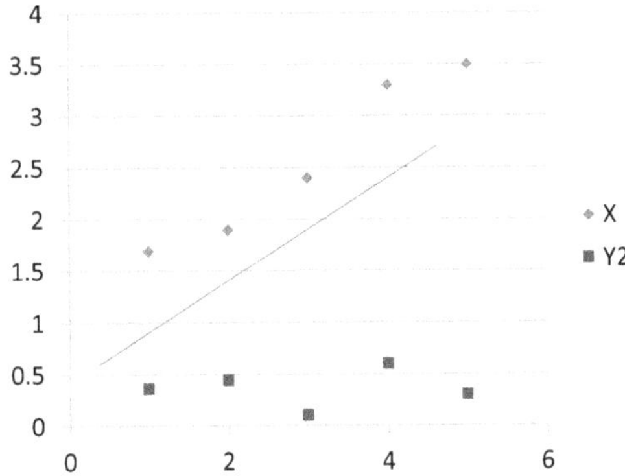

Figure 2.4 SVM Operation of Classification.

according to the dimensionality of the input) in an N-dimensional domain. It is also worth remembering that the prototype SVM classifier was created to solve binary classifier difficulties. Contrary to Logistic Regression, which utilizes maximum likelihood estimation to identify the best-fitting sigmoid curve, and linear regression, which uses the conception of the line of best fit, SVMs instead make use of the concept of Markov chains. It is crucial to understand the SVM method's historical background before learning how to apply it to classification and regression issues. SVM was developed in the 1970s by Vladimir Vapnik [21].

It was allegedly created as part of a bet in which Vapnik asserted that creating a decision boundary that aims to increase the margin between the two classes will produce excellent results and solve the over fitting issue. Everything changed, especially in the 1990s, when the kernel technique was developed and SVM could be used to tackle non-linear issues. Due to their tremendous complexity, this has been a significant impact on the prominence and advancement of neural networks for a time. SVM, on the other hand, was significantly more straightforward yet capable of addressing non-linear classification challenges rapidly and reliably [18]. The importance and reliance on SVM have not changed over time, despite the development of deep learning and neural networks more generally [19]. In a wide range of businesses that depend on machine learning, it is still highly regarded and applied.

2.5 WORKING OF SUPPORT VECTOR MACHINE ALGORITHM

It is simpler to comprehend the SVM approach by concentrating on its main type, the SVM classifier. The SVM classifier divides different data points into different classes in an N-dimensional space by creating a hyper-plane depicted in Figure 2.5. The hyper-plane with the maximum margin between the two classes, however, is the one that is selected for this hyper-plane based on margin. A particular kind of data point called a support vector is used to determine these margins. Support vectors are data points that can be used to align the hyper-plane as they are close to it.

If it is necessary to understand the SVM classifier's operation theoretically, one can do it in the following ways:

- The SVM algorithm can detect the classes in phase one. The number 1 designates one of the classes, whereas the number –1 designates the other.
- The organizational challenge is converted into a mathematical equation with unknown variables using all machine learning approaches. The work is then transformed into an optimal control

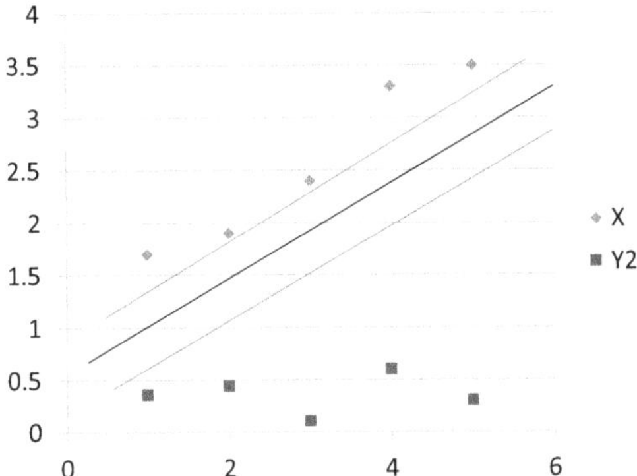

Figure 2.5 Classification Process of Hyper–Plane.

problem to find these unknowns. Throughout the case of the SVM classifier, the hinge loss function is used and changed to determine the greatest margin because optimization problems always seek to maximize or minimize something while evaluating and accounting for uncertainties.

- Hinge Loss Function

$$C\left(x,\ y,\ f\left(x\right)\right) = \{0 \ \text{if} \ y^{*}f\left(x\right) \geq 1\ 1 - y^{*}f\left(x\right) \ \text{else} \tag{2.1}$$

- Step 3: This loss function can alternatively be thought of as a cost function for the sake of clarity, with a cost of 0 for wrong predictions of no classes. In the absence of this, the error/loss is computed. The current predicament forces one to choose between maximizing margins and the significant loss that results in the process. These approaches are conceptually grounded by the addition of a regularization parameter.

Loss Function of SVM

$$\|w\|^{2} + \sum_{i=1}^{n}\left(1 - y;\ \langle x;w\rangle\right) \tag{2.2}$$

- Step 4: Weights are optimized by computing gradients through partial derivatives, a topic from intermediate calculus, as is the case for the bulk of optimization issues.

Gradients

$$\frac{\delta}{\delta w_k} \lambda \|w\|^2 = 2\lambda w_k \tag{2.3}$$

$$\frac{\delta}{\delta w_k}\left(1 - y;\ \langle x;w \rangle\right) = \{0\ \text{if}\ y_i\left(\langle x;w \rangle\right) \geq 1 - y;\ x_{ik}\ \text{else} \tag{2.4}$$

- Step 5: The gradients are updated only when there is no classification error using the regularization parameter, and when misclassification occurs using the loss function.

Updating of gradients when there is no misclassification-

$$w = w - \propto \left(2\lambda w\right) \tag{2.5}$$

Updating of gradients when there is misclassification-

$$w = w - \propto \left(2\lambda w\right) \cdot \left(y_i \cdot x_i - 2\lambda w\right) \tag{2.6}$$

- Step 6: The gradients are modified using the normalization parameter whenever there is no classification error; when misunderstanding happens, the loss function is also used.

2.6 PARTS OF SUPPORT VECTOR MACHINE

SVMs should be used in a machine learning context only; it is critical to comprehend some of the ideas and phenomena that result from their operation. The data points used to compute and maximize margins are support vectors shown in Figure 2.6. The quantity of support vectors or the extent of their influence is one of the hyper-parameters that must be managed [22].

Hard margin: The above concept is crucial for comprehending SVMs particularly and SVM classification in general. A "strict margin" judgement boundary assures that each data point is appropriately classified. This not only ensures that the SVM classifier is error-free but also has the ability to decrease the margins, defeating the very objective of utilizing an SVM technique [23].

Soft margin: A normalization parameter is also added to the loss function in the SVM classification method. The user can optimize margins at the risk of misclassification by combining the loss function and

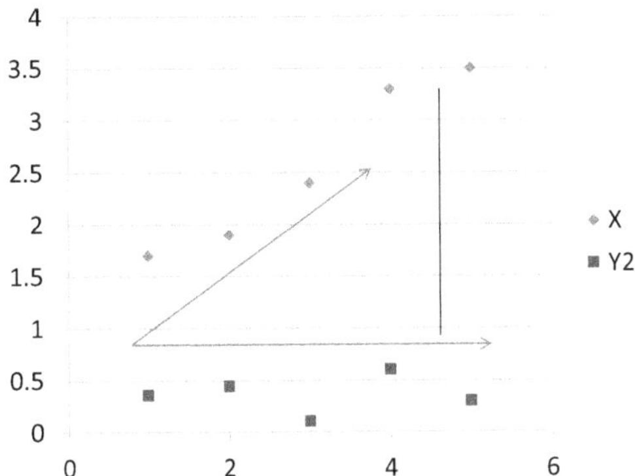

Figure 2.6 Data Margin Classification.

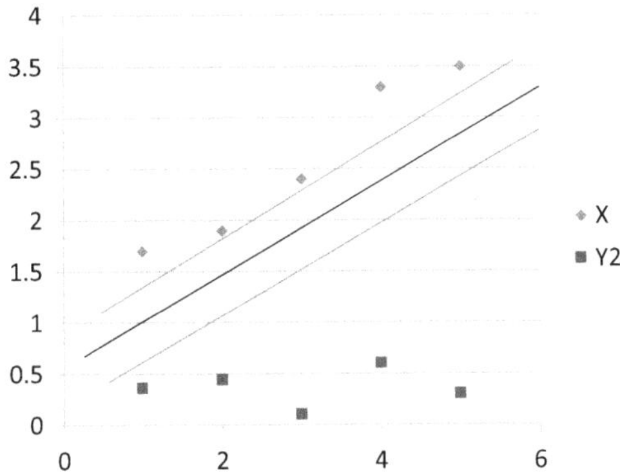

Figure 2.7 Margin Representation of Data.

the regularization parameter. However, this classification must be regulated, necessitating the modification of yet another hyper-parameter [24]. Hard margin classification and soft margin classification based on data are shown in Figure 2.7, whereas Figure 2.8 describes different kernel representations of data.

Different kernels: As it makes use of kernels, the SVM algorithm is a potent method for machine learning. The SVM algorithm, as demonstrated throughout this discussion, generates a linear hyper-plane. A non-linear problem, on the other hand, can occasionally lead a linear classifier to underperform [25]. The concept of kernel transformation is useful in this situation. SVM is primarily a non-linear classifier because kernel transformation transforms a low-dimensional space into a high-dimensional space where a linear

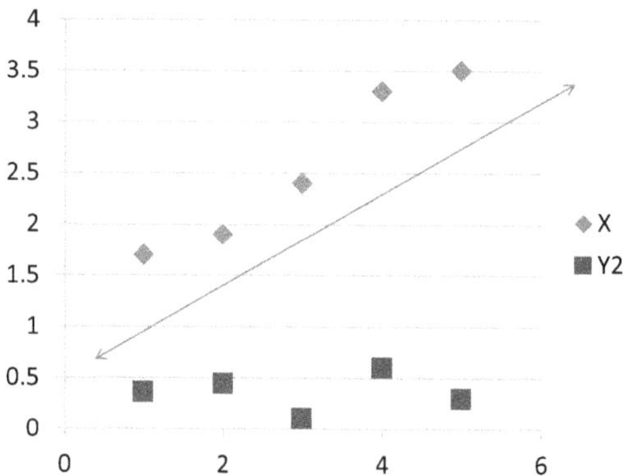

Figure 2.8 Kernel Representation of Data.

hyperplane can readily recognize the input points. Different kernels can be used to address multiple linear and nonlinear issues [26]. When these kernels are chosen, one more hyperparameter is added for consideration and optimal tuning.

2.7 IMPLEMENTATION OF SUPPORT VECTOR MACHINE

SVM is simply included in the vast majority of regularly used predictive modelling systems. SVMs: Using Python, an SVM classifier can be built with Python's learn package - This may be accomplished in languages like R and Python with a suitable support vector example. The SVM algorithm's steps are as follows:

Step 1: Load the important libraries
      >> import pandas as pd
      >> import numpy as np
      >> import sklearn
      >> from sklearn import svm
      >> from sklearn.model_selection import train_test_split
      >> from sklearn import metrics
Step 2: Import dataset and extract the X variables and Y separately.
      >> df = pd.read_csv("mydataset.csv")
      >> X = df.loc[:,['Var_X1','Var_X2','Var_X3','Var_X4']]
      >> Y = df[['Var_Y']]
Step 3: Divide the dataset into train and test
      >>X_train, X_test, y_train, y_test = train_test_split(X, Y, test_size = 0.3, random_state=123)
Step 4: Initializing the SVM classifier model

```
>>svm_clf = svm.SVC(kernel = 'linear')
```
Step 5: Fitting the SVM classifier model
```
>>svm_clf.fit(X_train, y_train)
```
Step 6: Coming up with predictions
```
>>y_pred_test = svm_clf.predict(X_test)
```
Step 7: Evaluating model's performance
```
>>metrics.accuracy(y_test, y_pred_test)
>>metrics.precision(y_test, y_pred_test)
>>metrics.recall(y_test, y_pred_test)
```

Similarly, an SVM classifier can be created in R also.
SVMs: Implementation in R

Step 1: Load the important libraries
```
>> require(e1071)
>> require(Metrics)
```
Step 2: Import the dataset
```
>>df<- read.csv("mydataset.csv")
```
Step 3: Divide the dataset into train and test
```
>>samp<- sample(1:nrow(df), floor(df)*0.7))
>> train <- df[samp,]
>> test <- df[-samp,]
```
Step 4: Fitting the SVM classifier model
```
>>svm_clf<- svm(Var_Y~., data = train, kernel = 'linear',
cost = 10, type = 'C-classification')
```
Step 5: Coming up with predictions
```
>>y_pred_test<- as.numeric(as.character(predict(svm_clf,
test))))
```
Step 6: Evaluating model's performance
```
>> accuracy(test$Var_Y, y_pred_test)
>> recall(test$Var_Y, y_pred_test)
>> precision(test$Var_Y, y_pred_test)
```

When there are more than three functions, it becomes challenging to imagine. Think of ×1 and ×2 as two independent variables, and a circle – either blue or red – as the dependent variable. The above illustration clearly demonstrates that there are multiple components that divide our data points or categorize those into red and blue circles (our hyper-plane in this case is a line because we are only looking at two input features, ×1, ×2). So how would we decide which line, or more generally, which hyper-plane, best divides our data points?

The hyper-plane that offers the greatest margin or separation between both the classes is the ideal hyper-plane. As a consequence, we identify the hyper-plane that can be reached from the closest data point upon every side

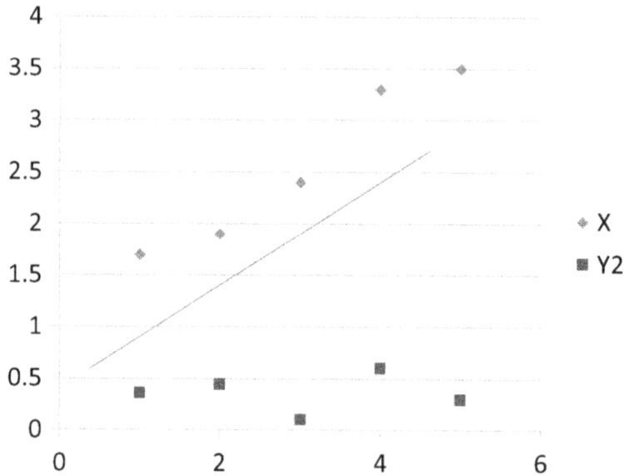

Figure 2.9 Defence Data Classification Representation.

in the shortest amount of time. If such a hyper-plane exists, then maybe the maximum-margin hyper-plane or hard margin occurs. So from Figure 2.9 we choose L2.

2.8 CONCLUSION

A supervised machine learning method called SVM can be applied to both classification and regression problems. It is more appropriate for classification even though we can classify it as a regression problem. The purpose of the SVM algorithm is to find a hyper-plane that can distinguish between different data points in an N-dimensional space. The hyper-plane's dimension depends on the number of features. When there are only two input characteristics, the hyper-plane is just a line. The hyper-plane is essentially a line if there are simply two input features. The hyper-plane changes into a two-dimensional plane when there are three input features.

REFERENCES

[1] A. J. Ramisa *et al.*, "Gene expression data classification and pattern analysis using data-driven approach," *Proc. – Int. Conf. Mach. Learn. Cybern.*, vol. 2021-December, 2021, doi: 10.1109/ICMLC54886.2021.9737248.

[2] H. Dai, "Research on SVM improved algorithm for large data classification," *2018 IEEE 3rd Int. Conf. Big Data Anal. ICBDA 2018*, pp. 181–185, May 2018, doi: 10.1109/ICBDA.2018.8367673.

[3] M. Nadeem *et al.*, "Multi-level hesitant fuzzy based model for usable-security assessment," *Intell. Autom. Soft Comput.*, vol. 31, no. 1, 2022, doi: 10.32604/IASC.2022.019624.

[4] F. A. Alzahrani, M. Ahmad, M. Nadeem, R. Kumar, and R. A. Khan, "Integrity assessment of medical devices for improving hospital services,"

Comput. Mater. Contin., vol. 67, no. 3, p. 3619, Mar. 2021, doi: 10.32604/ CMC.2021.014869.

[5] R. Tahboub and Y. Saleh, "Data leakage/loss prevention systems (DLP)," *2014 World Congr. Comput. Appl. Inf. Syst. WCCAIS 2014*, Oct. 2014, doi: 10.1109/WCCAIS.2014.6916624.

[6] W. Lusoli, M. Bacigalupo, F. Lupiáñez-Villanueva, N. N. G. de Andrade, S. Monteleone, and I. Maghiros, "Pan-European survey of practices, attitudes and policy preferences as regards personal identity data management." Jun. 04, 2012, Accessed: Jan. 21, 2023. [Online]. Available: https://papers.ssrn. com/abstract=2086579.

[7] A. Nadeem and M. Y. Javed, "A performance comparison of data encryption algorithms," *Proc. 1st Int. Conf. Inf. Commun. Technol. ICICT 2005*, vol. 2005, pp. 84–89, 2005, doi: 10.1109/ICICT.2005.1598556.

[8] S. Shorey and P. N. Howard, "Automation, big data and politics: a research review," *Int. J. Commun.*, vol. 10, pp. 5032–5055, Nov. 2016, Accessed: Jan. 21, 2023. [Online]. Available: http://ijoc.org.

[9] D. Quick and K. K. R. Choo, "Big forensic data reduction: digital forensic images and electronic evidence," *Cluster Comput.*, vol. 19, no. 2, pp. 723–740, Jun. 2016, doi: 10.1007/S10586-016-0553-1/TABLES/5.

[10] E. Yildirim, A. Aslan, and I. Ozturk, "Energy consumption and GDP in ASEAN countries: bootstrap-corrected panel and time series causality tests," *Singap. Econ. Rev.*, vol. 59, no. 02, p. 1450010, 2014, doi: 10.1142/s0217590814500106.

[11] C. C. Aggarwal, "Data classification: advanced concepts," *Data Min.*, pp. 345–387, 2015, doi: 10.1007/978-3-319-14142-8_11.

[12] C. C. Aggarwal, "Data classification," *Data Min.*, pp. 285–344, 2015, doi: 10.1007/978-3-319-14142-8_10/COVER.

[13] A. Chalumuri, R. Kune, and B. S. Manoj, "Training an artificial neural network using qubits as artificial neurons: a quantum computing approach," *Procedia Comput. Sci.*, vol. 171, pp. 568–575, 2020, doi: 10.1016/ j.procs.2020.04.061.

[14] C. C. Aggarwal, "Data classification," *Data Min.*, pp. 285–344, 2015, doi: 10.1007/978-3-319-14142-8_10.

[15] W. Wen and F. Mizoguchi, "Web security: authentication protocols and their analysis," *New Gener. Comput. 2001 193*, vol. 19, no. 3, pp. 283–299, 2001, doi: 10.1007/BF03037600.

[16] H. Alyami *et al.*, "Analyzing the data of software security life-span: quantum computing era," *Intell. Autom. Soft Comput.*, vol. 31, no. 2, 2022, doi: 10.32604/iasc.2022.020780.

[17] W. Alosaimi *et al.*, "Impact of tools and techniques for securing consultancy services," *Comput. Syst. Sci. Eng.*, vol. 37, no. 3, p. 347, Mar. 2021, doi: 10.32604/CSSE.2021.015284.

[18] R. Yoshida, M. Takamori, H. Matsumoto, and K. Miura, "Tropical support vector machines: evaluations and extension to function spaces," *Neural Networks*, vol. 157, pp. 77–89, Jan. 2023, doi: 10.1016/ J.NEUNET.2022.10.002.

[19] G. Chen, T. D. Bui, and A. Krzyzak, "Sparse support vector machine for pattern recognition," *Concurr. Comput. Pract. Exp.*, vol. 28, no. 7, pp. 2261–2273, May 2016, doi: 10.1002/CPE.3492.

[20] H. Luo and S. G. Paal, "A data-free, support vector machine-based physics-driven estimator for dynamic response computation," *Comput. Civ. Infrastruct. Eng.*, vol. 38, no. 1, pp. 26–48, Jan. 2023, doi: 10.1111/MICE.12823.

[21] K. D. Thangavel, U. Seerengasamy, S. Palaniappan, and R. Sekar, "Prediction of factors for Controlling of Green House Farming with Fuzzy based multiclass support vector machine," *Alexandria Eng. J.*, vol. 62, pp. 279–289, Jan. 2023, doi: 10.1016/J.AEJ.2022.07.016.

[22] U. S. Bist and N. Singh, "Analysis of recent advancements in support vector machine," *Concurr. Comput. Pract. Exp.*, vol. 34, no. 25, Nov. 2022, doi: 10.1002/CPE.7270.

[23] T. H. Le, C. P. Nguyen, T. D. Su, and B. Tran-Nam, "The Kuznets curve for export diversification and income inequality: evidence from a global sample," *Econ. Anal. Policy*, vol. 65, pp. 21–39, Mar. 2020, doi: 10.1016/j.eap.2019.11.004.

[24] I. K. Maji, "Impact of clean energy and inclusive development on CO_2 emissions in sub-Saharan Africa," *J. Clean Prod.*, vol. 240, p. 118186, Dec. 2019, doi: 10.1016/j.jclepro.2019.118186.

[25] K. Qi and H. Yang, "LS-GNHSVM: a novel joint geometrical nonparallel hyperplane support vector machine," *Expert Syst. Appl.*, vol. 215, p. 119413, Apr. 2023, doi: 10.1016/J.ESWA.2022.119413.

[26] J. Ren, Y. Yuping, and X. Deng, "Slack-factor-based fuzzy support vector machine for class imbalance problems," *ACM Trans. Knowl. Discov. Data*, Sep. 2022, doi: 10.1145/3579050.

Chapter 3

Securing Industry 5.0 data through unified emerging technologies

Archana Singh, Kavita Sahu, and Bineet Kumar Gupta

3.1 INTRODUCTION

Industry 4.0, sometimes referred to as the "fourth industrial revolution," is currently under development. This revolution simplified manufacturing structures and integrated critical technologies such as the Internet of Things (IoT), artificial intelligence (AI), cloud computing, and edge computing (EC) to establish smart manufacturing and increase efficiency. Industry 4.0 has fundamentally changed earlier iterations, when the only goals were to increase production and produce a large number of items at once.

Mechanical Industry 1.0, which began in the 1970s, is primarily driven by the production of steam and water for energy. Henry Ford invented the assembly line idea and used electricity to produce huge quantities of goods at low costs in 1870, which led to the creation of assembly line manufacturing in Industry 2.0. Industry 3.0 witnessed the transition from mechanical to digital manufacturing, and some industrial processes began to be partially automated. Large-scale computers and memory-programmable controller logic are used in industrial planning in this generation, which minimises the need for human labour.

There was a significant transition worldwide toward Industry 4.0 in 2011. Information and communication technologies (ICTs) are used in this generation to support remote manufacturing lines that are managed by means of assisted networks. The notion of the "smart factory" was introduced by the introduction of IoT-driven cyber–physical systems (CPSs), which allowed for self-organised manufacturing, self-monitoring, and networked logistics to maximise financial returns. The next revolution is Industry 5.0, which expands on Industry 4.0's complex cognitive process and increases interpersonal contact. As a consequence, it would be value-driven rather than process-driven. Figure 3.1 depicts a possible Industry 5.0.

Industry 5.0 is seen as a way to increase production quality by having robots handle tedious, repetitive tasks while freeing up people to focus on intelligent, critical-thinking tasks. As a result, a specialised market for skilled labourers would emerge. In Industry 5.0, robots will be operated by

DOI: 10.1201/9781003514312-3

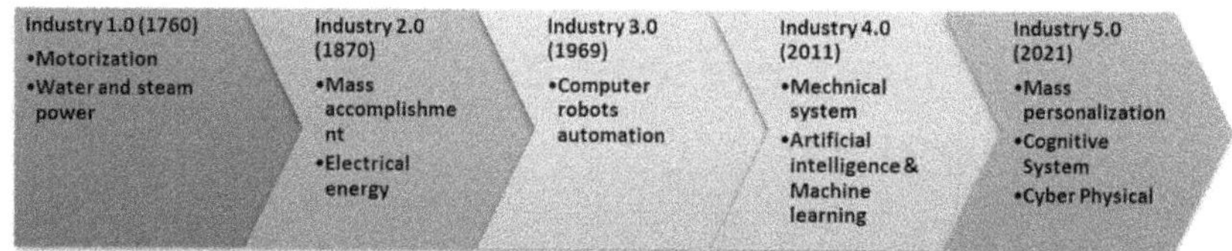

Figure 3.1 A Potential Transition towards the Industry 5.0 Environment.

humans, with a focus on personalisation. CPS and Industry 5.0 are closely related, because CPS envisions a collaborative interaction between people and robots. Collaborative robots (cobots) is a novel concept, which would serve as the process management industry's fundamental building block. Additionally, it would be more environmentally friendly, with an emphasis on sustainable development and manufacturing.

Automation is a must, and pipelines powered by AI would be backed by powerful reinforcement learning (RL), machine learning (ML), and deep learning (DL) models. However, it is essential to include the human aspect to manage the operations properly. Processes in an Industry 5.0 driven by CPS would exchange enormous amounts of data over the web via assisted wireless networking routes. To prevent hostile assaults that might compromise the concealment of conscious data, the processed data must be secured. Additionally, with specialised needs, communication nodes must maintain client information. Using equitable privacy policies and user-defined standards to obtain data with authorisation is a fundamental strategy for solving privacy challenges.

The supporting industrial technologies and issues of the previous industrial conventionality are discussed in a number of published books (i.e., Industry 4.0). The motivation of this research is to provide a review of emerging technologies and data security in these technologies of Industry 5.0. The remainder of the chapter is organised as follows: In Section 3.2, we describe the new aspects of Industry 5.0 in contrast to earlier industrial evolutions. Various data security issues in industries 4.0 and 5.0 are presented in Section 3.3. In Section 3.4, the prospective applications and data security in these applications of Industry 5.0 are covered. In Section 3.5, we highlight the discussions and future scope relating to the research. Finally, we conclude this chapter in Section 3.6.

3.2 INDUSTRY 5.0: EVOLUTION

The purpose of Industry 5.0 is to make industrial operations easier to understand by fusing human input with automated procedures. The autonomous

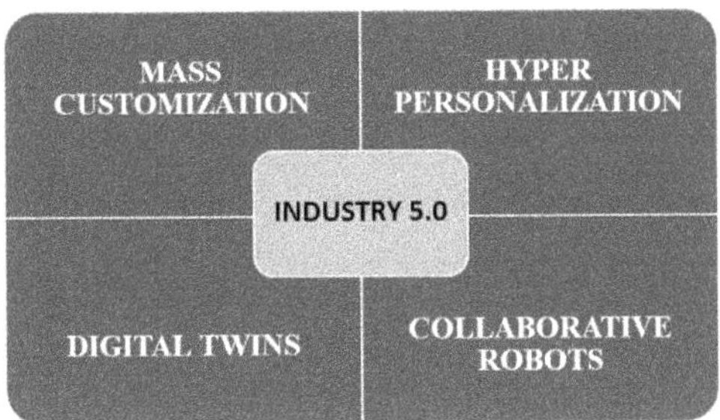

Figure 3.2 The Key Attributes of Industry 5.0.

workforce in Industry 5.0 is aware of human needs and intentions. As the former understands and engages with them, humans would be able to work closely with their robotic counterparts without feeling threatened. As a result, there would be reduced resource waste and a decrease in capital and operating costs as the process would be synergistic and in line with business models. Additionally, Industry 5.0 would help the green production movement by being self-sustaining and self-healing.

Industry 5.0 is built on simple yet powerful techniques like the 6R (recognise, reconsider, realise, recycle, reduce, and reuse) approach and the logic efficiency design (LED) philosophy, despite its complexity. The 3R (reuse, reduce, and recycle) approaches are a superstructure in the 6R methodology, which tries to include the non-functional approach into waste management design. The LED approach seeks to optimise operational efficiency by creating logic blueprints for industrial operations. The basic purpose of the LED concept is to increase supply chain profitability. The primary design technologies supporting Industry 5.0 aspirations are then covered. Our next discussion will focus on the major design technologies that will support Industry 5.0 aspirations. The key terms of industry 5.0 are mass customisation, digital twins (DT), hyper-personalisation, and collaborative robots, as discussed in Figure 3.2.

3.2.1 Mass customisation

The digital user experience for mass customisation would be at the heart of Industry 5.0 procedures. The parties develop the required set of essential functionalities for the product offering through extensive user interviews. The choice of parameters is influenced by user feedback, demographics, and design elements of rival goods. The demands of the end-user are met by these items. Therefore, mass personalisation enables businesses to reach decision-makers across sectors and improve client engagement.

3.2.2 Digital twins

Digital simulations of actual industrial settings or processes that are designed to precisely mimic the behaviour of physical objects are known as digital twins (DTs). Simulators of actual items operate similarly to DTs, but the key distinction between the two is the degree of industrial process scalability. In contrast to a simulator, which might only function with a restricted set of inputs, DT can perform several simulations to investigate various processes. Real-time data is sometimes important in simulations, in contrast to DTs, which are built on a two-way information dissemination where detectors input the system with information. With specified input sets, that data is processed using ML and DL models, and an estimation of the amount of inaccuracy is supplied with feedback to the controller. As DTs have richer and more current information than traditional simulations, they can examine complex themes and related relationships better.

3.2.3 Hyper-personalisation

The goal of hyper-personalisation is to create experiences, services, and products that are tailored to the interests of each individual client. To achieve hyper-personalisation, the industrial process must have a deep understanding of the generated goods, their readership, and the cost of the relevant technology.A tailored marketing plan is developed based on the knowledge gained, and digital technologies, particularly hyper-cognitive systems, ML and AI pipelines, and computer vision, which complement the effort by enabling industrial programs to choose a particular outcome for each consumer. Robots and human inputs work closely together during mass manufacturing to tailor the product. Priority should be given to the product's functional behaviour while also minimising costs. To expand manufacturing processes, the supply chain shifts from prototype design to agile manufacturing.

3.2.4 Collaborative robots

Cobots, also known as collaborative robots, are created to collaborate closely with people, adding a human touch to the production. Today's cobots are sensor-driven and respond to human input. Cobots are personalised in Industry 5.0 to provide a complete personalisation engine for a product. These are a vital component of the workforce in contemporary businesses, as they not only increase output but also have the ability to automate menial activities.

3.3 DATA SECURITY IN INDUSTRY 4.0 AND INDUSTRY 5.0

The European Union (EU) has announced a new industrial project in part as a reaction to coronavirus disease 2019 (COVID-19) and climate change issues. While the objectives of Industry 5.0, which can be summed up as the transformation of industry through digital technology, have been substantially adopted by the corporate sector, Industry 5.0 calls for rethinking industry around a human-centric approach. The Industry 5.0 initiative focuses on a number of important issues, mentioned as follows:

- Ensuring that digital technologies used in industry, such as robots and AI, are designed with human workers in mind
- Industry must be sustainable with a focus on enabling it.
- Industry needs to be built so that it can handle shocks like COVID-19 and climate change.

Industry 5.0, in contrast to Industry 4.0, is not only concerned with digital transformation. However, it is obvious that digital technology will play a significant role—and in places you would not expect. Decentralised digital technologies based on digital ledgers, such as these, are also being hailed as one key to a robust human-centred industrial economy. Distributed communications technologies gave rise to the internet, a long-term communication system.

Trust is essential for any industrial data flow, as is one clear reality. Trust is scarce in today's operational technology (OT) and information technology (IT) settings, as businesses struggle to combat the widespread epidemic of ransomware and other cybercrimes. For Industry 4.0, let alone Version 5.0, the current absence of a trust bridge between the OT and IT sectors must be urgently rectified. Workers must have confidence that any industrial AI application, such as robots, will not harm them and that the data fed into these processes is accurate and has not been manipulated or sourced incorrectly. The trust that this data will be handled securely will be essential as these technologies have the ability to capture potentially sensitive information about employees and citizens in general.

The majority of the data that will serve as the foundation for Industry 5.0 applications, as well as Industry 4.0, will originate from sensors and IoT devices. There are several cybersecurity methods in use today that can guarantee the reliability and security of this data. To prevent this, it is necessary to securely identify the devices that send data, and the Public Key Infrastructure, which powers most of the internet's current security, can be crucial in producing distinctive device IDs for authentication. To further safeguard the software operating on the devices, the devices and the software that runs on them may make use of a mix of technologies. Encryption and authentication can help ensure that the data stays the same as it moves through the different internet networks. Using this data will require cooperation from numerous partners for many Industry 5.0 applications. There are cloud solutions that can improve data trust and cooperation by monitoring data rights, controlling access, giving audit trails, and monitoring data and device claims. It is noteworthy, however, that Industry 5.0 is gaining ground at a time when widespread coverage, reduced latency, and data prices, made possible by and along with rising computing power, which is fuelled by cheaper data storage costs, will push many industrial uses toward EC. In fact, a key feature of allowing trustworthy Industry 5.0 applications is EC. Data journeys to the cloud will be dangerous, because many mission-critical applications will necessitate AI systems making decisions almost instantly. The aforementioned trust mechanisms will perform well in scenarios involving Industry 5.0 EC. However, as predicted by Industry 5.0, there will soon come a time when these applications will benefit from decentralised technology for performance, robustness, and efficacy. Web3 technologies and the security fixes for them are still in the early phases of evolution. One pattern stands out: many assaults target the software tools that work with the blockchain rather than the blockchain itself. For the time being, old, centrally managed apps and new, decentralised apps will have to coexist. Here too, the right application of the same security measures and other well-known technology should be helpful.

3.4 PRACTICAL UTILITY DATA SECURITY IN ENABLED TECHNOLOGIES OF INDUSTRY 5.0

During the deployments, Industry 5.0 will have serious security problems. A new threat vector will be created by Industry 5.0's usage of AI and supporting automation. For security reasons, trustworthy execution is crucial for AI/ML operations. For Industry 5.0 applications to function properly, the integrity of the dataset used for ML model development as well as the AI methodology must be maintained at all costs. Industry 5.0 will also need to fulfil security requirements discussed in this section including integrity, availability, authentication, and audit elements, similar to the traditional CPSs [1].

3.4.1 Authentication

To build mutual trust in the ecosystem, a huge variety of diverse stakeholders, including machines, IoT nodes, collaborative partner nodes, communication nodes, and fog nodes, must be authenticated. Industry 5.0 authentication methods need to be everywhere, easy to use for IoT nodes, and resistant to quantum computing so that they can be used with future quantum computing applications.

3.4.2 Integrity

As controlling instructions and analysing changes will be communicated via network interfaces, integrity is a crucial concern from the standpoint of data security in Industry 5.0. The integrity checks, however, must not have an impact on the system's performance aspects.

3.4.3 Access control

In the Industry 5.0 networks of the future, establishing authentication and authorization is an essential security measure to ensure that only authorised parties have access to delicate resources like intellectual property (IP) rights. In the majority of computer systems, it is difficult to set up access control measures as demand grows.

3.4.4 Audit

The ability to be audited is a key factor in determining if service operations are in line with regulatory compliance requirements. The audit records also mandate that you look into incidents of dispute settlement.

The massive interconnectivity foreseen in the subsequent Industry 5.0 systems must be supported by the data collection and management in Industry 5.0.

A multitude of enabling technological advancements, including DT, EC, Internet of Everything (IoE), blockchain, cobots, 6G and beyond, and big data analytics, may help businesses increase productivity and provide custom solutions more quickly when cognitive abilities and innovation are integrated. These enabling technologies have propelled the development of Industry 5.0, an enhanced manufacturing model that centres on human–robot interaction. As smart robots are designed to work alongside humans, human skills are now more productive and simpler to automate for both individuals and small businesses than ever before. Enabling technologies for Industry 5.0 are briefly covered in this section.

3.4.4.1 *Blockchain*

Industry 5.0 might greatly benefit from the utilisation of blockchain technology. A major issue in Industry 5.0 is the centralised operation of a sizable number of various or mixed networked devices. Building distributed and decentralised management platforms is possible using the blockchain, which facilitates distributed trust [2]. Blockchain technology offers a permanent ledger for recording information during secure peer-to-peer communication. The irreversible ledger also encourages operational responsibility and openness for the crucial things in Industry 5.0 applications. For the Industry 5.0 ecosystem's dispute resolution procedure in particular, transparency is essential. Future Industry 5.0 apps can employ smart contracts to ensure security through authentication and automated service-oriented operations. Additionally, a segmented and distributed strategy employing blockchains may be used to provide a better level of safety for data and transactions. Blockchain may also be used to facilitate data collection and receipt.

Blockchain might be used in Industry 5.0 to manage consumers more efficiently by giving different people and organisations digital identities. All industrial processes carried out via a public network necessitate access control and stakeholder identification. The management of assets, such as structures, products, and services, may also be done using these digital identities. Blockchain technology may be used to register and categorise IP rights. By automating the procedures for reaching agreements between various parties, blockchains and smart contracts can aid in the automation of the contractual process. Additionally, blockchain technology-based hardware connectivity and data exchange are made possible by cloud manufacturing provided by blockchain.

3.4.4.2 Edge computing

Edge computing, a revolutionary idea that allows data processing at the network edge, has emerged in response to the IoT's rapid expansion and the availability of multiple cloud services. Both the shift to Industry 4.0 and the subsequent Industry 5.0 may greatly benefit from EC. EC can meet your objectives for latency costs, data confidentiality, battery life limits, reaction time requirements, and privacy. It helps keep applications running well even in remote areas, while lowering the cost of connecting to the internet.

Furthermore, EC may handle data without sending it to a public cloud, which would ease security worries for critical Industry 5.0 events. A few of the useful jobs that EC may do are cache coherency, data processing, transferring, compute offloading, and request delivery.

To provide security, dependability, and privacy with all of these network processes, the edge must be properly constructed. Low latency, data security, and privacy are all guaranteed for Industry 5.0 applications by EC, which also provide effective services to users. EC enables future Industrial 5.0 applications to connect in real time, including autonomous vehicles, unmanned aerial vehicles (UAVs), and remote medical monitoring. Industry 5.0 can now access and share information with EC about their industrial sectors' use of technology and software that is becoming more widely used in the outside world.

Industries strive to routinely access data from local servers to manage massive volumes of data. The sheer volume of raw data generated by all of these technologies makes their analysis one of the most difficult tasks. Industry 5.0 may filter data because of EC as it reduces the amount of data that has to be transferred to a centralised server. Prophylactic analytics are allowed by EC in Industry 5.0, enabling the early detection of machine failure and its mitigation through the empowerment of human decision-making.

Edge computing brings processing close to datasets by utilising some of the most recent technological advancements, such as data transport, cloud computing, and privatisation. In this instance, data protection and privacy preservation have been the fundamental standards for safeguarding customers in the industry, economy, and actual activities. Additionally, we must acknowledge that security and protection will be discussed at every stage of designing EC systems.

3.4.4.3 Digital twins

A DT is a digital recreation of a tangible object or equipment. Real-world objects like wind farms, jet engines, factories, buildings, or even more complex systems like smart cities may be digitally simulated through the use of DT. Despite the fact that the concept of DT was initially proposed in 2002, it has only lately become a reality due to the growth of the IoT. Because of the IoT, DT is now affordable and available to many sectors [3].

Data from physical items is sent to their digital analogues via IoT devices for simulation. It is possible to avoid, analyse, and monitor issues before they materialize in the actual world through the digital mapping of legitimate devices and systems employing DT. DT has been able to maintain low maintenance costs and upgrade system execution because of the rapid growth of big data analytics, AI, and ML.

For the formation of specialised goods in the trades, improved business procedures, decreasing flaws, and quickly expanding creative business models to make money, DT may be very valuable in Industry 5.0. The DT may help Industry 5.0 address technical issues by more quickly identifying them, locating components that can be updated or modified depending on productivity, more accurately generating projections, identifying problems in the future, and removing significant financial losses. Organisations may now more rapidly and sequentially reap economic benefits because of this form of smart architectural design.

DT may be used in Industry 5.0 to create simulation models and access real-time computational data, allowing businesses to update and alter physical items from a distance. In Industry 5.0, DT is used to customise products in a way that improves the user's experience. This is done through a buying process that lets customers create virtual environments to try out different options.

3.4.4.4 Internet of Everything

IoE connects people, processes, information, and objects together. It can significantly assist in the formation of a new catalogue for Industry 5.0 applications. IoE advancements have the ability to bring forth new functionality, an enhanced user experience, and anticipated advantages for sectors of the economy and entire economies. The importance of IoE in Industry 5.0 is improving client satisfaction and loyalty and developing bespoke experiences based on IoE-generated data. By removing congestion from communication lines and lowering latency, the use of IoE in Industry 5.0 offers the chance to cut operational expenses. For Industry 5.0, efficiency in the logistics and supply chain is a difficult problem. IoE will streamline production procedures and cut down on supply chain waste. Since the IoE has advanced so much, humans now share information wirelessly, mostly with the aid of wireless sensors.

3.4.4.5 6G and beyond

6G will eventually be able to provide Industry 5.0 with considerable value-added services. With a densely packed network of tens of billions of sensors, embedded systems, and robots, radio industrialisation is complicated. It will be hard for existing networks to fulfil the fast-rising bandwidth needs due to the robust expansion of smart infrastructure and projected applications (such as 4G and 5G networks). In the Industry 5.0 revolution, employing 6G and even beyond may produce better latency, greater services, widespread IoT networks, and unified AI capabilities.

By offering intelligent spectrum management, mobile EC driven by AI, and smart mobility, 6G networks contribute to the efficient and effective improvement of application administration in Industry 5.0 applications. According to projections, 6G networks will be able to support Industry 5.0 applications that require ultra-low latency, ultra-high dependability, ultra-high data rates, high traffic capacity, high energy efficiency, etc. The biggest problems facing 6G networks in Industry 5.0 are handover management and mobility. The 6G networks will have several layers and be very dynamic, leading to frequent handovers. To ensure effective connectivity, AI approaches may be employed to generate the best mobility forecasts and handover solutions [4]. Providing a high bandwidth for various applications is the biggest difficulty facing Industry 5.0 services. These problems can be resolved via 6G free-space optical communication and quantum communication. Energy economy is a crucial issue in Industry 5.0 applications as there are many linked smart devices that consume excessive amounts of energy. Through the application of cutting-edge energy-expending tactics and energy harvesting techniques, 6G networks maximise energy management.

3.4.4.6 Collaborative robots

Robotics and automation technology advancements have made working with robots more important than ever. As a consequence of the incredibly swift developments in AI and smart technology, which have also given rise to a new field of research known as "cobots," it is evident that devices with computer potential have become more intelligent. Collaborative robots are those that have been designed to work with humans. Through this cooperation, human skills are becoming more effective and simpler to integrate for both individuals and modest enterprises. Robots frequently contain sensors that can stop on their own when human operators spot any misplaced objects in their path, as they are extremely responsive to the detection of an unplanned hit. Compared to traditional industrial robots, they are often far more dependable in terms of workplace safety.

Robots are far more tolerant than people and are better at producing large quantities of goods. In terms of critical thinking ability, robots fall short of humans. Customisation or personalisation of products may present a big challenge when utilising robots. As a result, it is critical to manage human interactions within production processes. Cobots have a lot to offer in Industry 5.0. Robots can accomplish their intended task while working in concert with humans, making it possible to quickly and accurately supply highly customised and personalised items to clients. In Industry 5.0, personalising cobots can take many various forms, such as providing smart applications, medical treatments that accurately summarise a patient's optimal health, and medical requirements to produce an entirely distinct workout regimen and health. One of the uses for cobots is in surgery, where a skilled surgeon and a robotic assistant collaborate to complete the procedure. Medical facilities currently profit from collaborative industrial processes assisted by robotics.

Cobots are used to boost productivity in Industry 5.0 and support the development of a new kind of interaction between people and machines. Cobots aid in enhancing performance and safety in Industry 5.0 applications while also giving human workers more inviting tasks and boosting productivity. In markets where there is a lot of competition, industries need to realise that cobots can help improve operational performance and save money on labour costs.

3.4.4.7 Big Data analytics

The importance of big data analytics is frequently highlighted in Industry 5.0. Some companies in Industry 5.0 may employ big data analytics to better understand consumer behaviour, which may subsequently be used to focus on improving production efficiency, optimising product costs, and helping to reduce overhead expenses. A vital problem is appreciating user behaviour as it exists now, social relationships, and human behavioural norms. The conquest of the Industry 5.0 ecosystem depends on highly tailored manufacturing processes, the injection of data, and intelligent automation in the production process. Big data analytics may be used to make decisions in real time to boost industrial competitive advantage, with an emphasis on delivering advice on predictive discoveries for major events in Industry 5.0 applications. Big data analytics supports Industry 5.0's mass customisation processes with resources that integrate flawlessly. Because real-time analytical data is exchanged with intelligent systems and data centres, manufacturers may produce and organise large amounts of data. The literature review is being carried out, revealing various enabling technologies and their model for data security in Industry 5.0, shown in Table 3.1.

Table 3.1 Enabling Technologies with Their Models for Data Security in Industry 5.0

S. no.	Author	Model	Enabled technologies
1	Sharma et al. [5]	The hybrid architectural scheme (the core and the edge)	Blockchain (scalability difficulties,bandwidth, latency, and security)
2	Biswas et al. [6]	Decentralised four-layer blockchain (BC) security strategy	Blockchain (database, interface, physical, and communication)
3	Rivera et al. [7]	A digital identification model	Blockchain: a digital centre that links the government, the business, and institutions of higher learning
4	Liao et al. [8]	Four light-weight procedures, three-layered, BC-based lottery system	Blockchain, initialisation, purchaseand validating winning number, and close time
5	Liang et al. [9]	Permissioned BCs and a mobile-controlled hyperledger fabric layout	Blockchain, Edge computing
6	Samaniego et al. [10]	Network latency issues and data transfer issues	Cloud-centric IoT systems
7	Khan et al. [11]	Using a model IoT security demands and barriers broken down into categories	IoT and Big Data analytics
8	Singh et al. [12]	Three distinct IoT security patterns based on BC	IoT and blockchain
9	Reyna et al. [13]	Scalability, storage capacity, data privacy, and anonymity	IoT and big data
10	Li et al. [14]	To promote confidence between customers and service providers and to enable secure service and data sharing, BV*mfg* developed a five-tier architecture.	Smart manufacturing, cobots
11	Castelló Ferrer [15]	New frameworks for decision-making, behaviour differentiation, security, and commerce	Cobots

Table 3.1 *(Cont.)*

S. no.	Author	Model	Enabled technologies
12	Bhattacharya et al. [16]	Combining security and management mechanisms (concept)	Edge computing
13	Gupta et al. [17]	Enormous data processing and support per second (isolation and independently)	5G and 6G services
14	Bodkhe et al. [18]	Method based on consensus (sensible network security transmit data using wireless networks.)	Cobots and IoT
15	Xiong et al. [19]	Intelligent greenhouse farms using BC-based architecture in which IoT sensors communicate data to a private BC that is managed by a single authority	IoT and blockchain
16	Tenopir et al. [20]	Methods to strike a balance between user experience and privacy	Big data Analytics
17	Kanagavelu et al. [21]	Using federated learning, it is possible to mine data that is scattered over several websites.	Big data and Edge computing
18	Shrestha et al. [22]	Tests showing Google's alarming data collection practices	Big data analytics
19	Iyilade et al. [23]	Multi-application life log exchange using a networked architecture	Big data and supply chain management
20	Dim et al. [24]	Utilising a distributed sharing model with a single independent agent	Big data management
21	Stoica et al. [25]	Chord, structured P2P system	Decentralised cloud and blockchain
22	Caro et al. [26]	Centralised cloud infrastructures	Supply chain management
23	Tian [27]	*AgriBlockIoT*, a decentralised SPM approach based in BC	Blockchain and IoT
24	Leong et al. [28]	Hazard analysis and critical control points (HACCP),	Supply chain management and big data

According to the state of the art, a range of applications benefit greatly from Industry 5.0's adoption of enabling technical developments, including DT, EC, IoE, cobots, big data analytics, blockchain, and 6G. All of these supporting technologies combine with creativity and cognitive abilities to help businesses boost productivity and provide customised goods more quickly.

3.5 DISCUSSION

Industry 5.0 research that integrates big data, AI, and IoT was outlined in [29]. The future of many fields, including a digital drug to track actual medication adherence, can be found in the construction of complex, safe, and highly interconnected networks. IoT, AI, and Industry 5.0 will all be helpful for the accumulation of big data across numerous digital settings. A value-sensitive design for smart factories was published by Aslam et al. [30]. The future industry that relies on human–machine synergy and its prospects are discussed. Furthermore, industrial systems engineering addressed the ethical implications of machines on human employees.

To meet the specifications of Industry 5.0 as well as the IoT, managers should develop a deeper awareness of business strategy, the innovation ecosystem, and design thinking, according to the Absolute Innovation Management Framework created by Javaid and Haleem [31]. However, the report does not discuss how the framework and safety measures would be implemented after they were integrated with business operations. Kotianová [32] conducted a study of Industry 5.0's essential elements and advantages over its predecessor. According to the report, Industry 5.0 enables smart production by utilising data intelligently and connecting various industrial data and cutting-edge technology to create more customised goods.

Because of a cognitively enabled manufacturing process, customers may have access to the most personalised services through Industry 5.0. Some of the implementation issues that may arise, which are covered in this section, must be resolved for seamless services. Data security is one of the biggest potential difficulties. As computing becomes more digitalised, we must double-check security flaws in data handling and the use of cloud services for diverse users and industrial system integration. When providing customers with customised and more predictive services, confidentiality in data transactions, privacy in content aggregation, and ethical concerns must all be taken into account. Human–robot coworking must take into account the challenges associated with upgrading users and manufacturing processes to provide individualised customer service. To minimise potential downsides and detrimental influence on society, the ethical concerns associated with AI adoption must also be taken into account.

A new threat vector will be created by Industry 5.0's use of supporting automation and AI. For security-related reasons, trusted execution is crucial for AI and ML functions. For Industry 5.0 applications to function as intended, it is essential to protect the authenticity of the content source used for ML model training and the AL approach. For instance, Industry 5.0 tenants should safely exchange observable data to train AI models or update models incrementally, as in federated learning. Deploying proactive security procedures [33] and mitigating zero-day attacks are two examples of the increased security needs that will result from Industry 5.0 applications' strong reliance on ICT systems. Additionally, Industry 5.0 may function in the quantum computing age as a result of the advancement of quantum computing. In such circumstances, Industry 5.0 systems should make use of post-quantum encryption or cryptography that is quantum-resistant to ensure the necessary level of security.

Privacy is a crucial need for Industry 5.0 applications as the entire eco-system relies on pricey IP, expensive manufacturing inputs, and subscription management. To connect people and machines, designers and other partners, and to provide monitoring and control information, data is delivered via the internet in Industry 5.0. To maintain the confidence of the cloud manufac-turing ecosystem, such data must not be accessible to hackers online. One of the privacy concerns is the concept of human data protection rights, or the idea that individuals own their data. They are entitled to compensation for any data theft that affects their personal or sensitive data. To protect the privacy of user data, it must be kept private while it is being used for cogni-tive analysis to predict maintenance needs.

The system's resilience, adaptability, and responsiveness in the face of dynamic workload changes are all examples of scalability. In Industry 5.0, scalability refers to a system's capacity to operate effectively under a range of settings as the number of hyperconnected, interconnected systems in the network grows or shrinks. Industry 5.0 aims to link and communicate with multiple systems from different factories as well as countless individuals. Scalability, a feature of Industry 5.0, which is an improvement over Industry 4.0, is one of its challenges, but it becomes more serious when humans and machines are made to work together as a team.

3.6 CONCLUSION

In this study, we provide a research walkthrough on Industry 5.0's enab-ling technologies and data security in these technologies. We started our study by outlining a few Industry 5.0 tenets from both the educational and commercial sectors' perspectives. Following is a description of the pri-mary Industry 5.0 enabling technologies. We started our study by outlining a few Industry 5.0 tenets from both the academic and business sectors' perspectives. Following the overview of the main Industry 5.0-enabling technologies (blockchain, cobots, IoE, EC, DT, etc.), in conclusion, the idea of "Industry 5.0" was developed to continually align the productivity of workers and machines. Industry 5.0 is expected to upgrade consumer con-tentment, industrial production, and data security. It is made possible by numerous innovative applications and supporting technologies. We also discussed a number of difficulties and unresolved problems that need to be addressed to fully actualise the idea of Industry 5.0 in the near future, including human–robot collaboration in a factory, skilled labour, reliability, adaptability, and confidentiality.

REFERENCES

[1] Porambage, P., Gür, G., Osorio, D. P. M., Liyanage, M., Gurtov, A., & Ylianttila, M. (2021).The roadmap to 6G security and privacy. IEEE Open Journal of the Communications Society, 2, 1094–1122.

[2] Arastouei, N. (2023). 6G technologies: key features, challenges, security and privacy issues. In: Skarmeta, A., Canavese, D., Lioy, A., Matheu, S. (eds.), Digital sovereignty in cyber security: New challenges in future vision. CyberSec4Europe 2022. Communications in computer and information science, vol 1807. Springer, Cham. https://doi.org/10.1007/978-3-031-36096-1_7

[3] Jiang, Z., Guo, Y., & Wang, Z. (2021). Digital twin to improve the virtual-real integration of industrial IoT. Journal of Industrial Information Integration, 22, 100196.

[4] Yang, H., Alphones, A., Xiong, Z., Niyato, D., Zhao, J., & Wu, K. (2020). Artificial-intelligence-enabled intelligent 6G networks. IEEE Network, 34(6), 272–280.

[5] Sharma, P. K., & Park, J. H. (2018). Blockchain based hybrid network architecture for the smart city. Future Generation Computer Systems, 86, 650–655.

[6] Biswas, K., & Muthukkumarasamy, V. (2016, December). Securing smart cities using blockchain technology. In 2016 IEEE 18th international conference on high performance computing and communications; IEEE 14th international conference on smart city; IEEE 2nd international conference on data science and systems (HPCC/SmartCity/DSS) (pp. 1392–1393). IEEE, Sydney, NSW, Australia,.

[7] Rivera, R., Robledo, J. G., Larios, V. M., & Avalos, J. M. (2017, September). How digital identity on blockchain can contribute in a smart city environment. In 2017 International smart cities conference (ISC2) (pp. 1–4). IEEE, Wuxi, China.

[8] Liao, D. Y., &Wang, X. (2017, October). Design of a blockchain-based lottery system for smart cities applications. In 2017 IEEE 3rd international conference on Collaboration and Internet Computing (CIC) (pp. 275–282). IEEE, San Jose, CA.

[9] Liang, X., Zhao, J., Shetty, S., Liu, J., & Li, D. (2017, October). Integrating blockchain for data sharing and collaboration in mobile healthcare applications. In 2017 IEEE 28th annual international symposium on personal, indoor, and mobile radio communications (PIMRC) (pp. 1–5). IEEE, Montreal, QC, Canada.

[10] Samaniego, M., & Deters, R. (2016, December). Hosting virtual IoT resources on edge-hosts with blockchain. In 2016 IEEE international conference on computer and information technology (CIT) (pp. 116–119). IEEE, Nadi, Fiji.

[11] Khan, M. A., & Salah, K. (2018). IoT security: Review, blockchain solutions, and open challenges. Future Generation Computer Systems, 82, 395–411.

[12] Singh, M., Singh, A., & Kim, S. (2018, February). Blockchain: A game changer for securing IoT data. In 2018 IEEE 4th World Forum on Internet of Things (WF-IoT) (pp. 51–55). IEEE, Singapore.

[13] Reyna, A., Martín, C., Chen, J., Soler, E., & Díaz, M. (2018). On blockchain and its integration with IoT. Challenges and opportunities. Future Generation Computer Systems, 88, 173–190.

[14] Li, Z., Liu, L., Barenji, A. V., & Wang, W. (2018). Cloud-based manufacturing blockchain: Secure knowledge sharing for injection mould redesign. Procedia Cirp, 72, 961–966.

[15] Castelló Ferrer, E. (2018, November). The blockchain: A new framework for robotic swarm systems. In Proceedings of the future technologies conference (pp. 1037–1058). Springer, Cham.

[16] Bhattacharya, P., Tanwar, S., Shah, R., & Ladha, A. (2020). Mobile edge computing-enabled blockchain framework—a survey. In Proceedings of ICRIC 2019 (pp. 797–809). Springer, Cham.

[17] Gupta, R., Reebadiya, D., & Tanwar, S. (2021). 6G-enabled edge intelligence for ultra-reliable low latency applications: Vision and mission. Computer Standards & Interfaces, 77, 103521.

[18] Bodkhe, U., Tanwar, S., Bhattacharya, P., & Kumar, N. (2022). Blockchain for precision irrigation: Opportunities and challenges. Transactions on Emerging Telecommunications Technologies, 33(10), e4059.

[19] Xiong, H., Dalhaus, T., Wang, P., & Huang, J. (2020). Blockchain technology for agriculture: Applications and rationale. Frontiers in Blockchain, 3, 7.

[20] Tenopir, C., Palmer, C. L., Metzer, L., van der Hoeven, J., & Malone, J. (2011). Sharing data: Practices, barriers, and incentives. Proceedings of the American Society for Information Science and Technology, 48(1), 1–4.

[21] Kanagavelu, R., Li, Z., Samsudin, J., Hussain, S., Yang, F., Yang, Y., ... Cheah, M. (2021). Federated learning for Advanced Manufacturing based on industrial IoT Data Analytics. In Implementing Industry 4.0 (pp. 143–176). Springer, Cham.

[22] Shrestha, A. K., Vassileva, J., &Deters, R. (2020). A blockchain platform for user data sharing ensuring user control and incentives. Frontiers in Blockchain, 3, 497985.

[23] Iyilade, J., &Vassileva, J. (2013). A decentralized architecture for sharing and reusing lifelogs. In UMAP Workshops. Springer-SBM. https://ceur-ws.org/Vol-997/llum2013_paper_1.pdf.

[24] Dim, E., &Kuflik, T. (2012). User models sharing and reusability: a component-based approach. In 20th Conference on User Modeling, Adaptation, and Personalization, UMAP 2012, 872. Montreal, QC, Canada.

[25] Stoica, I., Morris, R., Karger, D., Kaashoek, M. F., Balakrishnan, H., &Chord, A. (2001). A scalable peer-to-peer lookup service for internet applications. SIGCOMM Comput. Commun. Rev. 31(4), 149–160. https://doi.org/10.1145/964723.

[26] Caro, M. P., Ali, M. S., Vecchio, M., &Giaffreda, R. (2018, May). Blockchain-based traceability in Agri-Food supply chain management: A practical implementation. In 2018 IoT vertical and topical summit on agriculture—Tuscany (IOT Tuscany) (pp. 1–4). IEEE, Tuscany, Italy.

[27] Tian, F. (2017, June). A supply chain traceability system for food safety based on HACCP, blockchain & Internet of things. In 2017 International conference on service systems and service management (pp. 1–6). IEEE, Dalian. DOI: 10.1109/ICSSSM.2017.7996119

[28] Leong, Y. K., Tan, J. H., Chew, K. W., & Show, P. L. (2020). Significance of industry 5.0.In P. L. Show, K. W. Chew, & T. C. Ling (Eds.), The prospect of Industry 5.0 in biomanufacturing (Ch. 2.2, pp. 1–20). CRC Press, Boca Raton, FL.

[29] Longo, F., Padovano, A., & Umbrello, S. (2020). Value-oriented and ethical technology engineering in industry 5.0: A human-centric perspective for the design of the factory of the future. Applied Sciences, 10(12), 4182.

[30] Aslam, F., Aimin, W., Li, M., & Ur Rehman, K. (2020). Innovation in the era of IoT and industry 5.0: Absolute innovation management (AIM) framework. Information, 11(2), 124.

[31] Javaid, M., & Haleem, A. (2020). Critical components of Industry 5.0 towards a successful adoption in the field of manufacturing. Journal of Industrial Integration and Management, 5(3), 327–348.

[32] Kotianová, Z. (2019). Aspects of safety and security in Industry 4.0. Industry 4.0, 4(6), 319–321.

[33] Bilge, L., & Dumitraş, T. (2012, October). Before we knew it: An empirical study of zero-day attacks in the real world. In Proceedings of the 2012 ACM conference on computer and communications security (pp. 833–844). ACM, Raleigh, NC. https://doi.org/10.1145/2382196.2382284

Chapter 4

Blockchain technologies
Shaping futuristic smart countries

Abhay Kumar Yadav and Virendra P. Vishwakarma

4.1 INTRODUCTION

A blockchain is a distributed, decentralized network that records all peer-to-peer (P2P) transactions. In a blockchain, records are securely stored across many interconnected systems making blockchain technology more secure than a centralized system. A blockchain creates a permanent, verifiable, and unalterable ledger by validating transactions across a distributed network. Furthermore, every "block" in the blockchain is uniquely correlated with the previous block's hash function by means of digital signatures. Hence, making modifications in the chain difficult results in enhanced security of stored information.

A key characteristic of blockchain technology is its ability to permit users to securely transfer data assets in a decentralized manner, removing the necessity of centralized third parties [1]. Additionally, blockchain provides enhanced security features by using cryptographic methods of data exchange. This makes it ideal for storing sensitive data, such as financial transactions and medical records, that requires enhanced security measures.

Blockchain technology can be used as a distributed electronic database encrypted by different cryptography functions thus eliminating the limitations of traditional systems. Traditional systems use a centralized authentication system making it difficult for various users to access the database due to limited correct credential access and server capacity. Blockchain is similar to a distributed ledger, with different transactions encrypted together and, in the chain, and are permanent. The data is secured by using a linked list based on hash functions and not on pointers. The hash function generates a hash value which is used for encryption of input data. Each block is connected with previous blocks by a digitally verified signature causing difficulty in changing records, making blockchain immutable.

Governments are working on providing an ecosystem where blockchain would work as a central technology in providing all citizen-related services related to economic, social, legal, and environmental aspects. Various advantages of blockchain technologies are described in Figure 4.1,

DOI: 10.1201/9781003514312-4

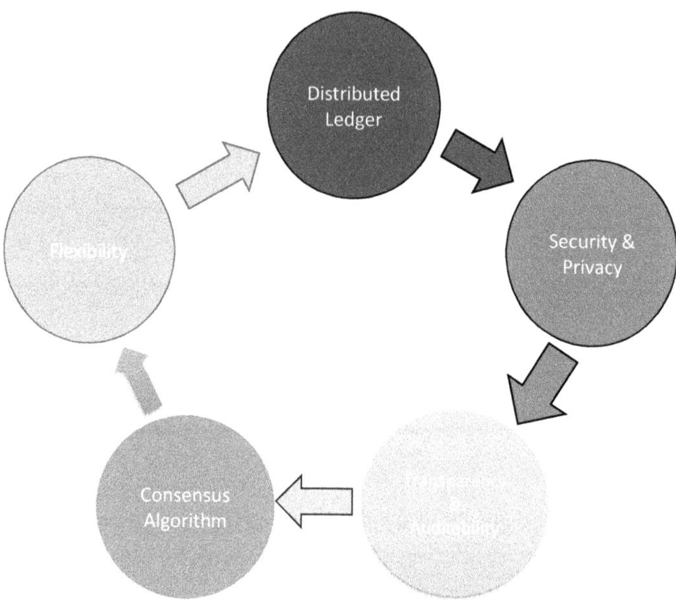

Figure 4.1 Various Advantages of Blockchain.

including flexibility, security and privacy, transparency and audibility as critical components. Anything having the possibility of being owned may be considered an asset. Any assets whether having substantial or insubstantial values can be converted easily into cash. A client in blockchain may include a customer, supplier, business network, government, or any regulatory body residing in an organization. A contract is a set of conditions using which any transaction can take place in blockchain.

4.1.1 Blockchain structure

Blockchain is a P2P-based network that uses cryptographic methods for providing security using digital signatures. The decentralized network allows smart devices to communicate securely among themselves. Each block of transactions on the blockchain is connected with the previous block's hash by a hash function and every block of transactions on blockchain is arranged in a linear manner.

A blockchain block consists of transaction data, digital signature of the previous block, and nonce. A miner searches for an eligible signature by frequently changing the nonce. Miners require computational power for continuously trying different nonces which is supplied by electricity consumptions. The higher the computational power; the faster their speed of inserting nonces randomly in turn increasing the probability of finding eligible signatures in less time. Hence, blockchain is an efficient combination of many reliable technologies implemented together for providing higher security to all digital data and assets over the network.

4.1.2 Blockchain components

In blockchain, various components work together to provide different services to its users.

4.1.2.1 Ledger

Basically, blockchain is just a collection of fixed digital ledgers somehow similar to a standard ledger. The ledger is useful in tracking and managing ownership of different assets. Blockchain is a record-keeping mechanism for different kinds of data. Even though blockchain is being presented as an innovative technology, it is just a combination of earlier present concepts, techniques, and methodologies in a much more innovative way [2].

4.1.2.2 Cryptography

In a blockchain, various advanced cryptographic techniques are used to maintain anonymity, enhancing immutability of the distributed ledger and validating claims that users claims against the tracked assets and are managed on the blockchain. The chain of different blocks is implemented together and executed in a linear fashion with the help of a particular function "cryptographic hash." These hash functions return a unique identifier or output for any given input. Any attempt to manipulate data will lead to the creation of a newer hash or ID that will be completely different from the original recorded value of the chain next block. Altering any data causes the creation of a completely new hash function and data associated with this newly created hash will be different than the hash function of the next block leading to blockchain breakage and all data linked with blockchain will be invalid making blockchain tamper proof and censorship-resistant [3].

4.1.2.3 Peer-to-peer network

Blockchain uses existing P2P network architectures for information sharing. P2P networks work as the underlying of the modern internet backbone. P2P network increases fault tolerance and redundancy by eliminating single points of vulnerability issues in centralized client/server architectures [1].

4.1.2.4 Assets

An asset is simply an item that is being tracked, a piece of information that is relevant to a particular use case or solution. Anything requiring the record of ownership is termed as assets. It can be either monetary or non-monetary in nature, or maybe a collection of data records such as healthcare, event tickets, or any patent information. Initially, blockchain began keeping records on any system by recording the flow of transactions with the help of tokens or coins. These tokens and coins require keeping a record of digital ownership leading to the creation of blockchain [4].

4.1.2.5 Consensus algorithms

Assuring nodes on the network accept transactions in line with the pattern and existence on a ledger is the concept of consensus. In cryptocurrency, consensus is very necessary for preventing issues like double spending issues or invalid data on the ledger database. Using consensus algorithms, there is a possibility of providing many solutions in different situations. While considering using any consensus algorithm, the calculation of opportunity cost is needed to be considered [5]. It is the way in which the delegation and rewards of validation are provided that makes different consensus mechanisms stand out from one another. Proof-of-work (PoW) and proof-of-stake (PoS) are mostly used consensus mechanisms. Many more consensus mechanisms, implemented in permissioned and private blockchains, such as hyperledger, which do not need high computational consensus mechanisms, are present. There are many more efficient consensus alternatives. Here, we have summarized a few consensus algorithms.

1. **Proof of Work:** Cryptocurrencies such as bitcoin execute Byzantine Fault Tolerance for providing a validation consensus called PoW. A network can be fault-tolerant by allowing two nodes to communicate with each other by using Byzantine Fault Tolerance by displaying similar types of data even though any other user in the network maliciously harms the network [6]. The Byzantine Generals Problem can be solved only when two-thirds of the nodes of the network in the blockchain are honest. In PoW consensus, whenever any single block is validated, each and every node in the network begins to compete to solve a guessing game problem with the purpose of providing validation to a data block. The problem is generally non-computational in nature and can easily be solved by random guesses efficiently. The nodes are termed as miners, and each miner tries to randomly predict the part of data called "nonce" for succeeding in block validation [6].

2. **Proof of Stack:** PoW had some limitations related to scalability and cost factors. A newer blockchain consensus system working similarly to PoW consensus for overcoming its drawbacks is PoS. PoS completely eliminates the need for guessing games for block validation removing the need for powerful and specialized hardware devices. PoS also reduces overall energy consumption. PoS works on "validator" nodes that can provide or accept a stake for validating transactions. In group consensus, all participating nodes lock up funds. Then, a node is randomly chosen, and the hash function of the node's is displayed to participating nodes. If a majority of nodes believe in the authenticity of a randomly selected node, that random node is remunerated along with all those who speculate on the random node [7].

3. **Proof of Activity:** This utilizes the abilities of PoW and PoS together for achieving consensus. By mining PoW templates, many template blocks that were initially empty can be filled by PoS transactions.

4. **Proof of Burn:** PoB is the method of elimination of cryptocurrencies. The coins are "burned" by moving them to a place from which they can't be accessed again. As the number of coins burned increases, the chances are higher for them to be again selected as miners for the next block.

5. **Proof of Capacity:** In PoC, all the hard drive space is completely used together for participation. As long as more spaces are "staked," the possibility of being selected for mining the next block increases. The consensus mechanism creates a large dataset termed as "plots" which result in higher storage consumption.

6. **Proof of Elapsed Time:** PET has characteristics similar to PoW but with more energy efficiency. It was created by Intel for running on Intel's own execution environment. This consensus requires high Intel trust and views Intel as the centralized permitting authority.

7. **Proof of Authority (PoA):** PoA works on the collection of "authorities" which are the general working nodes with the permission of creating new blocks and providing security to the Blockchain. PoA serves as a replacement for PoW but is restricted only to the private blockchain. The nodes involved need to secure permission for becoming a validator/authority. money and other items.

4.2 REVIEW OF LITERATURE

From last 10 years, many countries have switched over to blockchain for efficient management of government resources. This section discusses different countries' progress that has been made in implementing blockchain in smart governance. For simplicity, it has been arranged in order of continents.

4.2.1 Blockchain in African continents

The popularity of blockchain technology is increasing every day over the globe. Fortune Business Insights reports predict that the global blockchain market will grow from $27.84 billion in 2024 to $825.93 billion by 2032 [3, 5–7]. This study explores the usage of blockchain technology in African, American, and Asian continents, which are mentioned in Table 4.1, Table 4.2, and Table 4.3, respectively.

Rwanda has proposed a Land Title Registration model using blockchain, in partnership of Wise Key, Switzerland-based cybersecurity and IoT

Table 4.1 Blockchain Technology in African Continent

References	Country Name	Area of Application	Implementation
[8]	Rwanda	Land registry	The land registry would completely be digitized by blockchain with the use of WiseID apps of WiseKey's and Microsoft's Azure cloud. WiseKey has also helped in setting up Blockchain Center of Excellence in Kigali.
[9]	Tanzania	Unbanked	Humaniq provided a real-time access to unbanked transaction for Tanzania citizens. The Humaniq app can be accessed on any low-end mobile devices and support peer-to-peer(P2P) transactions
[10]	Uganda	Public services	Uganda government is promoting blockchain implementation in public sector enterprises.
[11]	Nigeria	Cryptography Development Initiative in Nigeria (CDIN) initiative	CDIN initiative was proposed by active blockchain community for consistent growth of blockchain. New blockchain startups are being established on large scale.
[12]	Kenya	BitHub Africa	BitHub provides consultancy services to industries interested in implementing blockchain solutions across African and Middle Eastern countries.
[13]	Sierra Leone	Elections	General elections of March 2018 were conducted by blockchain for ensuring transparency and fair counting of electoral votes.
[14]	Botswana	Daily agricultural production and stock	PLAAS is a mobile application which acts as a free platform that allows co-operatives as well as individual farmers real-time monitoring of their daily goods production and remaining stock on blockchain.

Table 4.2 Blockchain Technology in USA

References	Area of Application	Implementation
[15]	PokitDok, Healthcare	Easier identification and claims resolving, along with real-time verification and validation in medical treatment or medicines Distribution.
[16]	Patientory, Healthcare Solutions	Proper patient access management by connecting patients with the suitable care team for proper treatment eliminating the delay in choosing the doctors suitable for patients as per the symptoms.
[17]	Doc.AI, Healthcare	Healthcare data management: The medical data of patients are collected and are encrypted by the blockchain, Subsequently, the data is embedded with efficient machine learning algorithms. This would provide a medical forecast of the patient health condition and would enable doctors to start treating accordingly.
[18]	*Coral Health,* Healthcare	Build policies and platforms where different patients could upload their medical records along with their personal views with different stakeholders in a secure network providing personalized platform.
[19]	Akiri, Healthcare	Provided an architecture where all data would be temporarily stored till the time it is required. It helped in establishing trust by providing a fully secure platform with proper authentication and identification
[20]	Exochain, Healthcare	Provide individuals patient data with feature of controlling privacy over it. The public data shared by patients will serve as clinical trial data. The researchers may easily access patient medical data. This will lead to enhanced quality and quantity of patients accepted for clinical trials, with feature of giving individual's access control over his personal medical information.
[21]	Curisium, Healthcare	Patient-centric value-based contracts are enabled by the use of computation technologies secured by clients, service providers, and life science companies.
[22]	Healthcombix, Healthcare	Providing decentralized payment portals and providing confidentiality in data asset management in newer healthcare ecosystems.
[23]	Healthcare	Establishing trustworthy communication by linking patients and medical research together; patients can also permit researchers to access their data directly.
[13]	Stock Ownership	The American state Delaware in July 2017 permitted stock trading using blockchain. It officially passed a bill drafted for eliminating the gaps in the legislative framework, legalizing and regulating the blockchain in Delaware.

Table 4.3 Blockchain Technology in Asian Continent

References	Country Name	Area of Application	Implementation
[25]	China	Public Sector	Providing initiatives to startups, promoting all cryptocurrencies. It is the leading country in area-wise deployment of blockchain
[26]	Japan	Retail Industry	Many retail enterprises accept blockchain as a legal substitute to real currency. Bitcoin trading is highly popular among Japanese as a substitute to digital payment.
[27]	Russia	Capital Market	Secure trading and transactions among stakeholders. Provided digital interface for blockchain technologies where different applications be created and executed.
[28]	Mauritius	Regulatory Sandbox License (RSL)	RSL works for the legalization of blockchain. Creating environment for blockchain. Encourage investors to visit Mauritius and develop various blockchain-based applications with government permissions.
[29]	Singapore	Technological and financial sectors	Project "Ubin," a cross-currency inter-bank payment transactions system for establishing communication with banking industry of Canada and Europe was launched. Achieved interconnectivity by using blockchain.
[30]	South Korea	Banking	Blockchain ecosystem for banks. Secure transaction, increasing customer trust.
[31]	United Arab Emirates	360-degree, e-governance	Dubai has managed to drastically reduce the involved economic costs related to the present bureaucratic processes. Dubai expects to save 100 million document pages along with 25.1 million work hours every year.
[32]	United Arab Emirates	Bodyo, Healthcare Solutions	A network for measuring general checkup measures such as blood pressure, sugar, fat, height, muscle, weight, mass, etc. with a health-pod.

solution organization and Microsoft, Rwanda. The land registry would completely be digitized by using blockchain with the help of Wise ID apps of Wise Key's and Microsoft's Azure cloud. Wise Key has also helped in setting up the Blockchain Center of Excellence in Kigali, the capital city of Rwanda.

Sierra Leone has tested blockchain for ensuring fair elections in the country. General elections of March 2018 were conducted using blockchain with the help of Agora, a Switzerland-based Technology Company, for testing whether blockchain could help in ensuring transparency in national elections or not, and also a fair counting of electoral votes in the elections.

In Tanzania, the Humaniq project was proposed by a London-based FinTech organization that provided a real-time application for unbanked transactions in the last few years, The Humaniq application is now available in four more African countries. The perks of the Humaniq app include its accessibility on any low-end mobile devices and support for P2P transactions and also providing referral initiatives for increasing user count, Humaniq is helping establish connections between people who are unable to access traditional banking services properly without altering the structure of traditional banking systems.

Uganda is promoting blockchain implementation in public sector enterprises. The government has declared that the Uganda government will continue using blockchain technology wherever it would be a suitable technology for providing consistent services to the general public. Uganda's government has been continuously implementing blockchain in the public sector.

In Kenya there is Bit Hub Africa, which acts as a blockchain accelerator for local startups, situated in Nairobi Kenya. BitHub, Africa was founded in December 2015. The objective of BitHub is to provide consultancy services to industries that are eager to deploy blockchain solutions across African and Middle Eastern countries. Bit Hub provides help to local blockchain startups for their establishment in the competitive market. BitHub Africa interacts with local regulatory bodies to advocate blockchain adoption in Kenya's technological policy and promote more commendatory regulations for Initial Coin Offering (ICOs) and cryptocurrencies.

South African country Nigeria has a Cryptography Development Initiative in Nigeria (CDIN) initiative proposed by an active blockchain community for consistent growth of blockchain. The CDIN initiative has been active over the last 4 years. Moreover, a few newer blockchain startups including the CDIN initiative and Sure Remit are highly active, with an objective of promoting and informing Nigerians about the futuristic potential and benefits of blockchain and cryptocurrencies in the development of the country.

Botswana has implemented a blockchain network on a much smaller scale. The Satoshi Center, the only blockchain hub of Botswana has an interesting startup program called PLAAS (Afrikaander word for farm). PLAAS is a mobile application which acts as a free platform that allows operatives as well as individual farmer real-time monitoring of their daily goods production and remaining stock on a blockchain network.

4.2.2 Blockchain in American continent

Venezuelan economy has suffered a major economic downfall in recent years. The government of Venezuela has proposed its own digital currency—Petro [24]. Venezuela is the first country to issue its own digital currency as a substituting measure for recovering from the country's falling economy. The hyperinflation in Venezuela has encouraged the citizens to convert their saved money into cryptocurrencies.

In Chile, the government is already working on a pilot project for testing data entry and storage in the energy sector. The government is heavily investigating other potential areas where the government can implement blockchain. Ghana is a large producer of copper; the government has implemented blockchain in the mining industry. The government has initiated electricity production in government departments using blockchain.

Delaware, a North American state, permitted blockchain implementation for years. In July 2017, it was the first state in the US to permit stock trading by blockchain. It officially passed a bill drafted for eliminating the gaps in the legislative framework, legalizing and regulating the blockchain in Delaware. This bill is an exceptional attempt in encountering the regulatory challenges linked with various blockchain applications. Additionally, many industries are intended to be benefited from this new trade environment that would be created with the deregulating effect of blockchain.

4.2.3 Blockchain in Asian continent

Singapore is using blockchain in technological and financial areas; Singapore aims to achieve interconnectivity using blockchain. The Singaporean government has launched the project "Ubin," an inter-bank and cross-currency payment transactions system derived from the blockchain for establishing communication with the banking industry in Canada and Europe. Moreover, Singapore is also trying to build a secure and robust healthcare data system network for protecting patient's medical records.

South Korea has been using blockchain mainly in the banking system because of its strength in developing innovative reward and loyalty program. Blockchain resolves the security issues present in traditional banking systems. Korean banks are also investing in startups based on blockchain for maximizing the efficiency of their system with new ideas.

Mauritius has proposed a Regulatory Sandbox License (RSL) for the legalization of blockchain. The Mauritius island government has proposed a RSL intended to encourage investors to visit Mauritius and develop various blockchain-based applications under the guidance of the Board of Investment (BoI) of Mauritius.

The Chinese government is supporting the blockchain in a passive mode by providing initiatives to startups and promoting all cryptocurrency development on its soil. China is the leading country, in area-wise deployment of blockchain.

Japan has been among the earliest countries to set up the blockchain. Many retail units in the retail industry accept cryptocurrencies based on blockchain as a substitute for legal currencies. Bitcoin trading is highly popular among the Japanese as a substitute to digital payment. Bitcoin was created in Japan.

Dubai, the largest city of UAE, has intensively worked on blockchain. Dubai has a remarkable status in global blockchain applications. Dubai has managed to drastically reduce the involved economic costs related to the present rigid bureaucratic processes. Using blockchain, Dubai expects to save 100 million document pages along with 25.1 million work hours every year. The blockchain implementation would take place by a three-step verification process. First, increase the effectiveness of government transactions by increasing dependency on blockchain. Second, create new industries and business groups depending on blockchain. Third, promote the country as an international leader in blockchain adoption.

Russia is working extensively on blockchain implementation in capital market investment. It has allowed secure trading features among different stakeholders. Russia also provides a digital interface for blockchain technologies where different applications can be easily constructed and executed.

4.2.4 Blockchain in European continent

Switzerland is promoting blockchain in every sphere of e-governance. In 2016, Switzerland gained the status of being the world's first country to accept bitcoin as a legal medium of payment for taxes. In 2017, it announced the launching of its own government-owned decentralized Ethereum-dependent digital identification model. In 2018, it tested its initial decentralized voting system derived from local blockchain.

The government of Georgia had approved a private organization to create a blockchain network for the secure Registration of Land Titles working similarly to bitcoin. The objective was to reduce the fees on property registration and present a safe transaction system ensuring flexibility and transparency of the network. This reduced the delivery time of service from 1 to 3 business days to only a few seconds. Moreover, the operational costs were also reduced up to 90%. Blockchain also enabled real-time auditing capabilities.

Denmark is an electoral democracy. The Denmark Liberal Alliance (DLA) party has been in support of blockchain-based use cases since 2014. DLA has proposed a complete electronic e-voting system using blockchain that can be enforced in general elections for the welfare of the public.

Malta is implementing cryptocurrencies as a substitute for physical currencies. In Malta, cryptocurrencies are accepted as a legal currency in all areas. Virtual Financial Assets Act (VFA) provided provisions for registration in the future and managing accountability of online crypto service providers to customers. It will provide the necessary framework for regulating cryptocurrencies legalization.

Estonia is implementing blockchain in every aspect of e-governance. The government had created blockchain technology KSI which is specifically designed in Estonia and is intended to be used globally for making a secure network where user data are free from any alteration and ensuring data privacy to a large extent. KSI may be considered a test bed for all generalized government services for the people. So, it can work as a roadmap for blockchain applications elsewhere.

The United Kingdom has legalized blockchain implementation in the food sector. The Foods Standards Agency of the UK has approved blockchain technology to act as a regulatory method for managing data on slaughter houses meat such as their stock left, consumptions, sell, etc. Moreover, the government aims to provide blockchain in the voting and healthcare industry.

Various blockchain technologies in European continental are also highlighted in Table 4.4.

Sweden is working intensively on blockchain implementation in the real estate sector. For the last 4 years, Sweden has been working continuously for the complete digitization of real estate transactions. This is intended to make the land registry office Lantmäteriet future-ready and to reduce both fraud cases and transaction costs. Other advantages include a reduction of waiting time in committing land registry and transactions.

4.3 CONCLUSION

In this chapter, we have discussed the major advantages, structures, and components of blockchain and have seen how various organizations and governments across the globe are using it for people's welfare. This chapter has presented blockchain as a common platform using which governments and organizations can achieve their national and international development goals. The chapter notifies the current progress various countries' governments have made to date in blockchain implementation to ensure the welfare of their citizens. Since blockchain is a modern technology, its potential isn't completely revealed and understood. It still has many disadvantages. The government is somehow in support of blockchain but not in cryptocurrencies. Blockchain's full potential may completely replace the existing centralized digital networks with a decentralized blockchain network.

Table 4.4 Blockchain Technology in European Continental

References	Country	Area of Application	Implementation
[33]	Switzerland	All Areas	In 2016, Switzerland accepted bitcoin as a legal medium of tax payment. In 2017, it launched its own Ethereum-based identification system. In 2018, it successfully tested the first decentralized voting system.
[34]	Switzerland	Novartis, Healthcare	Implementing blockchain along with IoTs in recognition of false medicines and tracking the temperatures in real time in the medicine supply chain. This ensures secure distribution of drugs.
[35]	Switzerland	Medicalchain, Healthcare Solutions	Secure saving and sharing of patient medical records with other medical stakeholders in the network. Records are shared only after getting permission from the patients. Medicalchain architecture also supports telemedicine feature: it provides patients facilities of interacting with their doctors online, can have detailed conversations on the disease, and provision of providing patients health report digitally.
[36]	Estonia	360 governance	KSI specifically designed in Estonia and can be used globally for making a secure network were user data are free from any alteration and ensuring data privacy to large extents.
[32]	Estonia	Guardtime, Healthcare Solutions	Security and scalability of various enterprise solutions. Currently Guardtime is implemented in E-government of Estonia.

REFERENCES

[1] Nakamoto, Satoshi. "Bitcoin: A peer-to-peer electronic cash system." *Decentralized Business Review* (2008). Available at https://ssrn.com/abstract=3440802 or http://dx.doi.org/10.2139/ssrn.3440802

[2] Zheng, Zibin, et al. "An overview of blockchain technology: Architecture, consensus, and future trends." *2017 IEEE international congress on big data (BigData congress)*. IEEE, Honolulu, HI, 2017, pp. 557–564

[3] Guo, Xingcheng, and Xianglong Guo. "A research on blockchain technology: Urban intelligent transportation systems in developing countries." *IEEE Access* (2023).

[4] Swan, Melanie. *Blockchain: Blueprint for a new economy*. O'Reilly Media, Inc., 2015, vol. 11, pp. 40724–40740.

[5] Dirac, P. A. M. "The Lorentz transformation and absolute time." *Physica* 19.1–12 (1953): 888–896.

[6] Yadav, Abhay Kumar, and Virendra Prasad Vishwakarma. "Adoptation of Blockchain of Things (BCOT): Opportunities & Challenges." *2022 IEEE International Conference on Blockchain and Distributed Systems Security (ICBDS).* IEEE, 2022.

[7] Shalender, Kumar, Babita Singla, and Sandhir Sharma. "Emerging technologies and their game-changing potential: Lessons from Corporate World." *Contemporary Studies of Risks in Emerging Technology, Part A* (2023): 61–70.

[8] Mwanza, Kevin, and Henry Wilkins. "African startups bet on blockchain to tackle land fraud." *Reuters Media* 16 (2018): 2018.

[9] Farouk, Ahmed, et al. "Blockchain platform for industrial healthcare: Vision and future opportunities." *Computer Communications* 154 (2020): 223–235.

[10] Lubogo, Isaac Christopher. *The law of crypto currency and cryptography in Uganda.* Jescho Publishing House, 2022.

[11] Ebekozien, Andrew, Clinton Aigbavboa, and Mohamad Shaharudin Samsurijan. "An appraisal of blockchain technology relevance in the 21st century Nigerian construction industry: Perspective from the built environment professionals." *Journal of Global Operations and Strategic Sourcing* 16.1 (2023): 142–160.

[12] Lee, Nicole M., et al. "Digital trust substitution technologies to support smallholder livelihoods in Sub-Saharan Africa." *Global Food Security* 32 (2022): 100604.

[13] Sharma, Toshendra Kumar. "Top 10 countries leading blockchain technology in the world." *Blockchain Council.* Available at: www.blockchain-council.org/blockchain/top-10-countries-leadingblockchain-technology-in-the-world (2019).

[14] Ghosh, P.K., et al. "Blockchain application in healthcare systems: A review." Systems 11.1 (2023): 38.

[15] Farouk, Ahmed, et al. "Blockchain platform for industrial healthcare: Vision and future opportunities." *Computer Communications* 154 (2020): 223–235.

[16] McFarlane, Chrissa, et al. "Patientory: A healthcare peer-to-peer EMR storage network v1." *Entrust Inc.: Addison, TX, USA* 3 (2017): 19.

[17] Shabani, Mahsa. "Blockchain-based platforms for genomic data sharing: A de-centralized approach in response to the governance problems?" *Journal of the American Medical Informatics Association* 26.1 (2019): 76–80.

[18] Rattanawiboomsom, Vichayanan, et al. "Blockchain-enabled Internet of Things (IoT) applications in healthcare: A systematic review of current trends and future opportunities." *International Journal of Online & Biomedical Engineering* 19.10 (2023): 99–117.

[19] Alquwayzani, Alanoud, and MM Hafizur Rahman. "Short review on blockchain technology for smart city security." *Mobile Computing and Sustainable Informatics: Proceedings of ICMCSI 2023* (2023): 45–66.

[20] Exochain [Online]. Available: https://exochain.com/ (accessed on January 17, 2023).

[21] Curisium [Online]. Available: https://pitchbook.com/news/reports/q1-2024-healthcare-it-report (accessed July 21, 2023).

[22] Yaeger, Kurt, et al. "Emerging blockchain technology solutions for modern healthcare infrastructure." *Journal of Scientific Innovation in Medicine* 2.1 (2019). https://doi.org/10.29024/jsim.7

[23] Joshi, Parikshit, et al. "Blockchain technology for sustainable development: A systematic literature review." *Journal of Global Operations and Strategic Sourcing* 16.3 (2023): 683–717.

[24] Rosales, Antulio. "Unveiling the power behind cryptocurrency mining in Venezuela: A fragile energy infrastructure and precarious labor." *Energy Research & Social Science* 79 (2021): 102167.

[25] Wang, Yunwu. "Report on smart education in China." *Smart Education in China and Central & Eastern European Countries*. Singapore: Springer Nature Singapore, 2023. 11–50.

[26] Ahl, Amanda, et al. "Exploring blockchain for the energy transition: Opportunities and challenges based on a case study in Japan." *Renewable and Sustainable Energy Reviews* 117 (2020): 109488.

[27] Karapetyan, ME, et al. "The development of blockchain technology in Russia: Outlook and trends." *International Journal of Economics & Business Administration* 7.2 (2019): 279–289.

[28] Luchoomun, Keshav, Sameerchamd Pudaruth, and Somveer Kishnah. "Implementation of a proof of concept for a blockchain-based smart contract for the automotive industry in Mauritius." *International Journal of Advanced Computer Science and Applications* 11.3 (2020). http://dx.doi.org/10.14569/IJACSA.2020.0110309

[29] Zhou, Yusheng, et al. "The key challenges and critical success factors of blockchain implementation: Policy implications for Singapore's maritime industry." *Marine Policy* 122 (2020): 104265.

[30] Kim, Hwang. "An empirical analysis of navigation behaviors across stock and cryptocurrency trading platforms: Implications for targeting and segmentation strategies." *Electronic Commerce Research* (2022): 1–29.

[31] Madakam, Somayya. "Blockchain technology: Concepts, components, and cases." *Industry Use Cases on Blockchain Technology Applications in IoT and the Financial Sector*. IGI Global, 2021. 215–247.

[32] Bodyo [Online]. Available: https://bodyo.com/ (accessed on March 12, 2023).

[33] Witzig, Pascal, and Victoriya Salomon. *Cutting out the middleman: A case study of blockchain-induced reconfigurations in the Swiss Financial Services Industry*. Institut de Sociologie/GRET, Neuchâtel, 2018.

[34] Scorringe, Mark. "More than medicine: Pharmaceutical industry collaborations with the UK NHS." *Sustainable Entrepreneurship: The Role of Collaboration in the Global Economy* (2019): 111–137.

[35] Armstrong, Stephen. "Bitcoin technology could take a bite out of NHS data problem." *BMJ: British Medical Journal* 361 (2018). www.jstor.org/stable/26960400

[36] Semenzin, Silvia, David Rozas, and Samer Hassan. "Blockchain-based application at a governmental level: Disruption or illusion? The case of Estonia." *Policy and Society* 41.3 (2022): 386–401.

Chapter 5

Blockchain role in enhancing financial risk management in banking sector using correlation analysis

Abhay Kumar Yadav and Virendra P. Vishwakarma

5.1 INTRODUCTION

Users of blockchain have the ability to introduce decentralized improvements to the network. There is no need for a third party for monitoring activities that take place on blockchain networks. Blockchain may be used to store information, and the technology behind distributed ledgers makes it possible for that information to be more readily transferred. It is possible to utilize it to have one-on-one conversations with other people who are using the network. All of the transactions that take place on the blockchain network are safe and anonymous. The solid security features offered by blockchain technology make it desirable to a wide variety of various sorts of enterprises. The process of data reconciliation is time-consuming and resource-intensive because, in the modern business world, each organization is responsible for its own accounting obligations [1]. The use of blockchain technology might provide a solution to this problem by enabling the simultaneous recording of several sorts of information in real time inside a distributed ledger, such as transactional data, contractual terms, and other data. This suggests that evaluations of adherence to legal requirements will be carried out in an automated manner. It is anticipated that the actions of the organization will have a much higher effect. The experience of the consumer might perhaps be enhanced, which would result in increased safety of identification and data transactions [2]. The concept of a distributed ledger, which records every transaction and maintains the chronology and correctness of the data stored therein over a secure and unbreakable worldwide network, is the cornerstone around which blockchain was built.

5.1.1 Risk management in banking

Risk management is a critical undertaking within the banking sector, aimed at recognizing, evaluating, and mitigating potential risks that financial institutions may encounter. Its significance lies in upholding the stability and longevity of banks, while safeguarding the welfare of depositors,

DOI: 10.1201/9781003514312-5

shareholders, and other involved parties. Given the inherent exposure of the banking industry to a multitude of risks, such as credit, market, operational, liquidity, and regulatory risks, the implementation of effective risk management practices becomes indispensable. By employing robust risk management strategies, banks can proactively anticipate and tackle these risks, bolstering their resilience and capacity to navigate adverse circumstances [3].

Risk management in Indian banks is a relatively recent development, but it has become increasingly significant in response to rising competition and market volatility. The adoption of risk management practices has led to improved efficiency in managing Indian banks and has also promoted a culture of corporate governance. The primary objective of the risk management model is to minimize or mitigate the risks associated with the products and services offered by banks. To effectively address internal and external risks, there is a pressing requirement for a robust risk management framework. Different types of risk in banks are discussed in Figure 5.1.

Credit risk is the potential for a bank to incur losses when borrowers or counterparties are unable to fulfil their financial obligations. This risk arises from the loans, investments, or credit exposures that the bank holds. If borrowers default on their payments, it can lead to significant losses for the bank. Effective management of credit risk involves assessing the creditworthiness of borrowers, setting appropriate credit limits, and implementing risk mitigation measures such as requiring collateral and diversifying loan portfolios [4].

Market risk refers to the potential impact of adverse fluctuations in financial market variables, such as interest rates, exchange rates, and asset prices. The banking sector is susceptible to market risk due to its investments and

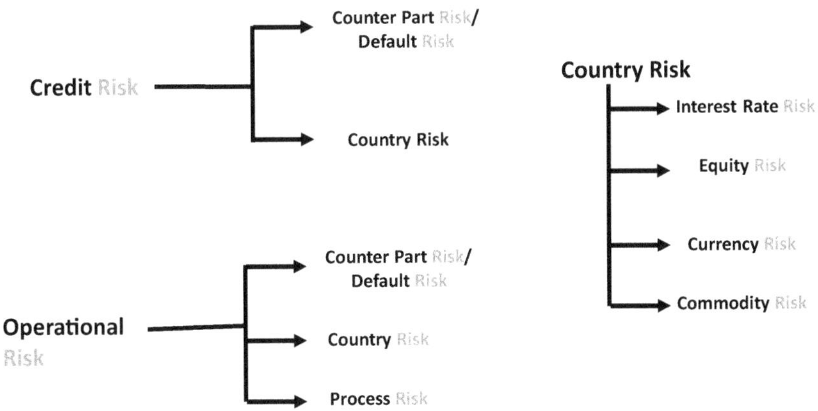

Figure 5.1 Types of Risks in Banks.

trading activities. Fluctuations in these market variables can result in losses in the value of financial instruments held by the bank. To protect their capital and earnings from adverse market movements, banks implement market risk management strategies. This involves assessing and monitoring the level of risk exposure, using hedging techniques, and setting risk limits to mitigate potential losses [5].

Operational risk refers to the possibility of experiencing losses due to deficiencies or failures in internal processes, personnel, and systems, as well as from external events. This category of risk covers a broad spectrum of potential issues, such as human errors, system malfunctions, fraudulent activities, compliance breaches, and external occurrences like natural calamities. Operational risk management aims to identify and address potential vulnerabilities within the bank's operations. It involves establishing strong internal controls, detecting and preventing operational weaknesses, and fostering a risk-aware and responsible culture throughout the organization [6].

5.1.2 Blockchain in banking

The implementation of blockchain technology is vital in enhancing the management of financial risks in banking companies by using correlation analysis. Correlation analysis is a statistical tool that assesses the association between two variables and it can be used to evaluate how blockchain technology impacts the management of financial risks in banking companies [7]. By examining the correlation, it can reveal the strength and direction of the relationship between these variables and help identify areas where blockchain technology can improve financial risk management. Furthermore, it can also reveal if there is a causal link between blockchain technology and financial risk management. To summarize, correlation analysis can provide important information on effect of blockchain technology on risk management in banking companies and aid in making decisions related to implementing and utilizing blockchain technology in this industry [8].

Smart contracts implementation in blockchain would make it possible to instantaneously purchase and run a vehicle, while simultaneously studying answers to problems like electromobility. This would be a huge step forward for the automotive industry. Businesses that use the technology for financial operations are able to post their invoices to the blockchain utilizing smart contracts. On a blockchain, it is possible to record not just billing schedules but also inventory counts and customer information. When a consumer fulfils their financial obligation to pay an invoice, the smart contract changes the status of the invoice to "paid" and alerts the respective companies that they have been paid. Before doing business with a consumer that uses financial services, blockchain technology might be used to confirm the customer's identity [9]. The near future heralds that a large number of financial transactions will be processed by means of blockchain technology.

Blockchain may assist to maintain harmony in the middle of the current digital revolution by mediating conflicts that arise between technology, user data, and privacy concerns. It is possible that the time and effort spent by auditors on reviewing a large volume of mundane transactions will be redirected towards the resolution of more complex and contentious issues. Therefore, the need for accountants and auditors has not been eliminated as a consequence of the automation of processes [10]. Both blockchain and artificial intelligence are incredibly unique technologies that have an astonishingly wide range of potential applications. On the other hand, artificial intelligence is a highly centralized service that is dependent on safe data that cannot be accessed or duplicated in any way. Their collaboration brings about a number of advantages, most notably in the form of monetary assistance. With the help of smart contracts, which are made possible by blockchain technology, all parties involved in a transaction are able to form legally enforceable agreements that will be executed when all conditions are met [11].

Blockchain technology enables continuous data transfer amongst parties engaged in transactional activities. Similar to present-day contracts, smart contracts ensure proper terms are followed in real time and without any room for interpretation on a blockchain [12]. This removes the need for a third party and increases accountability in a way that traditional contracts cannot. By having a decentralized network of computers perform the intermediate responsibilities using the internet, the distributed ledger system removes the necessity of having to rely on the services of a reliable third party. A public distributed ledger keeps a record of all transactions and sends them out across the network to all nodes. The system is more secure than the current centralized ledger system, because each node has a copy of ledger. It can also validate asset ownership and make transactions transparent [13].

The proliferation of digital technologies has made many unattainable forms of collaboration possible. Cloud-based software with analytics are customized for particular use cases. Cash transfer, cashier's cheques, and electronic transfers are the types of payments that are considered to have the highest level of reliability. However, a single-wire transfer cannot be used to transmit both the money and the time simultaneously [14]. These issues are resolved by payment systems that are based on blockchain technology, which also helps in boosting customer trust. The use of this technology makes it easier to monitor financial transactions and gives the possibility of automating previously manual processes. It is possible for banks and other types of financial organizations to utilize smart contracts to keep an eye on clients' transactions involving money and other items.

1. **Data Integrity:** Blockchain ensures the immutability and transparency of data. In stock market prediction, it can be used to securely

store and verify historical stock market data, financial reports, and trading records. By ensuring data integrity, blockchain can help build trust and accuracy in predictive models [15].

2. **Decentralized Data Sources:** Blockchain enables the aggregation of data from multiple sources in a decentralized manner. This can be beneficial in stock market prediction by allowing access to a wide range of data, such as social media sentiment, news articles, economic indicators, and company-specific information. Incorporating diverse data sources can enhance the accuracy of prediction models [16].

3. **Smart Contracts:** Self-executing contracts codes work on predefined regulations encoded on the blockchain. They can be utilized in stock market prediction by automating the execution of trading strategies based on certain predefined conditions. For example, when a specific price threshold is reached, a smart contract can automatically execute a buy or sell order [17].

4. **Tokenized Assets:** Blockchain can enable the tokenization of assets, including stocks and securities. This allows for fractional ownership and efficient transfer of ownership. Predictive models can leverage these tokenized assets to analyse market trends and make predictions based on trading patterns and investor behaviour.

5. **Peer-to-Peer Prediction Markets:** Blockchain-based prediction markets can provide a platform for users to make predictions about stock market movements. These markets incentivize participants to provide accurate predictions by using cryptocurrencies or tokens as rewards. Aggregating the collective wisdom of participants can potentially improve the accuracy of stock market predictions [18].

The chapter is structured in following order: Section 5.1 discusses the background and Section 5.2 follows literature review. Section 5.3 is responsible to discuss conceptual model and hypotheses development. Section 5.4 discusses research methods with emphasis on participant demographic profiling and validating the research questions, Sections 5.5 and 5.6 present the data analysis and results, with discussion and implications followed by conclusion, limitations, and future scope of research.

5.2 REVIEW OF LITERATURE

Within the expansive field of blockchain in the domain of financial services and its architecture, there are a select few tools and techniques that particularly stand out. It has been discovered during the course of time that the tools and procedures used in blockchain applications are rather unique. The tools and procedures are represented in Figure 5.1. These strategies and tools are able to swiftly adapt to rapidly shifting financial situations as blockchain principles underpin their operation. The following open-source software tools – Geth, Infura, Metamask, Parity, Truffle, and others – are featured here. With the help of these innovative and cutting-edge solutions, blockchain technology will continue to enhance financial services and linked industries [19].

Since the beginning of this decade, the financial and insurance sectors have been pondering the potential benefits brought about by blockchain technology. Blockchain ledger stores the time and date of each transaction. All transactions pertaining to money may be accurately documented in this manner [20]. Because there are several copies of the ledger, blockchain is nearly indestructible and very secure. This makes it extremely difficult, if not impossible, for hackers to modify or fabricate any part of the record. Businesses are now able to more readily and confidently place their faith in one another because of blockchain technology. As a direct consequence of this, deterministic smart contracts may be written and used in immutable systems to simplify business procedures, increase productivity, and build confidence [21]. It protects sensitive information at every step of the software development. Applications based on the blockchain might potentially make banking more accessible and less expensive.

The fact that the blockchain technology offers a high level of security is only one of the numerous benefits that contribute to the technology's growing popularity in the financial sector. Encryption is applied for reasons of safety to the distributed ledger that records blockchain transactions. Because of this, the information was unavailable to anybody who was not aware of the hidden code [22]. The banking industry currently provides a wide variety of possibilities within the realm of fintech [23]. As a result, it may be challenging for those who supply financial services to home in on the solution that will prove to be the most successful. It is possible that the answers to some of the financial services issues may be realized using blockchain. The administration of financial services is still performed in the traditional, centralized,and convoluted manner around the globe. Transparency has been reduced as a result of the majority of financial data being maintained in centralized systems and having to pass through a number of intermediaries. In addition, the protection of the data can only be compromised through intermediaries and database security [24].

However, even the most secure systems are susceptible to being hacked and having their data stolen. Lack of transparency generally results in complicated security issues as discrepancies are not brought to anyone's attention until it is discovered that a data breach or other system problem has occurred. When compared to the issue of physical securities, the process of digital securities may be said to be both quicker and more efficient [25]. Issuers of digital financial solutions have the ability to customize these solutions so that they meet the requirements of individual investors. Examples of this include fractional ownership of real objects, tokenized microeconomies, and asset transfers that are secure, scalable, and quick. These elements provide a number of advantages, including enhanced stakeholder incentive alignment, increased efficiency in corporate operations, and governance structures that are both more visible and responsible [26].

There is growing demand in specialized markets, venture capital, private equity, and real estate funds to modernize their methods for liability risk management, to build decision-making frameworks that are nimbler, and to cope with the complexities of ever-evolving rules. Blockchain technology, which functions as a distributed ledger, has the ability to completely transform stakeholder and asset management. The fact that digital money was the first sort of data to be stored on blockchains gives financial applications the potential to usher in a period of profound change across the sector [27].

An insurance company may make use of smart contracts to speed up the process of filing claims. An instantaneous evaluation of a client's claim will be performed by the programmes that are preinstalled in the blockchain. Payment to the customer will be provided according to terms of the smart contract only if the contract is found to be legally binding. In an effort to prevent activities of fraud and money laundering, the majority of banks and other companies that offer financial services now require their customers to go through some form of identity verification procedure. With addition of new block in blockchain, a digital ledger is created. Because blockchain ledgers provide several advantages over more traditional digital ledgers, there is a growing possibility that they may be used in financial industry. Now, using blockchain technology, we are able to establish our very own decentralized digital ledgers. As a result, there is no need to rely on a centralized organization to process or preserve information on transactions [28].

When adopting blockchain, there is a potential reduction in the risk of transaction data being hacked as there is no centralized repository for holding transaction data that uses its own distinct security mechanism. Applications based on the blockchain might potentially make banking more accessible and less expensive.

5.3 DATA ACQUISITION AND INPUT

5.3.1 Data gathering

The study follows an empirical approach and is limited with geographical area of Delhi National Capital Region (NCR), due to the presence of many investment banks working predominantly in the area. A well-defined questionnaire was presented for collecting participants' data. The questionnaire was divided into two sections: first section contains demographic profile of participants. The second section consists of validating some statements depending on service differentiation, disruptive technologies, and employee performance. The participants of presented study were mainly employees of banks particularly near southwest Delhi. Around 23 banks (12 private sector banks, 11 public sector banks) were included as participants for the study. The questionnaire was sent to nearly 450 potential respondents but were able to obtain fully completed response from 149 respondents only.

5.3.1.1 Demography profiling

Out of 450 participants intended to participate, only 149 participants answered the questionnaire. The demography proofing is done only of the respondents who completely filled the questionnaire as shown in Table 5.1.

5.3.2 Research questions

RQ 1: Is there any relation amongst the application of blockchain technology and enhanced safer clearing and settlements?

RQ 2: Is there any relation amongst blockchain application and real-time financial market tracking for effective management?

RQ 3: Is there any relation existing for blockchain application in creation of cost-effective tools for reporting credit?

5.4 METHODOLOGY

It is possible that developing solutions based on blockchain technology might be effective in the financial markets. Traditional trade finance methods have, for a very long time, been a source of annoyance for businesses. This is due to the lengthy procedures involved, which may sometimes cause delays in operations and make it difficult to manage liquidity. The use of blockchain technology may simplify the processes involved in international business transactions and company operations, facilitating secure corporate transactions. Use cases that need an immutable record, such as monitoring

Table 5.1 Demographic Profiling of Participants

Variables	Categories	Frequencies	Response rate
Gender	Male	97	65.10%
	Female	52	34.34%
Age	24–34	68	45.63%
	35–45	71	47.65%
	46 & above	10	06.71%
Designation	VP/branch manager	15	10.06%
	Analyst/managers	39	26.17%
	Clerical staff	76	51.00%
	Technical support	08	05.36%
	Others	11	07.38%

Table 5.2 Blockchain Supports in Fraud Prevention

Fraud prevention	*Frequency*	*in %*
Strongly agree	57	38.25
Agree	54	36.24
Neutral	10	6.71
Disagree	15	10.06
Strongly disagree	13	8.72
Total	149	100

things in real time as they are exchanged between multiple parties across the supply chain, are a good fit for blockchain because of its properties [26]. The distributed ledger technology, often known as blockchain, has the potential to completely transform industries that produce and sell a broad range of goods.

Assumptions

Hypothesis 1: There exists no association amongst the application of blockchain technology and enhanced safer clearing and settlements.

Hypothesis 2: There exists no association amongst blockchain application and real-time financial market tracking for effective management.

Hypothesis 3: There exists no association amongst blockchain application for creation of cost-effective tools for reporting credit.

5.4.1 Percentage error analysis

Table 5.2 states 38.25% of respondents strongly agreed and 36.2% agreed to statement that blockchain technology supports in preventing fraud in the financial transactions, thereby reducing the financial risk. Hence, it can be stated that blockchain supports in fraud prevention.

Table 5.3 states 37.58% of respondents strongly agreed and 34.22% agreed to statement that blockchain technology supports in preventing fraud in the financial transactions, thereby reducing the financial risk. Hence, it can be stated that blockchain supports in fraud prevention.

5.4.2 Correlation analysis

To study the relationship amongst different independent variables, correlation analysis is used. Correlation analysis is used for measuring association between different variables. In this study, we have considered three different independent variables such as risk management and domain of

Table 5.3 Efficacy in Handling Transactions

Efficacyin handling transactions	Frequency	in %
Strongly agree	56	37.58
Agree	51	34.22
Neutral	12	8.05
Disagree	14	9.39
Strongly disagree	16	10.73
Total	149	100

Table 5.4 Correlation Analysis

Coefficients	Safer cleating and settlements	Real-time tracking	Cost effective	Application of blockchain
Safer clearing and settlements	1	0.888	0.823	0.858
Real-time tracking	0.888	1	0.858	0.868
Cost effective	0.823	0.858	1	0.823
Application of blockchain	0.858	0.868	0.823	1

improving financial and cash management towards dependent variable for proper decision-making.

The analysis in Table 5.4 indicates a strong positive variable correlation, with a coefficient ranging from +0.823 to +0.888. The strongest correlation is found between real-time tracking of financial data and the use of blockchain technology at +0.868. Additionally, comparison between dependent and independent variable shows that highest correlation lies between real-time tracking and application of blockchain technology with the value of +0.868, also variables safer clearing and settlement and application of blockchain technology is +0.858, and the remaining variable cost effective and application of blockchain technology is +0.823.

5.4.3 Chi-square test analysis

The chi-square test is a hypothesis test for determining relation between two categorical variables as independent variable or not. It is used as the final step of data analysis shown in Table 5.5 and Table 5.6 is responsible to express a chi-square analysis between blockchain and real-time financial market tracking for effective management.

Hypothesis 1:
Null: There exists no association amongst application of blockchain technology and enhanced safer clearing and settlements.

Table 5.5　Chi-square Analysis between Application of Blockchain and Enhanced Safer Clearing and Settlements

Chi-square Test	Value	Degree of freedom	P value
Chi-square Data	290.180a	16	0.00
L ratio	197.414	16	0.00
Linear by linear	107.721	1	0.00

Table 5.6　Chi-square Analysis between Blockchain and Real-Time Financial Market Tracking for Effective Management

Chi-square test	Value	Degree of freedom	P value
Chi-square data	290.180a	16	0.00
L ratio	197.414	16	0.00
Linear by linear	107.721	1	0.00

Table 5.7　Chi-square Analysis between Blockchain and Creation of Cost-Effective Tools for Reporting Credit

Chi-square test	Value	Degree of freedom	P value
Chi-square data	298.677	16	0.00
L ratio	185.965	16	0.00
Linear by linear	96.969	1	0.00

Hypothesis 2:

Null: There exists no association amongst blockchain application and real-time financial market tracking for effective management.

Hypothesis 3:

Null: There exists no association amongst blockchain application for creation of cost-effective tools for reporting credit.

Depending on Table 5.7 analysis, the obtained value of p is 0.00 (less than value of significance). Therefore, alternate hypothesis is being accepted stating there is an association for blockchain application for creation of cost-effective tools for reporting of credit.

Hence, hypothesis statements are as follows:

Hypothesis	Decision
Alternate H1:There exists an association in blockchainapplication and improved safe clearing and settlement.	Accept
AlternateH2:Thereexistsanassociationamongstblockchain application and real-time financial market tracking for effective management.	Accept
Alternate H3:There exists an association amongst blockchain application for creation of cost-effective tools for reporting credit.	Accept

5.5 CONCLUSION

Blockchain technology is gaining popularity among businesses all over the world as more and more of their production facilities are networked together. The factories of the future will constitute a massive supply chain that will include a variety of different goods, businesses, and services. This innovation's primary objective is to provide a record for digital currency such as cryptocurrencies that is incorruptible and cannot be changed in any way. Information is protected by applications based on blockchain technology, which also enables companies to target particular audiences and ensures that artists are paid appropriately. The number of individuals who choose to conduct their financial dealings via the use of this technology is consistently growing. Payment processing is essential in today's society, as the vast majority of monetary transactions now take place between bank accounts. In exchange for safer transactions and the potential to create their own digital currencies, banks have embraced innovative technology and have been at the vanguard of the digital revolution. Financial organizations now have the ability to monitor all transactions in real time, because of the technology of blockchain. Transactions involving financial institutions will be able to be settled utilizing a public blockchain because of this innovation in technology. It will be necessary for banking executives to fulfil a number of conditions before the concept can be widely accepted by the banking industry. The blockchain ability of supporting distributed data sharing and having provision of temporary access to assets will fundamentally transform the manner in which we navigate our environments.

REFERENCES

1. Ning, Lianju, and Yaqin Yuan. "How blockchain impacts the supply chain finance platform business model reconfiguration." *International Journal of Logistics Research and Applications* 26.9 (2023): 1081–1101.
2. Sajid, Rabbia, et al. "The role of fintech on bank risk-taking: mediating role of bank's operating efficiency." *Human Behavior and Emerging Technologies* 2023 (2023).

3. Weber, Olaf. "Environmental credit risk management in banks and financial service institutions." *Business Strategy and the Environment* 21.4 (2012): 248–263.

4. Rao, M. Venkateswara, et al. "Credit investigation and comprehensive risk management system based Big Data analytics in commercial banking." 2023 9th International Conference on Advanced Computing and Communication Systems (ICACCS). Vol. 1. IEEE, 2023.

5. Adeabah, David, et al. "Reputational risks in banks: a review of research themes, frameworks, methods, and future research directions." *Journal of Economic Surveys* 37.2 (2023): 321–350.

6. Chowdhury, Debapriya, and Prasanna Kulkarni. "Application of data analytics in risk management of fintech companies." 2023 International Conference on Innovative Data Communication Technologies and Application (ICIDCA). IEEE, 2023.

7. Liu, Jingkuang, Lemei Yan, and Dong Wang. "A hybrid blockchain model for trusted data of supply chain finance." *Wireless Personal Communications* (2021): 1–25.

8. Boakye, Elijah Asante, Hongjiang Zhao, and Bright Nana Kwame Ahia. "Emerging research on blockchain technology in finance; conveyed evidence of bibliometric-based evaluations." *The Journal of High Technology Management Research* 33.2 (2022): 100437.

9. Ashima, Reem, et al. "Automation and manufacturing of smart materials in additive manufacturing technologies using Internet of Things towards the adoption of Industry 4.0." *Materials Today: Proceedings* 45 (2021): 5081–5088.

10. Rjoub, Husam, Tomiwa Sunday Adebayo, and Dervis Kirikkaleli. "Blockchain technology-based FinTech banking sector involvement using adaptive neuro-fuzzy-based K-nearest neighbors algorithm." *Financial Innovation* 9.1 (2023): 65.

11. Garg, Poonam, et al. "Examining the relationship between blockchain capabilities and organizational performance in the Indian banking sector." *Annals of Operations Research* (2023): 1–34.

12. Mosteanu, Narcisa Roxana, and Alessio Faccia. "Digital systems and new challenges of financial management—FinTech, XBRL, blockchain and cryptocurrencies." *Quality–Access to Success* 21.174 (2020): 159–166.

13. Duchenne, James. "Blockchain and smart contracts: complementing climate finance, legislative frameworks, and renewable energy projects." *Transforming Climate Finance and Green Investment with Blockchains.* Academic Press, 2018. 303–317.

14. Zheng, Kangning, et al. "Blockchain technology for enterprise credit information sharing in supply chain finance." *Journal of Innovation & Knowledge* 7.4 (2022): 100256.

15. Jena, Rabindra Kumar. "Examining the factors affecting the adoption of blockchain technology in the banking sector: an extended UTAUT model." *International Journal of Financial Studies* 10.4 (2022): 90.

16. Zhu, Yongjie, and Shanyue Jin. "How does the digital transformation of banks improve efficiency and environmental, social, and governance performance?" *Systems* 11.7 (2023): 328.

17. Kherbouche, Meriem, Galena Pisoni, and Bálint Molnár. "Model to program and blockchain approaches for business processes and workflows in finance." *Applied System Innovation* 5.1 (2022): 10.
18. Li, Dun, et al. "A blockchain-based secure storage and access control scheme for supply chain finance." *The Journal of Supercomputing* 79.1 (2023): 109–138.
19. Al Shanti, Ayman Mohammad, and Mohammad Salim Elessa. "The impact of digital transformation towards blockchain technology application in banks to improve accounting information quality and corporate governance effectiveness." *Cogent Economics & Finance* 11.1 (2023): 2161773.
20. Meng, Xue. "Risk assessment and analysis in supply chain finance based on blockchain technology." *Journal of Sensors* 2022 (2022).
21. Wang, Rui. "Blockchain and bank lending behavior: a theoretical analysis." *SAGE Open* 13.1 (2023): 21582440231164597.
22. Li, Gang, Ehsan Elahi, and Liangliang Zhao. "FinTech, bank risk-taking, and risk-warning for commercial banks in the era of digital technology." *Frontiers in Psychology* 13 (2022): 934053.
23. Gupta, Varun Chandra, Menita Agarwal, and Arundhati Mishra. "When trade finance meets blockchain technology." *International Journal of Innovative Science and Research Technology* 4.10 (2019): 342–346.
24. Bulut, Esra. "Blockchain-based entrepreneurial finance: success determinants of tourism initial coin offerings." *Current Issues in Tourism* 25.11 (2022): 1767–1781.
25. Chen, Yan, and Cristiano Bellavitis. "Decentralized finance: blockchain technology and the quest for an open financial system." Stevens Institute of Technology—School of Business Research Paper (2019).
26. Hashem, R. E. E. D. R., Al-Rifai Ibrahim Mubarak, and A. A. E. S. Abu-Musa. "The impact of blockchain technology on audit process quality: an empirical study on the banking sector." *International Journal of Auditing and Accounting Studies* 5.1 (2023): 87–118.
27. Gupta, Shivam, et al. "Influences of artificial intelligence and blockchain technology on financial resilience of supply chains." *International Journal of Production Economics* 261 (2023): 108868.
28. Dewey, Josias N., Robert Hill, and Rebecca Plasencia. "Blockchain and 5G-enabled Internet of Things (IoT) will redefine supply chains and trade finance." *Proc. Secured Lender* (2018): 43–45.

Chapter 6

Identification of security factors in cloud computing

Defence security perspective

Mohd Nadeem, Prabhash Chandra Pathak, Masood Ahmad, and Nafees Akhter Farooqui

6.1 INTRODUCTION

Protection for cloud computing is security for the technology. In plainer terms, cloud security aids in the security of the foundations, apps, and approaches while safeguarding data against trivial intrusions. The complex design of cloud computing infrastructure, which is all accessible via the internet and needs regular maintenance, gave rise to the concept of cloud security. Risk elements including outside information retention, reliance on the "public" the World Wide Web, lack of management, multiple tenancies, and coordination with internal security raise security concerns. The huge scale of the cloud and the simple fact that its service provider resources are completely dispersed, diversified, and virtualized are only two of the many distinguishing features that distinguish the cloud from conventional technology. In their current condition, identity, authentication, and authorization are insufficient for cloud security. Cloud computing is using the same security precautions as conventional IT environments in most cases. However, compared to traditional IT solutions, cloud computing may bring a different set of risks to a corporation due to its operational procedures, service provider model, and technology. Unfortunately, some individuals think that making these systems more secure makes them rigid [1]. Some of the security breaches are depicted in Figure 6.1 and listed below:

- Facebook elected not to notify its users that their personal data had been collected before August 2019 until April 2021, when it released the data set to the public. Additionally, contact information from client profiles is stored for the information, including full names, addresses, phone numbers, and email addresses. Even though the company eventually wrote about the hacking attempt on its blog, Facebook's reputation was still irreparably harmed. Mark Zuckerberg, the creator of Facebook, was impacted by the issue's cascading consequences; however, the company claims the issue was quickly identified and resolved. To settle a privacy issue with the

DOI: 10.1201/9781003514312-6

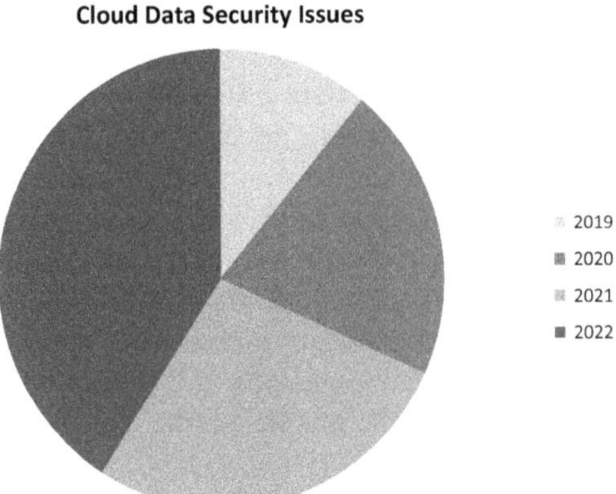

Figure 6.1 Pie Chart of Cloud Data Breaches from 2019 to 2022.

Federal Trade Commission, he had to testify before federal officials. The corporation also paid a $5 billion fine as part of the deal. Things only became worse after whistleblower Frances Haugen claimed in October 2021 that Facebook puts profits before safety [2].

- In November 2019, an attack on Alibaba's Chinese e-commerce platform Taobao resulted in the exposure of more than 1.1 billion user records. Before Alibaba recognized what was happening, a Chinese software developer spent eight months scouring the website and stealthily scraping user data. Data that was stolen included customer reviews, user IDs, and mobile phone numbers. Even though the programmer did not come close to passwords or other heavily encoded data, the breach was serious enough for the firm to call the authorities. It is unlikely that the full effects of this attack, which took place in China, will ever be known. A model does, however, highlight areas of strength for a more thorough evaluation of organizational and structural frameworks [3].

- Similar to Alibaba, LinkedIn also suffered a data breach in 2021. About 700 million LinkedIn profiles were impacted by the material, which was largely made public. The details from the hack are published on a dark web forum in June 2021, however. No private or sensitive information was revealed, according to LinkedIn. The company also claimed that the occurrence just broke the terms of service. Nevertheless, a scraped information test in an obscure blog post had email addresses, phone numbers, geo-location information, sexual orientations, and other nuances of online pleasure. A skilled hacker might utilize all of that knowledge in social engineering attacks. Despite LinkedIn's denial of guilt for the hack, it is undeniable that social media's potential for data security breaches has become better known [4].

- Sina Weibo is one of the most well-liked social networking sites in China. In June 2020, more than 538 million users' details, including real names, site usernames, gender, and locations, as well as 172 million users' phone numbers, were published on the dark web and other websites. Weibo's data was sold by the hacker for $250, most likely because passwords were not included in it, although the reason for the occurrence is unknown. Weibo is nevertheless occasionally used to share uncensored news from all around the country, although it is now tightly regulated and monitored. Because of this, anonymous Weibo users may be in the greatest danger from the intrusion [5].

- Hackers connected to the Lock Bit ransomware group attacked Accenture in August 2021. Even worse, the group stole and disclosed confidential company information by hacking into the clients' systems. The hackers claimed to have stolen six terabytes of data and wanted a $50 million ransom. Nevertheless, Accenture told a publication that due to the complete restoration of all affected systems from backups, neither Accenture's operations nor the systems of its clients were impacted [6].

- More than 500,000 Marriott division Starwood guests' private information was compromised following an assault in September 2018. In 2014, before Marriott acquired Starwood, the business learned through a forensic study that the Starwood network had been compromised. The IT infrastructure that Starwood had inherited from Marriott was still in use, and it is most likely that Marriott's usage of antiquated technology is what led to the hack. Although the attack did not force the company to stop operating, it did damage its reputation [7].

- In June 2021, the cyber analytics firm Cognyte exposed 5 billion records describing prior data breaches due to a failure to secure its database. No password or other kind of authentication was required to access the records because they were made available online. The number of passwords contained in the database after it was made public for four days is unknown, but each one includes names, email addresses, and the location of the data source. Hackers could utilize the info for a long time to come [8].

Even though you will not be able to block every attack, you must make your best effort. This calls for the creation of a disaster recovery plan, the building of a secure infrastructure, and the investment in technologies for detection and prevention of attack. The purchase of cyber security insurance might also aid in recuperation.

6.2 LITERATURE REVIEW

Distributed computing security guides the use of such fragile personal computer (PC) assets to prevent client information from being retrieved or divulged by other clients as all client information is stored on a single server [9]. The phrase "distributed computing" describes the persistent availability of framework resources, particularly capacity and processing power, without the client being directly engaged or effectively controlling them. Tasks are often divided across several locations in big mists, each of which is concealed behind a central information hub. Asset sharing is achieved using distributed computing by applying the "pay as you go" principle [10]. The purpose of the cloud computing system is to enable users to benefit from several technologies without having to have a thorough understanding of each one. Virtualization is the primary underlying technology for cloud computing. The front end, which involves the client machine device that provides access, web browsers, the network, and internet-based applications, is connected to the rear end, which is comprised of machines, databases, and servers [11]. All the information that the end client accesses is stored on the secondary end, which serves as a repository. A centralized server manages the scheduling of connections across the front end as well as the back end. Protocols enable the central server to facilitate data exchange. To control the communication between several devices on clients and cloud servers, the central server makes use of middleware and software. Typically, each job or programme uses a distinct server. Cloud computing relies on techniques like virtualization and automation. Users may easily distinguish between service and the foundational cloud architecture through virtualization and access services as logical entities. Without the direct intervention of the cloud provider's IT personnel, customers may provide resources, connect solutions, and deploy workloads using automation and orchestration features.

What other types of distributed computing administrations are there?
Three general help conveyance classifications or sorts of distributed computing can be distinguished in Figure 6.2:

> *Infrastructure as a service (IaaS):* IaaS providers, like Amazon Web Services (AWS), give its clients access to resources, a virtual server environment, and Application Programming Connection points (APIs) that enable them to transfer duties to Virtual Machines (VMs). Depending on the circumstance, clients can start, stop, access, and alter the VM and capacity. Each client has a set allocable stockpiling limit. For a variety of duty requirements, IaaS providers provide small, medium, large, extra-large, memory, or figure-boosted occasions in addition to supplying example customization. For commercial clients, the IaaS cloud model is the one that most closely resembles a far-off server farm [12].

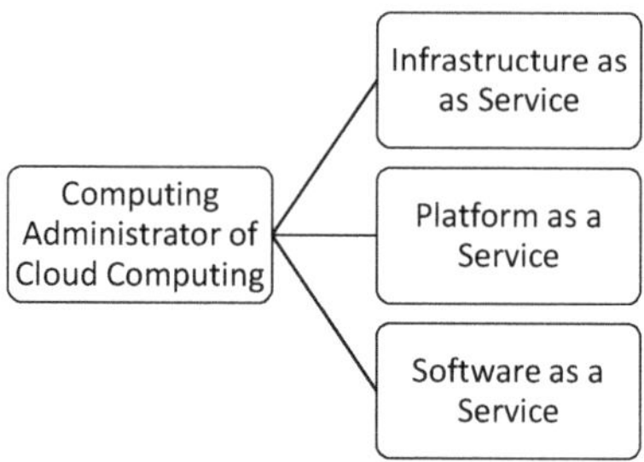

Figure 6.2 Different Administrators of Cloud Computing.

Platform as a service (PaaS): According to the PaaS concept, cloud-specialised companies have development tools on their systems. To access these devices over the internet, users use APIs, online points of interface, or entrance programming. PaaS is typically used for programming enhancement, and many PaaS companies have the product as soon as it is released. Popular PaaS options include Google Application Motor, AWS Flexible Beanstalk, and Lightning Stage from Business Power [13].

Software as a service (SaaS): The phrase "web administrations" is frequently used to describe this programme, which is delivered as software as a service (SaaS) through the internet. Customers may access SaaS apps and services from any location by using a PC or mobile device with an internet connection. Clients approach data sets and application development in the SaaS paradigm. The effectiveness and email management of a standard SaaS solution 365 by Microsoft [14].

6.3 FRAMEWORK

The flow diagram in Figure 6.3 that depicts the chapter's frameworks explains how cloud computing security is studied step by step in the context of contemporary computing in the twenty-first century.

6.4 SECURITY FACTORS OF CLOUD COMPUTING

The most important aspects come from mining the digital platform's library of various cloud-based research articles. These are the ten elements that influence cloud computing: the specialized components that affect distributed computing reception were highlighted by specialists. Examples of these components are comparative advantage, ambiguity, similarity, difficulty, and tradability. For instance, [15] mentioned that global IT neighbourliness

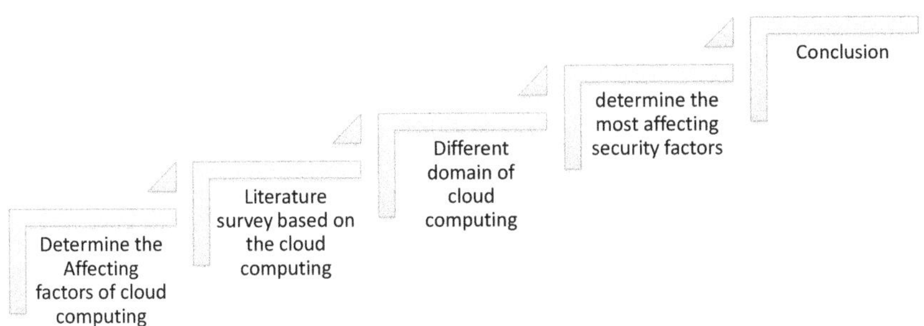

Figure 6.3 Framework of the Study on Cloud Computing.

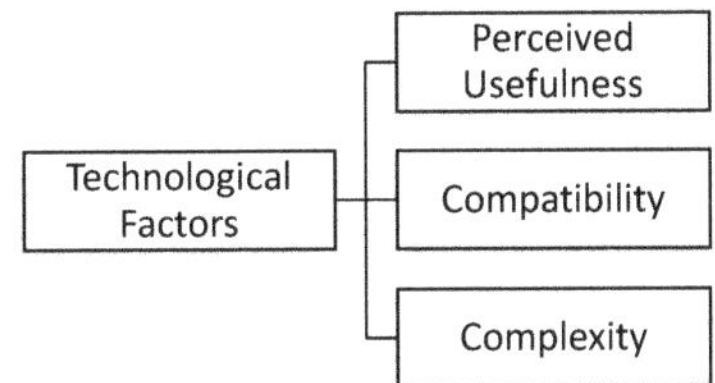

Figure 6.4 Technological Factors of Cloud Computing Security.

influences cloud reception, which might result in significantly different rates of cloud acceptance in some countries. Distributed computing features innovative components that include high accessibility, decreased complexity, the lessened total price of possession, minimal capital speculation, and adaptable adaptability, in contrast to the aspects of conventional processing models that are even better execution. Furthermore, one previous study [11] emphasized the importance of mechanical aspects like security and protection in the selection of the reception. It emphasized that while progress has been made, there are still a number of unique requirements that organizations must meet to properly handle people's personal information and ensure information security.

6.4.1 Technological factors

The technical parameters that affect cloud computing were emphasized by the researchers which include relative advantage, compatibility, and difficulty, as well as others are shown in Figure 6.4. The distinctive qualities of cloud computing are high availability, lesser complexity, reduction in total cost, low capital investment, and elastic scalability, in addition to the high-performance features of traditional computing models. Along with technology, privacy and security concerns are key in the adoption choice [15].

There are still several regulations that allow businesses to properly handle customers' data and safeguard data privacy. The cloud computing paradigm prevents businesses from breaking the law as it offers sufficient data security. Connectivity is a key facilitator for an organization's extensive adoption of cloud computing. A new industrial revolution which is rising demand for high-tech consumer goods is being made possible by the ability to access information through the cloud. Additionally, organizations must offer 24/7 support by using solid software. To have the least amount of disturbance possible in the event of failure, businesses must conduct disaster and emergency preparedness, efficiently explain that when it comes to small- and medium-sized enterprises (SMEs), the primary elements, such as relative benefit, uncertainty, compatibility, and tri-ability, play a crucial role in the adoption of cloud services. According to research, firms that identify the advantage of innovation are more likely to adopt it. SMEs must have a clear vision and understand the advantages of using cloud services [16]. The level of ambiguity of SMEs about cloud computing may have an impact on their decision to use it. It is also clear that the majority of early adopters and prospectors place a high value on having faith in the service providers.

It has been discovered that for SMEs to adopt cloud computing in their sector, they must test the service before deployment. This is because SMEs anticipate that cloud services will be user-friendly and compatible, which will influence their decision to adopt cloud computing. Cloud service providers, on the other hand, assert that the majority of adopters' expectations for consistency and added business value are met by cloud services. SMEs anticipate that the services offered should be comprehensive and user-friendly to utilize cloud computing [17]. Considering the difficulties encountered, it is advantageous for the customer to try the assistance before deciding whether or not to purchase it, believe that one of the biggest obstacles to the development of distributed computing is security concerns. Businesses should be concerned about inadequate security when entrusting their data to a third party, and service providers should demonstrate their ability to protect clients' data from serious threats like phishing and botnets (an organization of remote web-connected machines running continuously). Additionally, businesses are concerned about hosting their goods on servers controlled by suppliers. Many SMEs are hesitant to adopt new technologies, especially those that have not yet undergone thorough investigation and evaluation. This is due to the "assumed cost of obtaining" and the "apparent expense of disappointment" [18]. Except for African countries, most developed nations have carried out surveys on cloud computing. However, SMEs do not trust cloud computing when necessary information on how to use cloud services is needed due to the lack of references in this field.

6.4.1.1 *Perceived usefulness*

It is the extent to which a technical feature is thought to benefit businesses more. It makes sense for businesses to evaluate the benefits of implementing innovations. ERP software can be replaced or supplemented by cloud computing services, which enable operations to be mobilized and generalized through online transactions. The following are some of the anticipated advantages of embedded cloud computing services: rapid corporate communication, effective company coordination, improved customer communications, and market information mobilization [19]. As cloud computing is still relatively new, businesses might not have faith in it. Users may need a lot of time to learn and operate the new system. As a result, the intricacy of an idea can be a barrier to its implementation Complexity is typically negatively impacted by technology. The innovation model's tendency to spread is to look into how new technology is adopted. A model that attempts to broaden this outlook not only aids in determining whether a company is suitable for cloud computing by clearly tracing all the factors but also attempts to give a certain profitability valuation of the benefits associated with cloud computing has been proposed as a trade-off equation in previous studies, which indicates which technology can result in higher profits [20].

6.4.1.2 *Compatibility*

The degree to which an invention is compatible with the existing values, established habits, and present needs of a potential adopter. It has been said that compatibility is a crucial component of innovation. Businesses are typically inclined to consider the adoption of new technology when it is acknowledged that it is compatible with work application systems. When technology is thought to be fundamentally incompatible, big changes in processes that need a lot of learning are necessary. The researchers determined that relative advantages, complexity, and compatibility are the most widely used independent technological factors for IT adoption among enterprises from their thorough research of diffusion studies [21]. In this study, perceived utility—rather than proportional advantages—is seen as a key technology quality.

6.4.1.3 *Complexity*

The widespread availability of high-speed internet connectivity and the steadily falling cost of storage have finally allowed service providers to satisfy customers' demands for simplicity, affordability, and flexibility, which has led to the rise of cloud services. Consumers' recent influx of smart mobile devices—handheld wireless computers that are mobile—has sped up the creation of cloud services that give them application functionality. The ability of cloud computing to instantaneously deliver simple, user-friendly, smart, and high-powered computer programs and information that consumers could not otherwise access is one reason consumers have embraced it so quickly. Consumers' recent influx of smart mobile devices—handheld wireless computers that are mobile—has sped up the creation of cloud services that give them application functionality [22]. The ability of cloud computing to instantaneously deliver simple, user-friendly, smart, and high-powered computer programs and information that consumers could not otherwise access is one reason consumers have embraced it so quickly. A cloud computing setup can be implemented for an organization in a matter of minutes.

The IT resources of cloud computing small businesses may now compete on an equal footing with major corporations. The idea of "renting" essential IT services as opposed to purchasing software and hardware significantly lowers their cost. For instance, Think Grid provides SMEs with enterprise technology services for a low monthly charge that would otherwise cost hundreds of thousands of pounds. The adoption of cloud computing services provides users with the benefit of the economies of the usage of large-scale data centres, which are far more effective and use multi-tenant architecture to share resources across several clients. With the help of this paradigm, cloud computing companies can pass on cost savings to their customers. The usage of cloud computing also has the benefit of scalability and flexibility; clients can make necessary adjustments by IT needs, such as reducing capacity and users as necessary and being able to adapt to actual rather than anticipated requirements. Customers can benefit from more resource elasticity without having to pay a premium because cloud computing services are also provided in terms of actual consumption. Through the internet, cloud-based services enable secure access to apps and data from any place [23]. Costs are greatly reduced and resource consumption is maximized only when necessary because of the combination of resources in the form of big clouds. Considering the use of cloud computing. Organizations do not have to worry about over- or under-provisioning for services that do not live up to expectations or services that quickly gain popularity. Moving more infrastructure and apps to the cloud frees up time, effort, and financial resources so that they may be used to focus on the more important task of employing technology to enhance the enterprise's core business. As servers are continually in use and not idle, sharing computing resources among several tenants improves utilization rates and lowers costs while accelerating the development of applications.

However, a side effect is that when consumers are unable to plan for peak loads, computer capacity increases. Customers can save money on the acquisition and installation of IT infrastructure and/or applications, as was previously described. As a result, organizations can use this capital investment to fund simpler operational costs that can be planned for every month. Further, as IT solutions can be remotely patched and upgraded by the provider, deployed very fast, and managed and maintained, clients do not need to pay for resources to fulfil variable demands. Similarly, affordable service providers like Think Grid offer technical help around the clock, easing the pressure on IT workers [24]. As a result, they can concentrate on duties that are essential to the operation of the company, saving the company money on hiring and training new employees. IBM, a leading provider of information technology, has highlighted how cloud computing enables businesses to expedite their procurement procedures and does away with the need for duplicating certain computer administrative skills related to setup, configuration, and support.

6.4.2 Organizational factors

Prior studies concentrated on the organizational elements that influence business decisions to adopt cloud computing is depicted in Figure 6.5. Discusses additional concerns such as the Service Level Agreement (SLA), data migration, and the issue of cloud scalability, that hinder the adoption of cloud computing.

The report also emphasizes the importance of exercising caution when using the cloud. One needs to comprehend the issues and difficulties surrounding the use of these technical breakthroughs [5]. In other words, businesses should be technically capable of implementing such a computer paradigm successfully. Additionally, it makes the case that developments in the IT industry should be taken into account when a company uses cloud technology. Before making the decision, the designer advised the organization to understand its cycles and the benefits of applying distributed computing. It states that a number of factors, such as the company's size, top management support, and prior experience, have an influence on how SMEs use cloud services. Small associations feel that because of their ability to modify their vision and mission quickly and adaptably, their size makes

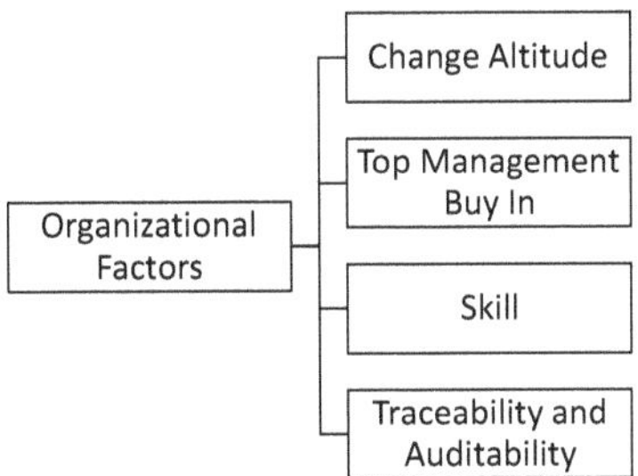

Figure 6.5 Organizational Factors of Cloud Computing Security.

it easier for them to retain new developments. The hierarchical scale is considered to be a crucial factor for start-ups and small organizations [25]. It is also generally agreed that start-up businesses are more likely to utilize cloud computing because it can help them avoid capital expenditure.

6.4.2.1 Change attitude

An organization should be aware of the dangers and other impacts cloud computing can have on its operational environment and consider those in its regular business operations. In some circumstances, cloud computing can infiltrate a company with ease while eluding normal management oversight mechanisms. Conventional processes and controls demand management's involvement and permission when a company commits considerable resources to a project that could take months or years to accomplish. Risk analyses, audits, and steering committees are all likely to be used to focus top management's attention on such initiatives [19]. Some cloud solutions can be quickly and simply implemented with only a little outlay of cash and the involvement of a small number of staff members. With cloud computing, a modest investment can have a great impact instead of the traditional equation of large investment = large impact. It may seem paradoxical to have to put in a lot of work to analyse the risks associated with cloud computing and carry out the necessary due diligence. As a result, management might forego taking time-consuming actions like verifying adherence to legal or regulatory standards or assessing the potential effects of the communications service providers on the organization's operations and risk profile. 90% of firms, according to a recent study, do not have sufficient disaster recovery or business continuity strategies, leaving them open to potential disruptions [26]. Numerous disaster recovery options, including cloud backup (which enables you to save essential files), are offered by providers.

6.4.2.2 Top management buy in

The significance of top management is in fostering the adoption of new technology by providing sufficient resources. The senior management of the organization commits to innovation, has a vision for it, and cultivates settings that support it. Cloud computing demands senior management support as it entails numerous resource integrations and process reengineering [27]. Additionally, research demonstrates that the adoption of new technologies is strongly positively correlated with the level of top management support and that the size of the organization is a major predictor of IT innovation. Large companies have been observed to adopt innovations more frequently due to their flexibility and capacity for taking calculated risks. As a result, the study postulates that a crucial element influencing cloud computing's anticipated value in technological growth is perceived. SMEs often have fewer slack resources to cushion the blow of a failed information system (IS) adoption investment. Due to the special characteristics of an SME, it is necessary to investigate if cloud adoption strategies created for large businesses can be similarly applicable to small- and medium-sized businesses [28]. Although many of the same limitations affect both large and small organizations, the impact on the latter is greater. Although the time, resources, and skills required for planning are not significant obstacles in large organizations, they do account for the majority of the challenges in SMEs. SMEs often have fewer slack resources to cushion the blow of a failed IS adoption investment. Due to the special characteristics of an SME, it is necessary to investigate if cloud adoption strategies created for large businesses can be similarly applicable to small and medium-sized businesses. Larger companies have more infrastructure and resources to support the adoption of innovation. SMEs experience a unique problem known as resource poverty. Resource poverty is a result of several SME-specific factors, such as operating in a highly competitive environment, facing financial constraints, lacking in-depth industry knowledge, and being vulnerable to outside forces [29].

This suggests that there are two clear motivations for weakening the favourable correlation between top administration support and the technical acceptability of modernization. For the best recognition and execution of an IT development, strong top administration support can initially provide a distribution of administrative resources (financial, specialized, and human) that is more than appropriate. Second, senior management may provide long-term vision, direction, support, and responsibility to create an environment that is beneficial for IT development to lessen hierarchical conflicts around the adoption of an IT innovation. According to this, there are two main reasons why the favourable correlation between top administration support and the acceptability of modernization may be mitigated. According to this, there are two main reasons why the favourable correlation between top administration support and the acceptability of modernization may be mitigated [28]. Second, top management can provide long-term vision, recommendations, support, and responsibilities to create a supportive atmosphere for IT innovation to lessen organizational conflicts around the adoption of an IT innovation.

6.4.2.3 Skill

The availability of technological infrastructure and the capacity of human resources have both been linked to an organization's ability and capacity to adopt new technologies. Installed network technologies and business systems are referred to as technological infrastructure because they provide a platform on which cloud computing applications can be used.IT human resources are equipped with the knowledge and abilities to implement IT applications connected to cloud computing. Services for cloud computing can only if companies possess the necessary technical know-how and infrastructure to participate in value chain operations. As a result, technologically advanced businesses are better equipped to use cloud computing. According to the Commission (2010), there are numerous reasons why businesses of all sizes and types are implementing this IT architecture. With the use of cloud computing, it is possible to enhance features or boost capacity instantly without spending money on new infrastructure, hiring additional staff, or licensing new software [30]. In the end, it can help businesses save a sizable sum of money. Because of this, small businesses are less aware of the advantages of cloud computing and lack both technical and IS knowledge. Due to their extremely competitive environment, SMEs also have limited financial resources and are prone to short-term planning.

6.4.2.4 Traceability and auditability

Increased auditability and traceability were other organizational elements that affected adoption. It is possible to track how each information service is used within an organization because of cloud capabilities like traceability. To make sure that businesses adhere to internal and external restrictions, it is essential to be able to trace the history, location, or application of an item using documented documentation. Internal compliance regulations may mandate that businesses examine how their global data is used [31].

6.4.3 Environmental factors

Previous studies underlined the significance of environmental factors as predictors of cloud computing adoption in addition to technological and organizational considerations. The size of the market is one factor influencing cloud computing adoption globally. The findings demonstrated that businesses that serve a large market likely to use cloud services, which could boost their productivity. The industry in which a corporation operates has an impact on how quickly technology is adopted. Adopters and prospectors, for instance, think that industries with high computing needs are likely to employ more cloud services [32]. The study also highlights the importance of supplier computing support, or external computing support, in the decision-making process. Most participants can be classified as providers, adopters, or prospectors. The environmental factors are classified in Figure 6.6 including competition, trends, and industry.

6.4.3.1 Competition

Competitive pressure is the amount of pressure experienced by the company from competitors in the same industry. It has been asserted that one

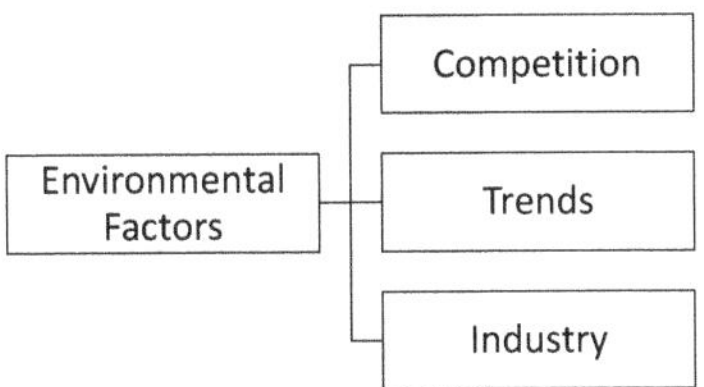

Figure 6.6 Environmental Factors of Cloud Computing Security.

key factor influencing IT adoption is exposure to fierce competition. Due to the high-tech industry's propensity for rapid change, businesses are under pressure to adopt new technology as quickly as their rivals do. Businesses gain a lot from implementing cloud computing, including improved market visibility, increased operational efficiency, and more precise data collection. Additionally, a lot of businesses depend on their trading partners to help with IT design and implementation. Trading partner pressure has been identified as a significant influencer of IT adoption and use, according to some empirical research studies [33]. Budgets for greenfield projects are constrained. To increase customer satisfaction and increase market share, most firms frequently provide new products and services to their clientele. Making investments in such initiatives is never easy due to tight IT expenditures. The initial start-up investment for these new ventures becomes minimal by using cloud services, and there are no costs associated with infrastructure breakdowns. Additionally, cloud services enable speedy on-demand infrastructure and service deployment [34]. Therefore, cutting-edge pilot projects can readily utilize Cloud services and deploy them with no risk. The cost of managing IT operations and systems, as well as maintenance, is significant and complicated: Cloud computing offers low-cost, simple-to-manage operational solutions and minimizes capital expenditure when managing IT operations and systems is complex and expensive. Some cloud service providers offer monitoring services and automatic fail-over systems, which lower overall management and operating expenses. In many cases, systems are designed to handle large-scale operations, but in practice, the demand for such computing needs rarely reaches its peak. Business processes have unpredictable needs for IT services. The demand for computing varies with time, which results in many systems being underutilized and computing resources being squandered [9]. As a result, infrastructure (hardware and software) is used inefficiently. These problems are addressed by cloud computing's on-demand elastic services, which are easily scalable as computing needs increase.

6.4.3.2 Trends

Commercially, organizations are putting more effort into developing internet-based trade partner solutions and integrating business processes into their current IS applications. Data transformation processes are widely used in high-tech enterprises, which is one of the most crucial variables for boosting operational efficiency. Building cloud computing competency is essential for boosting competitive advantage because it is quickly changing how organizations acquire, sell, and engage with consumers as well as becoming an even more essential part of businesses. After all, it allows businesses to carry out data transactions along value chain activities (such as production, financing, distribution, sales, customer support, information exchange, and collaboration with trading partners), the proliferation of cloud computing has become a key research area. Non-core IT operations are becoming commodities [6]. Many commoditized IT services, like email, archiving, storage, etc., are excellent candidates for cloud services. Outsourcing these services to a cloud service provider can frequently result in lower operational costs for the upkeep and management of these services as well as lower IT infrastructure costs. Although cloud computing has been regarded as a new technology that can offer its users several benefits, both operationally and strategically, the adoption rate is not increasing as quickly as anticipated [35]. In reality, a survey of numerous businesses from various sectors that had developed unique cloud-based apps was conducted, and the effects of cloud computing on business operations in the areas of security integration were examined.

6.4.3.3 Industry

Technology has a significant impact on the financial services sector. It has the power to completely change the insurance industry. The elusive value of IT in the insurance sector is shown by the rising focus of insurers on IT-driven transformation initiatives to set them apart. However, unless insurers—and all financial services organizations—invest the required time, resources, and experience, realizing value from IT investments can be difficult. The risk of cybercrime is one drawback of a technologically advanced industry like financial services. Cybercrime, also known as computer crime, is an economic crime committed using computers and the internet. If financial services organizations do not put in place the proper controls to protect their consumers and enhance information security, cybercrime in the financial services sector will increase due to the variety of product lines and mobile banking solutions.

6.4.4 Country-specific factors

According to the contextual theory, a greater knowledge of IT adoption, including the adoption of cloud computing, may be obtained by studying the phenomena and the conditions that are present in a particular environment. A nation like Lebanon has significant obstacles on the levels of the economy, infrastructure, and politics. As a result, it is impossible to overlook such difficulties when investing in a new computing paradigm [4]. Though scarce, earlier studies on cloud computing in the the Middle East and North Africa region took the environment's influence on adoption decisions into account. Demonstrates how a variety of factors, such as material prosperity, governmental stability, and geographic location, have an impact on cloud adoption globally. As it is unclear where the real data is stored, accessible from, and processed, and because each region may have its privacy laws and regulations when it comes to keeping and accessing data, it is discovered that political stability has a significant impact on the adoption of the cloud. By extension, politics influences the adoption of cloud computing and should be kept apart from it for cloud computing to become a borderless and international instrument [36]. Additionally, the World Bank's Global Competitiveness Index reveals that Lebanon's difficulties in adopting technology and innovation are caused by:

- Low levels of industry collaboration between universities and industry,
- Low levels of investment in research and development
- New start-ups and SMEs rely little on sophisticated technology,

According to a poll conducted by the Ministry of Economy and Trade in Lebanon, major obstacles to investments include the macroeconomic environment, corporate rules, infrastructure readiness, and the political status quo. The preparedness and interest of developing countries, including Lebanon, in adopting cloud computing technology services are also slow and disheartening, according to a report on the economy of the cloud in developing countries. This is illustrated by the numerous power outages, inadequate connectivity, subpar infrastructure, and other economic and political difficulties that these nations have. Contextual elements differ between studies and places. Several context-related aspects, including context management, supporting context (such as infrastructure), and situational circumstances, have been emphasized in prior studies (such as government policies, exchange rates, political situation, and national security).

6.5 INFLUENCE OF CLOUD COMPUTING ON BUSINESS OPERATIONS

The efficient use of manufacturing capability can be considered as operations strategic planning. Basically cloud computing business operations are directly related to cloud services including SaaS, PaaS, and IaaS shown in Figure 6.7. Using technology to achieve organizational and business objectives like

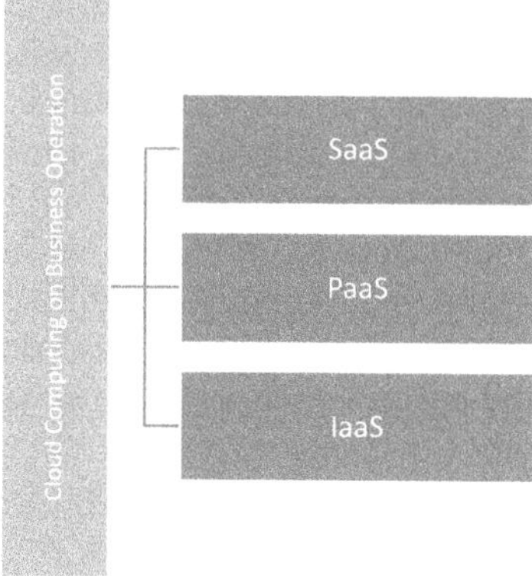

Figure 6.7 Cloud Computing on Business Operation.

Profit, innovation, personalization, product adaptability, product dependability, quality, and reaction are some of these objectives. Organizations have often installed applications by placing the application's software on one or more physical servers. When an application handles a large number of transactions or when there needs to be a high degree of certainty that the failure of a single server won't result in an application failing, many servers are frequently employed. The student registration process at a university is a well-known illustration of this kind of application. Transaction volumes will be very high during peak registration periods, and the university has to guarantee that a hardware malfunction will not cause the program to stop working. However, the application will be little used in the middle of the semester, thus an outage is probably not going to have a significant impact.

6.5.1 Software as a service

The applications are supplied via the World Wide Web as a service, and this is the Cloud's greatest degree of abstraction. Applications ranging from productivity tools to business tools like email hosting, supply chain management, or enterprise resource planning are available through this layer of cloud services.

6.5.2 Platform as a service

This is the second level of cloud abstraction, which also includes fundamental application infrastructure services like processing, messaging, networking, access control, etc. in addition to technical abstraction. A team of network, database, and system management experts is required to maintain everything operational in the traditional in-house computer approach. With the advent of cloud computing, these services are now delivered remotely by cloud service providers.

6.5.3 Infrastructure as a service

This is cloud computing's base layer. IaaS vendors offer services that abstract IT infrastructure resources like storage and memory. A cloud service provider oversees the physical infrastructure and provides end users with virtualized operating systems infrastructure. The virtual image, in this case, belongs entirely to the user and can be configured according to the needs. The remote distribution and support of an entire computer infrastructure (through the World Wide Web) is one of the products provided by this layer.

6.6 CHALLENGES OF CLOUD COMPUTING SECURITY

- Cloud security architecture and cloud security architect jobs are crucial to the success of cloud computing projects because the cloud requires a unique set of security design concepts, methods, and technology.
- The cloud risk assessment must be automated to meet corporate objectives.
- Tier 1 cloud providers typically offer the most secure foundation for all workload types. Because many firms have embraced a multi-cloud strategy, the usage of third-party security products is essential for consistent policy and governance across a multi-cloud landscape.
- End users, who increasingly employ hundreds of software service apps and numerous infrastructures as service offerings, are using the platform as a service technology to construct new services. As a result, cloud security initiatives must take into account the diverse cloud "form factors" used by numerous enterprises [11].

The challenges should be addressed in terms of security and how the quantitative assessment will be evaluated by soft computing techniques. The phrase "soft computing" refers to a collection of methods that work together to offer flexible information processing skills for handling confusing real-world scenarios. The computer benefits from the tolerability of the computing technique's capacity to impreciseness, uncertainty, approximation, and partial truth are all elements that contribute to the computing technique's capacity: the basic goal of soft computing is to provide a computation technique that will offer a workable resolution at a reasonable cost to a problem that can be expressed either accurately or imprecisely. Robustness, tractability, and low-cost solutions are other important ideas combining several disciplines such as fuzzy logic, fuzzy computing, genetics, and neuroscience.

6.7 CONCLUSION

One advantage of using a cloud does not need to comprehend the mechanics of the infrastructure, yet organizations typically desire to understand where the software's cloud "lives." The main problem with clouds hosted abroad is that data is subject to the laws and regulations of both the programme in question and the host nation. For instance, several Canadian provinces have made it illegal for apps to be hosted in the USA as the data would then be susceptible to the Patriot Act's provisions. Some American programmes cannot be hosted overseas due to similar restrictions on the transfer of computer system technology. Because of this, it is important to consider where a cloud is hosted when talking about difficulties with legal compliance auditability, and e-discovery. A corporation must have a lot of confidence in the security procedures and protocols employed by its cloud programming provider as cloud processing is a subgenre of computer outsourcing. With a variety of factors, including the type of software used and data stored in the cloud, a company might be confronted with issues connected to Health Insurance Portability and Accountability Act, Family Educational Rights and Privacy Act, Payment Card Industry Security Standards Council, or the Gramm-Leach-Bliley Act, as well as other statutes or regulatory authorities. Many companies have long contracted with a third-party supplier for IT services. However, the IT services industry, including outsourcing, is undergoing a fast transformation as a result of the increased usage of cloud computing. The worldwide computer technology outsourcing market has seen a significant transformation as more firms embrace cloud services. The full or a part of the lifecycle of building information systems is covered by a contract with a low-cost international service provider that handles global offshore outsourcing. Privacy and security issues are the main challenges the information technology business faces. To assess the risks and advantages of cloud computing to traditional on-site computing, they offer a threshold method, and they only partly switch to cloud computing if the advantages surpass the dangers. According to certain experts, there are still some hazards and uncertainties related to cloud computing, but it will take time for these concerns to be overcome. Decision-makers in some businesses prioritize data security when looking at cloud computing as a challenge when delivering cloud computation services to healthcare facilities like hospitals, therefore protecting confidential data is essential. Participants have recognized a problem with government policy as a danger to the effective adoption of cloud computing. Additionally, for security and privacy concerns, several countries imposed restrictions on cross-border data transfers. A participant in the conversation said, "Cloud computing" has emerged as a brand-new phenomenon used by all Saudi Arabian organizations. In this case, the government must undoubtedly plan and regulate its usage. How rapidly cloud computing is being embraced is directly related to the level of competence and experience in the field.

REFERENCES

[1] K. Hashizume, D. G. Rosado, E. Fernández-Medina, and E. B. Fernandez, "An analysis of security issues for cloud computing," *J. Internet Serv. Appl.*, vol. 4, no. 1, pp. 1–13, Feb. 2013, doi: 10.1186/1869-0238-4-5/TABLES/4.

[2] A. Alharbi *et al.*, "A link analysis algorithm for identification of key hidden services," *Comput. Mater. Contin.*, vol. 68, no. 1, 2021, doi: 10.32604/cmc.2021.016887.

[3] F. Alassery, A. Alzahrani, A. I. Khan, A. Khan, M. Nadeem, and M. T. J. Ansari, "Quantitative evaluation of mental-health in type-2 diabetes patients through computational model," *Intell. Autom. Soft Comput.*, vol. 32, no. 3, 2022, doi: 10.32604/IASC.2022.023314.

[4] S. A. Khan, M. Nadeem, A. Agrawal, R. A. Khan, and R. Kumar, "Quantitative analysis of software security through fuzzy PROMETHEE-II methodology: A design perspective," *Int. J. Mod. Educ. Comput. Sci.*, vol. 13, no. 6, 2021, doi: 10.5815/ijmecs.2021.06.04.

[5] M. Ahmad *et al.*, "Healthcare device security assessment through computational methodology," *Comput. Syst. Sci. Eng.*, vol. 41, no. 2, 2022, doi: 10.32604/csse.2022.020097.

[6] M. Alenezi, M. Nadeem, A. Agrawal, R. Kumar, and R. A. Khan, "Fuzzy multi criteria decision analysis method for assessing security design tactics for web applications," *Int. J. Intell. Eng. Syst.*, vol. 13, no. 5, 2020, doi: 10.22266/ijies2020.1031.17.

[7] H. Alyami *et al.*, "The evaluation of software security through quantum computing techniques: A durability perspective," *Appl. Sci.*, vol. 11, no. 24, 2021, doi: 10.3390/app112411784.

[8] A. Alharbi *et al.*, "Managing software security risks through an integrated computational method," *Intell. Autom. Soft Comput.*, vol. 28, no. 1, p. 179, Mar. 2021, doi: 10.32604/IASC.2021.016646.

[9] Y. Lu, "Cyber physical system (CPS)-based Industry 4.0: A survey," *J. Ind. Integr. Manag.*, vol. 02, no. 03, p. 1750014, Nov. 2017, doi: 10.1142/S2424862217500142.

[10] Y. Zhao, H. Liu, Y. Wang, Z. Zhang, and D. Zuo, "Reducing the upfront cost of private clouds with clairvoyant virtual machine placement," *J. Supercomput.*, vol. 75, no. 1, pp. 340–369, Feb. 2018, doi: 10.1007/s11227-018-02730-4.

[11] I. V. Pustokhina, D. A. Pustokhin, D. Gupta, A. Khanna, K. Shankar, and G. N. Nguyen, "An effective training scheme for deep neural network in edge computing enabled internet of medical things (IoMT) systems," *IEEE Access*, vol. 8, pp. 107112–107123, 2020, doi: 10.1109/ACCESS.2020.3000322.

[12] S. Shahzadi, M. Iqbal, Z. U. Qayyum, and T. Dagiuklas, "Infrastructure as a service (IaaS): A comparative performance analysis of open-source cloud platforms," IEEE Int. Work. Comput. Aided Model. Des. Commun. Links Networks, CAMAD, vol. 2017-June, Sep. 2017, doi: 10.1109/CAMAD.2017.8031522.

[13] P. R. Palos-Sanchez, F. J. Arenas-Marquez, and M. Aguayo-Camacho, "Cloud computing (SaaS) adoption as a strategic technology: Results of an

empirical study," *Mob. Inf. Syst.*, vol. 2017, Jun. 2017, doi: 10.1155/2017/ 2536040.

[14] J. M. Del Alamo, R. Trapero, Y. S. Martin, J. C. Yelmo, and N. Suri, "Assessing privacy capabilities of cloud service providers," *IEEE Lat. Am. Trans.*, vol. 13, no. 11, pp. 3634–3641, Nov. 2015, doi: 10.1109/TLA.2015.7387942.

[15] M. Armbrust *et al.*, "A view of cloud computing," *Commun. ACM*, vol. 53, no. 4, pp. 50–58, Apr. 2010, doi: 10.1145/1721654.1721672.

[16] C. M. Messerschmidt and O. Hinz, "Explaining the adoption of grid computing: An integrated institutional theory and organizational capability approach," *J. Strateg. Inf. Syst.*, vol. 22, no. 2, pp. 137–156, Jun. 2013, doi: 10.1016/J.JSIS.2012.10.005.

[17] G. Garrison, S. Kim, and R. L. Wakefield, "Success factors for deploying cloud computing," *Commun. ACM*, vol. 55, no. 9, pp. 62–68, Sep. 2012, doi: 10.1145/2330667.2330685.

[18] A. Rashid and A. Chaturvedi, "Cloud computing characteristics and services: A brief review proposing an innovative approach for dynamic resource scaling especially in multi-tenancy cases on cloud networks view project face recognition and artificial intelligence view project cloud computing characteristics and services: A brief review," *Artic. Int. J. Comput. Sci. Eng.*, 2019, doi: 10.26438/ijcse/v7i2.421426.

[19] W. K. Lee, H. Seo, Z. Zhang, and S. O. Hwang, "TensorCrypto: High throughput acceleration of lattice-based cryptography using tensor core on GPU," *IEEE Access*, vol. 10, pp. 20616–20632, 2022, doi: 10.1109/ ACCESS.2022.3152217.

[20] Y. Shingu *et al.*, "Variational secure cloud quantum computing," *Phys. Rev. A*, vol. 105, no. 2, Jun. 2021, doi: 10.1103/PhysRevA.105.022603.

[21] H. Gangwar, H. Date, and R. Ramaswamy, "Understanding determinants of cloud computing adoption using an integrated TAM-TOE model," *J. Enterp. Inf. Manag.*, vol. 28, no. 1, pp. 107–130, Feb. 2015, doi: 10.1108/ JEIM-08-2013-0065.

[22] M. D. Ryan, "Viewpoint cloud computing privacy concerns on our doorstep," *Commun. ACM*, vol. 54, no. 1, pp. 36–38, Jan. 2011, doi: 10.1145/ 1866739.1866751.

[23] Y. Wang, J. Li, and H. H. Wang, "Cluster and cloud computing framework for scientific metrology in flow control," *Cluster Comput.*, vol. 22, no. 1, pp. 1189–1198, Jan. 2019, doi: 10.1007/S10586-017-1199-3/TABLES/3.

[24] M. Ali, S. U. Khan, and A. V. Vasilakos, "Security in cloud computing: Opportunities and challenges," *Inf. Sci. (Ny)*, vol. 305, pp. 357–383, Jun. 2015, doi: 10.1016/J.INS.2015.01.025.

[25] S. Abidin, A. Swami, E. Ramirez-Asís, J. Alvarado-Tolentino, R. K. Maurya, and N. Hussain, "Quantum cryptography technique: A way to improve security challenges in mobile cloud computing (MCC)," *Mater. Today Proc.*, vol. 51, pp. 508–514, 2022, doi: 10.1016/j.matpr.2021.05.593.

[26] A. El Azzaoui, P. K. Sharma, and J. H. Park, "Blockchain-based delegated Quantum Cloud architecture for medical big data security," *J. Netw. Comput. Appl.*, vol. 198, p. 103304, 2022, doi: 10.1016/j.jnca.2021.103304.

[27] R. Pratap Singh, M. Javaid, A. Haleem, R. Vaishya, and S. Ali, "Internet of Medical Things (IoMT) for orthopaedic in COVID-19 pandemic: Roles,

challenges, and applications," *J. Clin. Orthop. Trauma*, vol. 11, no. 4, pp. 713–717, Jul. 2020, doi: 10.1016/J.JCOT.2020.05.011.

[28] M. Saraswat and R. C. Tripathi, "Cloud Computing: Analysis of Top 5 CSPs in SaaS, PaaS and IaaS platforms," *Proc. 2020 9th Int. Conf. Syst. Model. Adv. Res. Trends, SMART 2020*, pp. 300–305, Dec. 2020, doi: 10.1109/SMART50582.2020.9337157.

[29] N. S. Aldahwan and M. S. Ramzan, "Descriptive literature review and classification of community cloud computing research," *Sci. Program.*, vol. 2022, 2022, doi: 10.1155/2022/8194140.

[30] T. S. Behrend, E. N. Wiebe, J. E. London, and E. C. Johnson, "Cloud computing adoption and usage in community colleges," *Behav. Inf. Technol.*, vol. 30, no. 2, pp. 231–240, Mar. 2011, doi: 10.1080/0144929X.2010.489118.

[31] Q. Wang, C. Wang, K. Ren, W. Lou, and J. Li, "Enabling public auditability and data dynamics for storage security in cloud computing," *IEEE Trans. Parallel Distrib. Syst.*, vol. 22, no. 5, pp. 847–859, 2011, doi: 10.1109/TPDS.2010.183.

[32] A. Aguado *et al.*, "Hybrid conventional and quantum security for software defined and virtualized networks," *J. Opt. Commun. Networking*, vol. 9, no. 10, pp. 819–825, Oct. 2017, doi: 10.1364/JOCN.9.000819.

[33] K. Zhu, K. L. Kraemer, and S. Xu, "The process of innovation assimilation by firms in different countries: A technology diffusion perspective on e-business," *Manage. Sci.*, vol. 52, no. 10, pp. 1557–1576, 2006, doi: 10.1287/MNSC.1050.0487.

[34] R. K. Gupta, K. K. Almuzaini, R. K. Pateriya, K. Shah, P. K. Shukla, and R. Akwafo, "An improved secure key generation using enhanced identity-based encryption for cloud computing in large-scale 5G," *Wirel. Commun. Mob. Comput.*, vol. 2022, 2022, doi: 10.1155/2022/7291250.

[35] H. Alyami *et al.*, "Analyzing the data of software security life-span: Quantum computing era," *Intell. Autom. Soft Comput.*, vol. 31, no. 2, 2022, doi: 10.32604/iasc.2022.020780.

[36] I. Ion, N. Sachdeva, P. Kumaraguru, and S. Čapkun, "Home is safer than the cloud! Privacy concerns for consumer cloud storage," *SOUPS 2011—Proc. 7th Symp. Usable Priv. Secur.*, 2011, doi: 10.1145/2078827.2078845.

Chapter 7

A review on evolving technology of IoT and OT

Security threats, attacks, and defenses

Aditya Pratap Singh and Kavita Sahu

7.1 INTRODUCTION

The Internet of Things (IoT) is a crucial and evolving subject in the realms of technology, business, and society in general. A network of material things, called the IoT, is capable of recognizing, collecting, organizing, and transmitting data use internet protocols. IoT has fundamentally altered how people live. Recently, the use of the internet has expanded beyond computers to encompass smartphones, business machines, household products, and more [1,2]. Some examples of IoT in action are linked appliances and voice assistants like Fitbits, Google Home, Amazon Echo, and Alexa are all examples of activity monitors.

In today's modern civilization, operational technology (OT) is crucial as it powers a group of devices that are intended to function as a single, integrated system [3]. OT is a term used to describe software and hardware that is employed to keep an eye on and manage physical processes, equipment, and infrastructure to influence changes in industrial operations [4]. IoT and OT systems are spreading quickly and being increasingly used in everyday life, which might have negative effects that should be considered.

Due to a 2017 CNN report, the Food and Drug Administration (FDA) confirmed that St. Jude Medical's embedded cardiac devices had flaws that may allow an attacker to hack into a device. Once inside, they may exhaust the battery or provide unsuitable pacing or events, the FDA cautioned. The system, which comprises cardiac arrest devices and cardiac devices, is used to keep track of a patient's cardiovascular health, control it, and fend off heart attacks.

The transmitter that reads the device's data and wirelessly communicates it with doctors included vulnerability, the story added. The FDA said that by gaining access to a device's transmitter, the hackers may control it.

The Owlet WiFi neonatal heart monitor is situated behind the St. Jude cardiac devices. Cesare Garlati, the prpl Foundation's senior security strategist, claims that this most recent event is simply another example

DOI: 10.1201/9781003514312-7

of how even gadgets with the greatest of intentions may cause problems, like those that warn parents when their newborns experience cardiac problems, may put people in danger if used improperly by an unintentional person. Unfortunately, this is often the case when embedded computer is used in so-called smart gadgets. We will regretfully keep on hearing stories like this if manufactures and coders fail to take this into account and implement additional security precautions to safeguard devices at the hardware layer.

TechNewsWorld claims that TRENDnet promoted the safety of their SecurView cameras and said they could be used for a number of things, including home security and baby monitoring. However, due to faulty software, anyone with the IP address of a camera could convey user login credentials in plain, readable text via the internet and keep customers' login information in plain text. From at least April 2010 (until around January 2012), legible language was available on their mobile devices. As protecting IP addresses from hackers and encrypting or at the very least password protecting login credentials are typical security practices, TRENDnet's failure to do so was unexpected. Through integration, the organization's processes may be improved and capabilities expanded [5].

This work is focused on bringing out the issues of cybersecurity in IoT and OT devices. A review of literature related to these areas is done in this work and the problems associated are gathered along with possible solutions as a future work. The remainder of the chapter is structured as follows: second section focuses on review on recent literature. Next section is focused on background of famous Purdue model. Risk versus future trends and problems of IoT are discussed in the next two sections of this chapter. Conclusion and future work are discussed in the last section.

7.2 LITERATURE REVIEW

Two of the main issues in industrial cybersecurity research are protecting information technologies (IT) and OT systems from all conceivable risk situations and determining their security degree. The Jeep attack was mentioned on the IBM intelligence collection website a few years ago, with an announcement, It was just one, but it was many. In July [2015], a group of researchers entirely seized possession of the Jeep SUV via the CAN system. They used the Sprint cellular network to take control of the car by using a weakness in a firmware upgrade. The car's velocity, quickly, along with steering it off the road were all things they realized they could manage. For the just found IoT, it acts as a proof-of-concept assaults, highlighting the grave repercussions of businesses frequently putting their networks' or auxiliary devices' security at risk. If the IoT revolution is to continue benefiting people, we must create more reliable security protocols, methods, and standards. For diving more deeply into the topic, we have reviewed the recent research done in the field with its accuracy and results.

From the literature review above, it has been ruled out that there is no specific framework available simultaneously for IoT as well as OT security and risk associated with it. In the next sections, we will focus more on different types of models and attacks on IoT and OT devices.

7.3 BACKGROUND

Industrial networks are now so linked due to the spread of IoT and cloud adoption throughout the industrial value chain that the conventional air gap is ineffective. The technological operations, detectors, monitoring oversight, tasks, and logistics are divided into several categories and are the key components of the Purdue model, an organizational structure for industrial control system (ICS) security. It was developed at Purdue University. The paradigm has been around for a long time and is being used today despite the growth of edge computing and direct connectivity to the cloud. Its primary purpose is to safeguard OT against malware and other forms of assault. The Purdue model, developed from the ISA-99 model of the Purdue Enterprise Reference Architecture (PERA), serves as a conceptual representation of internal network segmentation. The demilitarized zone, the enterprise zone, and the industrial zone are the three zones that make up the Purdue model (DMZ) [6,7] as shown in Figure 7.1. The following web link (https://docs.aws.amazon.com/AmazonCloudFront/latest/DeveloperGuide/WhatsNew.html) describes the important changes made to Cloud Front documentation.

7.3.1 Level 4/5 enterprise zone

When doing a timing analysis, the accompanying timestamps for each event are examined. An adversary may employ specific attack techniques to gather data on network events like packet transit. This attack works by taking advantage of the varied execution times for different branches in the ecosystem.

7.3.2 Level 3.5 demilitarized zone

Security tools like firewalls and proxies are employed in this zone to try to stop lateral threat transfer between IT and OT. This IT–OT convergence layer can provide firms a competitive edge, but it can also raise their cyber risk, because the advent of automation has increased the requirement for bidirectional data flows between OT and IT systems.

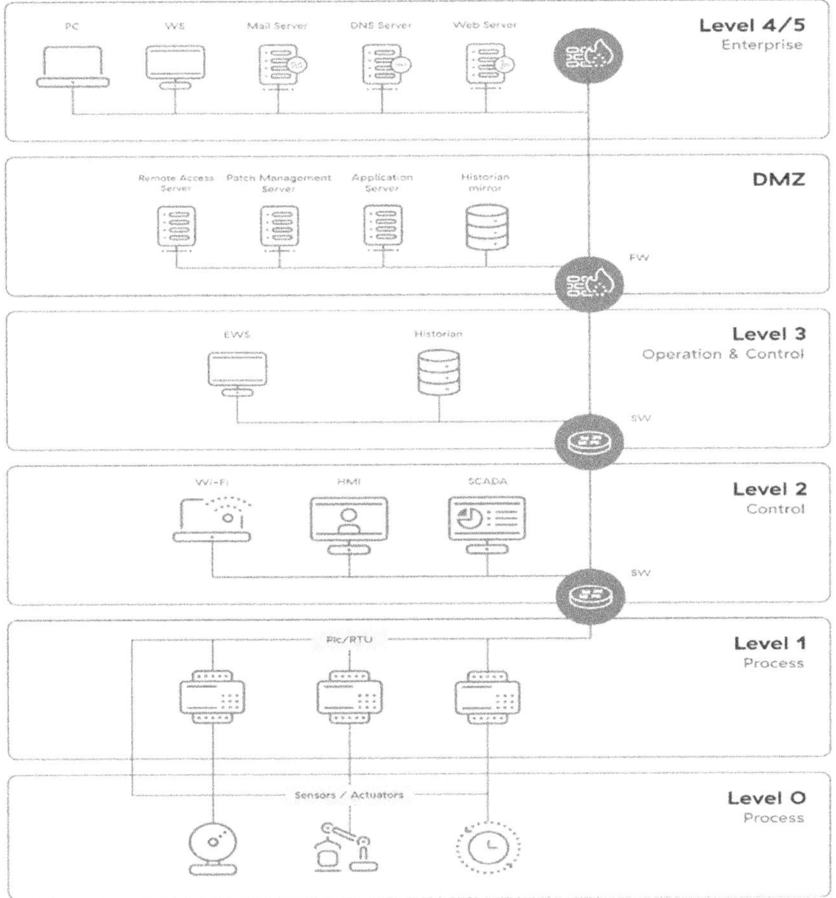

Figure 7.1 The Purdue Model.

7.3.3 Level 3 manufacturing operations system zone

This area features specialized OT equipment that controls shop floor production workflows:

Production activities are managed using manufacturing operations management (MOM) systems.

Real-time data are gathered by manufacturing execution systems (MES) to aid in production optimization.

Data historians archive process data and carry out contextual analysis (in contemporary solutions).

Similar to Levels 4 and 5, disturbances here may result in financial loss, breakdown of vital infrastructure, or a reduction in income.

7.3.4 Level 2 control system

Systems that direct, keep an eye on, and regulate physical processes are found in this zone.

Supervisory Control and Data Acquisition (SCADA) software gathers data for transmission to historians, while monitoring and controlling physical processes locally or remotely.

Distributed control systems (DCS) are implemented locally but fulfill SCADA functionality.

Basic controls and monitoring are made possible by connecting human–machine interfaces (HMIs) to DCS and programmable logic controllers (PLCs).

7.3.5 Level 1 intelligent devices

Instruments in this area transmit commands to Level 0 devices.

Programmable logic controllers keep track of manual or automated input in industrial operations and change output accordingly.

Remote terminal units (RTUs) link systems at Level 2 to hardware at Level 0.

7.3.6 Level 0 physical process zone

Sensors, actuators, and other equipment that is directly in charge of assembly, lubrication, and other physical processes are located in this area. Many contemporary sensors use cellular networks to connect directly to monitoring software on the cloud.

7.4 RISKS VERSUS FUTURE TRENDS

Even while IoT is expanding quickly and offers many benefits, some devices still lack security updates or fixes, making them susceptible and giving them only a few features [6]. Security, privacy, and safety are the three main areas into which the dangers to IoT may be divided. Given that IoT devices are expanding, prevalent than smartphones and other technological devices in our daily lives, it is evident how important these categories are. The most private and delicate data, including bank records, personal information, and social security numbers, will be available to it.

For instance, there are fewer issues when it comes to smartphones or laptops; however, with IoT devices, the issues multiply fast [9,10]. Some risks are given below:

- Inadequate access controls
- An excessively wide assault surface
- Out-of-date software
- No encryption
- Application flaws
- An execution environment that is not trusted
- The level of vendor security
- Inadequate privacy safeguards
- Ignorance of intrusion

In the future, we will see the finest of IoT and artificial intelligence (AI) combined in lethal ways. Data gathered from IoT devices are analyzed with the aid of AI algorithms, which in turn produce beneficial findings that are further deployed utilizing IoT devices [11]. They both function in this cycle together. As virtual user interface and the miniaturization of things (smart items) produce many benefits for consumers, development in this area is ongoing. Where work is continually being done, power consumption reduction or appropriate utilization of existing power sources is a crucial factor. IoT will have such a broad impact on society, encompassing important industries like transportation, manufacturing, and agriculture. The major issue with the emergence of OT is its security component [12,13]. Therefore, if we pay close attention to potential dangers and utilize the necessary strategies to resolve the problems, we may enhance communication, reduce the risk of cyber-attacks, increase efficiency, and improve user friendliness.

7.5 PROBLEMS OF INTERNET OF THINGS

As was already noted, security was not considered when developing IoT devices. As a result, there are a number of IoT security problems that might have serious repercussions which are shown in Figure 7.2. IoT security is governed by a very small number of standards and laws compared to other technology solutions [14]. The majority of people are also ignorant of the risks associated with IoT systems. Furthermore, they are not mindful of the magnitude of IoT safety hazards [15,16]. The basic security issues and their solutions are shown in Table 7.1. The following are some of the numerous IoT security concerns.

7.5.1 Lack of visibility

IoT device deployments by users typically occur without the IT departments' knowledge, making it challenging to have an accurate inventory of what has to be protected and tracked.

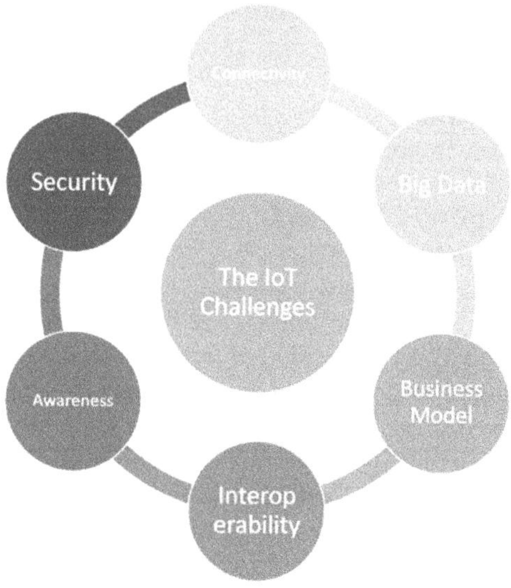

Figure 7.2 Problems of IoT.

7.5.2 Limited participation concerning safety

The IoT devices communicating with security systems can prove challenging or impractical due to their diversity and size.

7.5.3 Poor testing

Most IoT developers do not place a lot of emphasis on security, which prevents them from performing testing for vulnerabilities to detect problems with their systems.

7.5.4 Unpatched vulnerabilities

For multiple reasons, such as an absence of solutions and the difficulty in locating and deploying them, plenty of IoT devices have weaknesses that are not yet fixed.

7.5.5 Vulnerable APIs

Application programming interfaces (APIs) are frequently utilized as gateways by controlling and command centers to initiate tactics like injection of Structured Query Language (SQL) and widespread lack disruption

Table 7.1 Reviews of IoT Objectives, Datasets, Result, and Techniques

Author	Year	Objectives	Datasets	Results and Accuracy	Technique
Bagui et al. [53]	2021	IoT Botnet recognizing the penetration and attack.	Machine learning collection at UCI	The accuracy of many ML distribution was greater than 99% in several instances and 100% in others.	RF, SVM, LR
Thamaraiselvi and Mary [54]	2020	Problems with ML's serviceability in identifying abnormalities in IoT networks	IoT-23	The RF also produced the optimum outcomes, with a 99.5% accuracy rate.	Decision Tree, Naive Bayes, SVM, RF
Susilo and Sari [55]	2020	Utilizing machine learning techniques to increase IoT security	BoT-IoT	The highest accuracy rates were achieved by CNN and random forests.	CNN, Random Forest, and MLP
Hasan et al. [56]	2019	To forecast attacks and abnormalities on IoT networks, the effectiveness of several machine learning models has been rigorously examined.	NSL-KDD Real Traffic, DS2OS	The highest accuracy, 99.4%, was achieved by Random Forest.	DT, RF, SVM, ANN, and LR
Elmrabit et al. [57]	2020	Recognize unusual activity that might suggest an online assault	CICIDS-2017, UNSW-NB15,ICS cyber-attack	For the CICIDS-2017 dataset, the Random Forest The effectiveness of technique, which is 99.9%, is the most popular.	DT, KNN, LSTM, DNN, RF,AdaB, Simple RNN, GRU, CNN-LTSM, LR, GNB
Liu et al. [58]	2020	Improve IoT security by employing smart home invasion datasets to evaluate different machine learning methods.	IoT Network Intrusion dataset	Employing KNN algorithms, the precision was 99%	LR, SVM, RF, KNN, XGBoost

(continued)

Table 7.1 (Cont.)

Author	Year	Objectives	Datasets	Results and Accuracy	Technique
Aysa et al. [59]	2020	IoT device recognition of incidents and unusual events	Data from three IoT devices that UCI collected that was both normal and problematic	High accuracy was achieved by combining decision trees with random forests.	Random Forest, Decision Tree, SVM, and Neural Network
Al-Akhras et al. [60]	2020	Analyze the efficiency of different ML algorithms for spotting risks and irregularities in connected devices	UNSW-NB15	Without noise injection, the optimum accuracy for RF and KNN classifiers is 100%, while with 10% noise filtering, it is 99%.	KNN, RF, and Naive Bayes
Rani and Kaushal [61]	2020	Enhance the security and precision of the IDS (system for identifying breaches)	The KDDCUP99 and NSL and KDD	With minimal time and energy expended, the suggested simulation offers a 99.9% accuracy to recognize breaches.	Decision Tree, Logistics Regression, RF, KNN, NB
Stoian [62]	2020	Attacks and anomaly detection in IoT networks.	IoT-23	With 99.5% reliability, the RF algorithm generated the best performance.	NB, MLP, SVM, AdaBoost, and RF

ML: Machine Learning; UCI: University of California Irvine Machine Learning Repository; IoT: Internet of Things; RF: Random Forest; SVM: Support Vector Machine; LR: Linear Regression; CNN: Convolutional Neural Network; MLP: Multilayer Perception; KDD: knowledge discovery in a database; NSL: Neural Structured Learning; DT: Decision Tree; DS2OS: Distributed Smart Space Orchestration System (DS2OS) dataset; CICIDS-2017 Dataset: The Canadian Institute for Cybersecurity Intrusion Detection System 2017; UNSW-NB15: A network intrusion dataset; LSTM: Long short-term memory; DNN: Deep Neural Networks; GRU: Gated recurrent unit; AdaBoost algorithm, short for Adaptive Boosting; GNB: Gaussian Naive Bayes.

Table 7.2 Top IoT Vulnerabilities and Solutions

Sr. No	Vulnerabilities	Solutions
1.	Hardcoded, insecure, or weak passwords	Use a password management system and create complicated passphrases or passwords.
2.	Services on insecure networks	Use IDS and a firewall. Close any unwanted open ports and use the services' encrypted versions.
3.	Interfaces for unsecure ecosystems	Use systems for multi-factor authentication. Regularly assess the interfaces
4.	There are not any secure updated mechanisms	Use encrypted communications for secure delivery. To confirm the accuracy of changes, use checksum and hash.
5.	Use of outmoded or insecure components	Eliminate any dependencies or libraries for unsafe software. Avoid employing hardware or software from unreliable third parties.
6.	Insufficient privacy protection	Use the CIA triad. Anonymize user information gathered.
7.	Unsecure transfer and data storage	Use encrypted data transmission methods, and effectively implement access control mechanisms.
8.	Insufficient device management	Keep track of runtime settings. Block suspicious-looking devices.
9.	The default settings are unsafe	Replace any default usernames and passwords, and stay away from the remote access option.
10.	Insufficient physical hardening	Set a BIOS password. Use external connectors as little as possible.

of service [distributed denial-of-service (DDoS)], MITM attack, and intrusion into networks.

7.5.6 Weak passwords

The default passwords on many consumers' IoT devices are not changed. The top ten vulnerabilities and their solutions are mentioned in Table 7.2 for learning and research purposes.

7.6 SOME IOT ATTACKS THAT ARE PERFORMED ON A GLOBAL LEVEL

The IoT attacks can now find smart connected equipment everywhere, and they are permeating both our homes and workplace networks. The positive effects are numerous, ranging from mobile device-controlled thermostats to autonomous safety devices, from cutting-edge innovations like autonomous automobiles that will decrease traffic jams and roadway crashes to controlled by voice equipment that allow the less able to live on their own

[4]. There have been many different types of attacks for a very long time. Millions of issues with IoT connation might be exposed to common attacks via the internet, but they are considerably more widespread and frequently involve little to no security. This makes the scope and relative ease of IoT attacks relatively new. Even if attack techniques remain usually unchanged from the past, the results of each attack may vary significantly based on the ecology, the device and surrounds, the amount of resistance available, and many other aspects. Unfortunately, many IoT devices are sent from manufacturers without a secure configuration by default, and as these devices are often embedded, they are not frequently patched and protected after they are in use. Any flaws in a network are susceptible to attack by malicious software and can be easily exploited by hackers. As a result, companies have a responsibility to guarantee that they have implemented comprehensive security and compliance control across all IoT touch points. Some global attacks are described hereafter:

7.6.1 BlueBorne attack

A malicious person uses a BlueBorne attack to fully access via the use of Bluetooth to link to the device in question [17]; it is necessary for the attacked device to be associated with the target or just put to accessible state a device, which enables the execution of a wide variety of crimes, such as remote code execution (RCE) and MITM attacks. On a variety of IoT devices, including those running Android, Linux, Windows, and other operating systems [18], the BlueBorne attack may be carried out by following these steps:

- Attacker searches for all nearby, switched-on Bluetooth devices.
- They then get the device's media access control (MAC) address.
- Now, they keep probing the target device in an effort to discover the operating system (OS). After determining the OS, the attacker uses a flaw to gain access to the target device via the Bluetooth protocol.
- An adversary may now launch an RCE or MITM attack, because they have complete access to the device.

7.6.2 Rolling code attack

The majority of modern smart automobiles employ smart locking systems, which operate by transmitting an radio frequency (RF) signal in the form of a key's code to lock or unlock the vehicle [19,20]. Whether an auto gets a single rating twice, it is denied as it can only be used once. Replay attacks are prevented by doing this. A rolling code, sometimes known as a hopping code, locks or unlocks an automobile. The attacker now prevents the signal from being sent to get the rolling code. A jamming device is used in this attack to simultaneously jam the signal and sniff the car unlocking code, which the attacker may utilize afterward [21,22]. The rolling code attack may be carried out by the following steps:

- The victim unlocks the automobile by pressing a remote control.
- Attacker jams the car's receptor device by using the jammer to learn the initial code.
- As vehicle would not open the first time, the victim attempts transmitting the code again using the car remote button.
- Now the attacker may unlock the automobile using the second code that was recorded.

The usage of RF protocols makes it very hard to defend against rolling code attacks, as nothing can stop anyone from recording, replaying, and analyzing the RF signals that are transmitted. However, there are some actions that may be modified to improve protection, such as avoiding the use of a remote dongle to lock or unlock a car in favor of a push button on the door handle [23]. One can get theft insurance; in this case, financial defense is preferable than bodily defense.

7.6.3 Software-defined radio-based attacks

A radio communication system is a Software Defined Radio (SDR) system that generates radio communications and performs signal processing using software (or firmware) rather than hardware [24]. IoT devices' use of wireless physical communication creates previously unheard-of options for attackers, such as the ability to monitor IoT networks and use communication signals to distribute exploits to linked devices [25].

7.6.3.1 Reconnaissance of a target

The most crucial component of an IoT device is its operating system, which is occasionally listed with the Federal Communications Commission (FCC) ID or on the item's website [26]. SDR tools, such as the HackRF one, monitor a large spectrum of frequencies and identify the average frequency at which the device operates [27].

7.6.3.2 Decode data unknown RF protocol

The data from the transmission are decoded using the GNU Radio Companion program. To get the original signal, extra processes like reverse engineering the protocol are taken. The signal transmitted by the transmitter is captured in wav format using the HackRF One [28]. The wav file is loaded in Audacity, which is a program to examine and alter raw collected audio files. Finally, the signal is divided in 8-bit blocks and converted to write.

7.6.3.3 Replay attacks

The primary SDR attack is the replay attack. This attack uses signal capture and retransmission. Consequently, the receiver circuit continues its normal operation after replaying the signal [29]. The following are some procedures for a replay attack.

Once the transmission between the linked devices has started, keep an eye on the device's operational frequency, which was discovered during reconnaissance. Using programs like Universal Radio Hacker, the command sequence is separated and added to the transmission.

The device's action is then replayed when this frequency with a separate command sequence is transmitted.

7.6.3.4 Jamming attack

A sort of assault known as jamming RF involves interfering with transmitter and receiver communication [30] as shown in Figure 7.3. This is accomplished by conducting a denial-of-service (DoS) attack by transmitting a high-power signal at the device's operating frequency [31]. The endpoints are unable to communicate with one another as a result. This attack can succeed on any wireless access point.

7.6.4 DDoS attack

The amount of traffic on servers, online systems, or networks is saturating during a DDoS attack to use up bandwidth and other resources as shown in Figure 7.4. It does this by using a number of hacked computers [32,33].

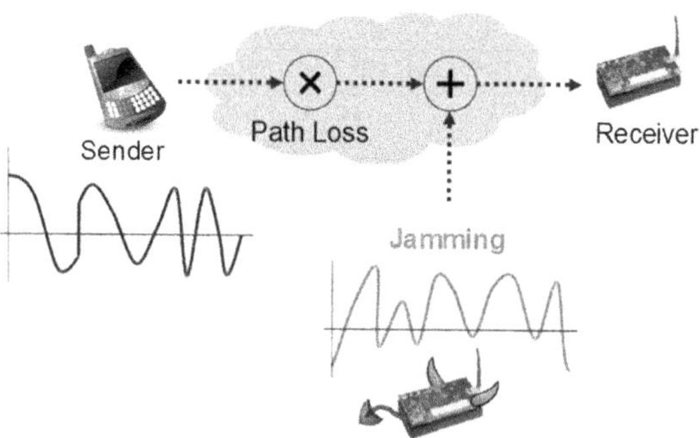

Figure 7.3 Jamming Attack.

As a result, systems become unresponsive to legitimate requests or run slowly. If an IoT DoS or DDoS assault is launched, the device may be compromised or turned into a botnet [34].

Attackers initiate the assault by introducing harmful software into their operating systems after first exploiting the device's security flaws.

Target systems experience a high amount of requests from different IoT devices located in diverse locations, which can cause the target become sluggish or even shut down [35].

7.6.4.1 Resource depletion attack

The resources needed by an IoT, such as RAM, CPU, and socket context, are immediately impacted by this attack [36]. This assault can be carried out via sending distorted packets, such as in the Ping-of-Death attack, or by taking advantage of network vulnerabilities, flaws in transport or application layer protocols, or both.

7.6.4.2 Bandwidth depletion attack

The bandwidth of the IoT network is intended to be totally consumed by this assault. The attack's density can be increased by broadcasting the corrupted packets. User Datagram Protocol flood attacks as well as Internet Control Message Protocol flood assaults are forms of bandwidth depletion attacks [37].

7.6.4.3 Infrastructure attack

The IoT gadget and its parts are immediately impacted by this assault, as the users can no longer access the bandwidth and resources [38].

7.6.4.4 Zero day attack

- A zero-day attack is a type of software attack that takes use of a vulnerability that a vendor or developer was unaware of [39]; after the attack, a patch is made available for these kinds of vulnerabilities.

7.6.4.5 Side-channel attack

Nearly all IoT devices create emitting side channels, which are signals that provide details about they operate inside. An intrusive party can launch a side-channel attack (SCA) by extracting information about encryption keys by monitoring these signals [40,41]. SCAs are designed with a constant data leak that attackers can take advantage of by using power or electromagnetic emissions.

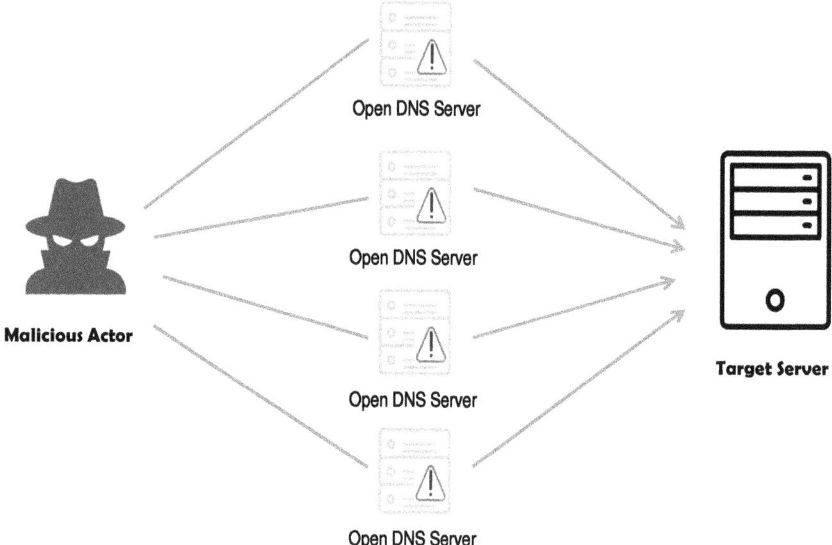

Figure 7.4 DDoS Attack.

- Timing analyses look at the timestamps associated with each occurrence. An adversary may use particular attack strategies to collect information about network events such as packet transit [42]. This attack exploits the different execution times of the ecosystem's branches..
- An attacker monitors the devices' power use during power analysis assaults [43]. An attacker must be near to the sensor node to estimate its power usage. Differential power analysis (DPA) and simple power analysis (SPA) are the two categories of power analysis [44]. SPA is a method for analyzing the power used by cryptographic activities, whereas DPA analyzes the power used by both computational and nondeterministic processes.
- Fault analysis attacks are possible when a cryptosystem flaw results in the leakage of relevant information [45]. These flaws might develop naturally or can be deliberately introduced by an enemy in two different ways. One is to give programs incorrect inputs or to flip certain memory bits using tools like a laser pointer.
- Critical nodes, like a sensor network's aggregator node, are described in traffic flow. The sensor nodes known as aggregator nodes are utilized to transport communications between nodes and the base station. To obtain topological data, attacks using traffic analysis are launched by examining these traffic flows [46].

- Analyzing the accompanying acoustic oscillations generated by gadgets can help an attacker learn sensitive information [47].

While it is impossible to cover all security risks and IoT-related threats in one chapter, we have tried to cover the most significant ones. The operating technology is now introduced. Various sectors have explored security issues and defense mechanisms, but contemporary OT research has not viewed them in depth.

7.7 PROBLEMS OF CYBER SECURITY IN OT

Critical infrastructure industries including health care, power plants, and water utilities all depend heavily on OT. Unfortunately, the majority of OT systems use outdated software and hardware, leaving them open to many vulnerabilities including eavesdropping, phishing, ransomware attacks, etc. [48,49] as shown in Figure 7.5. OT faces a number of difficulties that leave it open to several dangers and hazards, including:

- Convergence with IT
- Vulnerable communication protocols

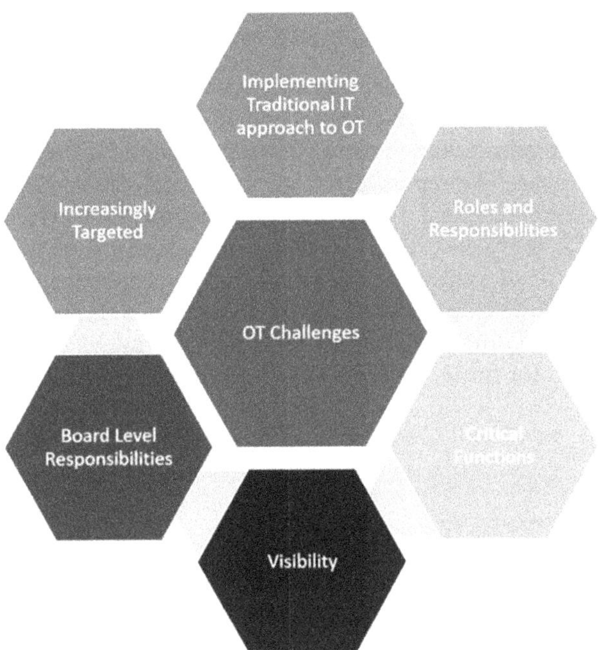

Figure 7.5 OT Attack Challenges.

- Lack of skilled professionals
- Outdated systems
- Lack of antivirus protection
- Insecure connections

7.8 IT/OT CONVERGENCE

The merging of IT computer systems with OT methods for tracking is known as IT/OT convergence [50, 51]. IT and OT convergence combines teams, operations, and technologies in addition to technology as shown in Figure 7.6.

Industrial IoT systems are made up of smart gadgets that are connected to monitor, analyze, and control the physical objects, equipment such as control systems, network modules, sensors, and other tools are used. These systems differ from more traditional ICS in terms of their variety and breadth because of their extensive connection with other systems and people [52]. An entitled document "Information Technology (IT) & Operational Technology (OT) Convergence: How Does It Benefits Digital Manufacturing" followed by a URL (www.tanand.com.my/it-ot-converge nce-how-does-it-benefits-digital-manufacturing) having a deeper impact to understand the concept and generate new efficiencies by using the flexibility and connectivity expertise.

7.9 CONCLUSION AND FUTURE WORK

This study covered the security concerns and risks that systems using OT and the IoT face. To clearly comprehend the vulnerabilities and assaults, techniques to replicate various attacks were suggested during investigation of various attacks. The report also suggests several risk-reduction methods as well as a few theoretical models that explain how these systems function. This research's study seeks to advance our understanding of security threats and methods for preventing attacks in OT and IoT systems. Some of them are put out following analysis of the most recent developments in security communities, and others are formed from generalizations of the findings of earlier research for future in terms of IoT an OT.

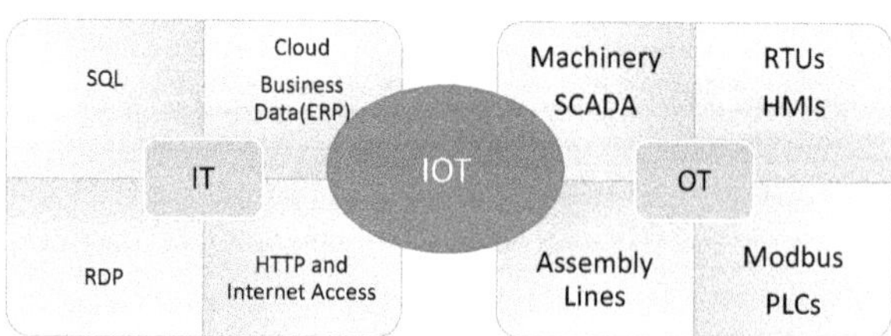

Figure 7.6 IT/OT Convergence.

- IoT for vehicles will develop.
- The uptake of IoT technologies will remain gradual.
- Equipment for 5G will expand.
- Consumer IoT will implement hardware firewalls.
- Cybersecurity for smart homes will increase.

REFERENCES

1. Chegini, H., Naha, R. K., Mahanti, A., Thulasiraman, P. (2021). Process automation in an IoT–Fog–Cloud ecosystem: a survey and taxonomy. IoT, 2(1), 92–118.
2. Abir, S. A. A., Islam, S. N., Anwar, A., Mahmood, A. N., Oo, A. M. T. (2020). Building resilience against COVID-19 pandemic using artificial intelligence, machine learning, and IoT: a survey of recent progress. IoT, 1(2), 506–528.
3. Hahn, A. (2016). Operational technology and information technology in industrial control systems. In Cyber-Security of SCADA and Other Industrial Control Systems (pp. 51–68). Springer, Cham.
4. Sudit, E. F. (1995). Productivity measurement in industrial operations. European Journal of Operational Research, 85(3), 435–453.
5. Piggin, R. (2014). Industrial systems: cyber-security's new battlefront [information technology operational technology]. Engineering Technology, 9(8), 70–74.
6. Kutlu, N., Gökdere, M. (2015). The effect of Purdue model based science teaching on creative thinking. International Journal of Education and Research, 3(3), 589–599.
7. Şener, N., Taş, E. (2017). Improving of students' creative thinking through Purdue model in science education. Journal of Baltic Science Education, 16(3), 350.
8. Silva, A. M., Matias, C. N., Nunes, C. L., Santos, D. A., Marini, E., Lukaski, H. C., Sardinha, L. B. (2019). Lack of agreement of in vivo raw bioimpedance measurements obtained from two single and multi-frequency bioelectrical impedance devices. European Journal of Clinical Nutrition, 73(7), 1077–1083.
9. Blasimme, A., Vayena, E., Van Hoyweghen, I. (2019). Big Data, precision medicine and private insurance: a delicate balancing act. Big Data Society, 6(1), 2053951719830111.
10. Chamola, V., Hassija, V., Gupta, V., Guizani, M. (2020). A comprehensive review of the COVID-19 pandemic and the role of IoT, Drones, AI, Blockchain, and 5G in managing its impact. IEEE Access, 8, 90225–90265.
11. Thames, L., Schaefer, D. (2017). Industry 4.0: an overview of key benefits, technologies, and challenges. In Thames, L., Schaefer, D. (eds.) Cybersecurity for Industry 4.0. Springer Series in Advanced Manufacturing. Springer, Cham. https://doi.org/10.1007/978-3-319-50660-9_1.
12. Ray, P. D., Harnoor, R., Hentea, M. (2010, October). Smart power grid security: a unified risk management approach. In 44th Annual 2010 IEEE

International Carnahan Conference on Security Technology (pp. 276–285). IEEE, San Jose, CA.

13. Tudosa, I., Picariello, F., Balestrieri, E., De Vito, L., Lamonaca, F. (2019, June). Hardware security in IoT era: the role of measurements and instrumentation. In 2019 II Workshop on Metrology for Industry 4.0 and IoT (MetroInd4.0&IoT) (pp. 285–290). IEEE.

14. Zhang, Z. K., Cho, M. C. Y., Wang, C. W., Hsu, C. W., Chen, C. K., Shieh, S. (2014, November). IoT security: ongoing challenges and research opportunities. In 2014 IEEE 7th International Conference on Service-Oriented Computing and Applications (pp. 230–234). IEEE.

15. Mahmoud, R., Yousuf, T., Aloul, F., Zualkernan, I. (2015, December). Internet of things (IoT) security: current status, challenges and prospective measures. In 2015 10th International Conference for Internet Technology and Secured Transactions (ICITST) (pp. 336–341). IEEE.

16. Almiani, M., Razaque, A., Yimu, L., Minjie, T., Alweshah, M., Atiewi, S. (2019, June). Bluetooth application-layer packet-filtering for blueborne attack defending. In 2019 Fourth International Conference on Fog and Mobile Edge Computing (FMEC) (pp. 142–148). IEEE.

17. Malallah, H., Zeebaree, S. R., Zebari, R. R., Sadeeq, M. A., Ageed, Z. S., Ibrahim, I. M., ... Merceedi, K. J. (2021). A comprehensive study of kernel (issues and concepts) in different operating systems. Asian Journal of Research in Computer Science, 8(3), 16–31.

18. Li, L., Yu, L., Yang, C., Gou, J., Yin, J., Gong, X. (2020, November). Rolling attack: an efficient way to reduce armors of office automation devices. In Australasian Conference on Information Security and Privacy (pp. 479–504). Springer, Cham.

19. Garcia, F. D., Oswald, D., Kasper, T., Pavlidès, P. (2016). Lock it and still lose it—on the (in)security of automotive remote keyless entry systems. In 25th USENIX Security Symposium (USENIX Security 16).

20. Thing, V. L., Wu, J. (2016, December). Autonomous vehicle security: a taxonomy of attacks and defences. In 2016 IEEE International Conference on Internet of Things (iThings) and IEEE Green Computing and Communications (GreenCom) and IEEE Cyber, Physical and Social Computing (CPSCom) and IEEE Smart Data (Smartdata)(pp. 164–170). IEEE.

21. Csikor, L., Lim, H. W., Wong, J. W., Ramesh, S., Parameswarath, R. P., Chan, M. C. (2022). RollBack: A New Time-Agnostic Replay Attack Against the Automotive Remote Keyless Entry Systems. arXiv preprint arXiv:2210.11923.

22. Oka, D. K., Furue, T., Langenhop, L., Nishimura, T. (2014, November). Survey of vehicle IoT Bluetooth devices. In 2014 IEEE 7th International Conference on Service-Oriented Computing and Applications (pp. 260–264). IEEE.

23. Le Roy, F., Roland, C., Le Jeune, D., Diguet, J. P. (2019, August). Risk assessment of SDR-based attacks with UAVs. In 2019 16th International Symposium on Wireless Communication Systems (ISWCS) (pp. 222–226). IEEE.

24. Alshouiliy, K., Agrawal, D. P. (2021). Confluence of 4G LTE, 5G, fog, and cloud computing and understanding security issues. In Fog/Edge Computing for Security, Privacy, and Applications (pp. 3–32). Springer, Cham.

25. Brown, J. S. (2017). Broadband privacy within network neutrality: the FCC's application expansion of the CPNI rules. University of St. Thomas Journal of Law and Public Policy, 11, 45.

26. Gummineni, M., Polipalli, T. R. (2020). Implementation of reconfigurable transceiver using GNU Radio and HackRF One. Wireless Personal Communications, 112(2), 889–905.

27. Hung, P. D., Vinh, B. T. (2019, February). Vulnerabilities in IoT devices with software-defined radio. In 2019 IEEE 4th International Conference on Computer and Communication Systems (ICCCS) (pp. 664–668). IEEE.

28. Mo, Y., Sinopoli, B. (2009, September). Secure control against replay attacks. In 2009 47th Annual Allerton Conference on Communication, Control, and Computing (Allerton) (pp. 911–918). IEEE.

29. Li, M., Koutsopoulos, I., Poovendran, R. (2010). Optimal jamming attack strategies and network defense policies in wireless sensor networks. IEEE Transactions on Mobile Computing, 9(8), 1119–1133.

30. Caparra, G., Ceccato, S., Formaggio, F., Laurenti, N., Tomasin, S. (2018, September). Low power selective denial of service attacks against GNSS. In Proceedings of the 31st International Technical Meeting of the Satellite Division of the Institute of Navigation (ION GNSS+ 2018) (pp. 3028–3041).

31. Geng, X., Whinston, A. B. (2000). Defeating distributed denial of service attacks. IT Professional, 2(4), 36–42.

32. Specht, S., Lee, R. (2003). Taxonomies of distributed denial of service networks, attacks, tools and countermeasures. CEL2003-03, Princeton University, Princeton, NJ, USA.

33. ushir, B., Sehgal, H., Nair, R., Dezfouli, B., Liu, Y. (2021). The impact of DoS attacks onResource-constrained IoT devices: a study on the Mirai attack. arXiv preprint arXiv:2104.09041.

34. Goolsbee, A., Syverson, C. (2021). Fear, lockdown, and diversion: Comparing drivers of pandemic economic decline 2020. Journal of Public Economics,193, 104311.

35. Cao, X., Shila, D. M., Cheng, Y., Yang, Z., Zhou, Y., Chen, J. (2016). Ghost-in-ZigBee: energy depletion attack on zigbee-based wireless networks. IEEE Internet of Things Journal, 3(5), 816–829.

36. Deshmukh, R. V., Devadkar, K. K. (2015). Understanding DDoS attack &its effect in cloud environment. Procedia Computer Science, 49, 202–210.

37. Sisalem, D., Kuthan, J., Ehlert, S. (2006). Denial of service attacks targeting a SIP VoIP infrastructure: attack scenarios and prevention mechanisms. IEEE Network, 20(5), 26–31.

38. Bilge, L., Dumitras, T. (2012, October). Before we knew it: an empirical study of zero-day attacks in the real world. In Proceedings of the 2012 ACM Conference on Computer and Communications Security (pp. 833–844).

39. Lerman, L., Bontempi, G., Markowitch, O. (2011). Side channel attack: an approach based on machine learning. In International Workshop on Constructive Side-Channel Analysis and Security Design. Center for Advanced Security Research Darmstadt, 29.

40. Zhang, T., Zhang, Y., Lee, R. B. (2016, September). Cloudradar: A real-time side-channel attack detection system in clouds. In International Symposium on Research in Attacks, Intrusions, and Defenses (pp. 118–140). Springer, Cham.

41. Song, D. X., Wagner, D., Tian, X. (2001). Timing analysis of keystrokes and timing attacks on SSH. In 10th USENIX Security Symposium (USENIX Security 01).

42. Kocher, P., Jaffe, J., Jun, B. (1998). Introduction to differential power analysis and related attacks. Cryptography Research, San Francisco. www.cryptography.com.

43. Mangard, S., Oswald, E., Popp, T. (2008). Power Analysis Attacks: Revealing the Secrets of Smart Cards (Vol. 31). Springer Science Business Media.

44. Biham, E., Shamir, A. (1997, August). Differential fault analysis of secret key cryptosystems. In Annual International Cryptology Conference (pp. 513–525). Springer, Berlin, Heidelberg.

45. Deng, J., Han, R., Mishra, S. (2005, September). Countermeasures against traffic analysis attacks in wireless sensor networks. In First International Conference on Security and Privacy for Emerging Areas in Communications Networks (SECURECOMM'05) (pp. 113–126). IEEE.

46. Kuo, B. C., Sarigul-Klijn, N. (2010). Conceptual study of micro-tab device in airframe noise reduction:(I) 2D computation. Aerospace Science and Technology, 14(5), 307–315.

47. Hahn, A. (2016). Operational technology and information technology in industrial control systems. In Cyber-Security of SCADA and Other Industrial Control Systems (pp. 51–68). Springer, Cham.

48. Moteff, J., Parfomak, P. (2004, October). Critical Infrastructure and Key Assets: Definition and Identification. Congressional Research Service, Library of Congress, Washington, DC.

49. Felser, M., Rentschler, M., Kleineberg, O. (2019). Coexistence standardization of operation technology and information technology. Proceedings of the IEEE, 107(6), 962–976.

50. Titu, A. M., Stanciu, A. (2020, June). Merging operations technology with information technology. In 2020 12th International Conference on Electronics, Computers and Artificial Intelligence (ECAI) (pp. 1–6). IEEE.

51. Stouffer, K., Falco, J., Scarfone, K. (2011). Guide to Industrial Control Systems (ICS) Security. NIST Special Publication 800-82 (pp.16–16). National Institute of Standards and Technology.

52. Williams, T. J. (1994). The Purdue enterprise reference architecture. Computers in Industry, 24(2–3), 141–158.

53. Bagui, S., Wang, X., Bagui, S. (2021). Machine learning based intrusion detection for IoT botnet. International Journal of Machine Learning and Computing, 11(6), 399–406.

54. Thamaraiselvi, D., Mary, S. (2020). Attack and anomaly detection in IoT networks using machine learning. International Journal of Computer Science and Mobile Computing, 9, 95–103.

55. Susilo, B., Sari, R. F. (2020). Intrusion detection in IoT networks using deep learning algorithm. Information, 11(5), 279.

56. Haji, S. H., Ameen, S. Y. (2021). Attack and anomaly detection in IoT networks using machine learning techniques: a review. Asian Journal of Research in Computer Science, 9(2), 30–46.

57. Elmrabit, N., Zhou, F., Li, F., Zhou, H. (2020, June).Evaluation of machine learning algorithms for anomaly detection. In 2020 International Conference on Cyber Security and Protection of Digital Services (Cyber Security) (pp. 1–8). IEEE.

58. Liu, Z., Thapa, N., Shaver, A., Roy, K., Yuan, X., Khorsandroo, S. (2020, August). Anomaly detection on IoT network intrusion using machine learning. In 2020 International Conference on Artificial Intelligence, Big Data, Computing and Data Communication Systems (icABCD) (pp. 1–5). IEEE.

59. Aysa, M. H., Ibrahim, A. A., Mohammed, A. H. (2020, October). IoT DDoS attack detection using machine learning. In 2020 4th International Symposium on Multidisciplinary Studies and Innovative Technologies (ISMSIT) (pp. 1–7). IEEE.

60. Al-Akhras, M., Alawairdhi, M., Alkoudari, A., Atawneh, S. (2020). Using machine learning to build a classification model for IoT networks to detect attack signatures. International Journal of Computer Networks & Communications (IJCNC), 12, 99–116.

61. Rani, D., Kaushal, N. C. (2020, July). Supervised machine learning based network intrusion detection system for Internet of Things. In 2020 11th International Conference on Computing, Communication and Networking Technologies (ICCCNT) (pp. 1–7). IEEE.

62. Stoian, N. A. (2020). Machine learning for anomaly detection in IoT networks: malware analysis on the IoT-23 data set (Bachelor's thesis, University of Twente).

Chapter 8

Managing big data integrity for IoT healthcare information systems

Sarita Shukla, Bineet Kumar Gupta, Pallavi Somvanshi, and Mukesh Mishra

8.1 INTRODUCTION

The rapid growth and integration of developing technology into virtually all aspects of our lives and society has the potential to produce amazing opportunities, but it also poses unique obstacles. In the recent digital era, data is the most precious asset and data integrity management is a complicated and challenging task for every security expert and researcher. Data integrity also defines the manner of assuring data quality, and efficiency throughout its lifecycle. It might comprise patient information, health reports, diagnostic reports, laboratory test reports, and other reports in the healthcare industry. It is very difficult to secure the patients' private reports and data for the healthcare sector and scientists, the target of the attacker is to manipulate patients' data in healthcare subdomains. Hence, the protection of data integrity in healthcare sector is the most prioritized issue and it is important when the larger ramifications of the data integrity breach go unidentified like attackers use these data breaches to exploit other attacks. Each country is following to be a digitized healthcare for well experience and infrastructure requirements. The trend of digitization in the healthcare sector, on the other hand, presents several complex issues for security specialists. Attacks on confidentiality, privacy violations, information breach believability, and many other issues are all becoming increasingly problematic for digitalization techniques and experts.

In the health sector, data integrity is also a major concern and its breach in healthcare organizations might have major ramifications. Incidents of cybersecurity are now often regarded as the most serious threat to healthcare. Maintaining data integrity has become a difficult task in healthcare companies due to their organizational structure, which combines high-end point complexity and legal constraints. The various security breaches have demonstrated that the healthcare business is still falling behind other industries in endeavours to protect its stakeholders' data integrity and confirming the organization's product image and the clients' expectations. Some data breaches can result in a significant loss of revenue as well as a loss of

DOI: 10.1201/9781003514312-8

customers' trust in the companies' legitimacy. Organizations are more vulnerable to this type of threat than they are to threats to privacy and availability. The importance of critical data integrity becomes even more serious when numerous data integrity attacks go unnoticed or unidentified and the attackers employ incorrect information or data in various sorts of attacks.

The main objective of this literature review is to demonstrate the present data integrity techniques that are explained by the researchers to secure healthcare information. The literature review undertaken in the work builds a database of data integrity tactics used by attackers and also informs users about those that need to be improved. This literature is divided into two sections. In the first section, the literature review delivers brief information about the prior attacks on data integrity related to healthcare sectors that provides an overview of recent data integrity scenarios in healthcare. In the second section, the literature review offers a review of prior research initiatives correlated to data integrity in healthcare sectors.

8.2 RISKS TO DATA INTEGRITY: CURRENT TRENDS

The issue of data integrity is one of the most challenging issues for the global healthcare business, and data integrity management is a difficult and challenging task for security professionals in the healthcare industry. Various problems associated with healthcare data management create several opportunities for attackers to take advantage of an industry [11, 12, 38].

As a result, the researcher developed a novel approach that uses attack statistics to present an overview of the current healthcare data integrity plot. According to multiple reports, the amount of data breaches affecting the healthcare business is on the rise. During 2009–22, a study on data breach assaults on healthcare companies was published in the online survey journal "HIPPA". According to research, compared to 2009, the latest data breach attack on the healthcare industry is at its worst [2]. The graph of attacks in Figure 8.1 shows that the healthcare industry requires strong malware protection to maintain data integrity, privacy, and accessibility.

According to HIPPA's study, the healthcare industry has seen 25 major data breaches in the previous ten years and the first half of 2021. With the aid of that record, researchers were able to classify the most common attacks in healthcare companies. Medical data breaches are caused commonly by hackers obtaining confidential information. It should be emphasized, however, that healthcare companies are considerably better at detecting cyber incidents. It is conceivable that the low ratio of hacking/IT attacks is related to the fact that prior years' hacking incidents and malware attacks were not detected. Healthcare companies enhance the screening of insider breaches and disclosure of such damages to the Office of Civil Rights (OCR), much as they do with hacking. Figure 8.2 demonstrates that IT events account for 62 percent of large healthcare assaults, which is a significant percentage for

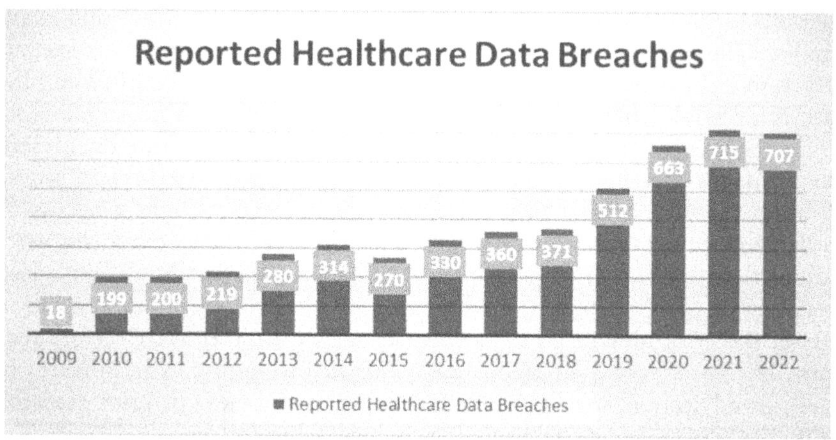

Figure 8.1 Reported data breaches in healthcare (2009–2021).

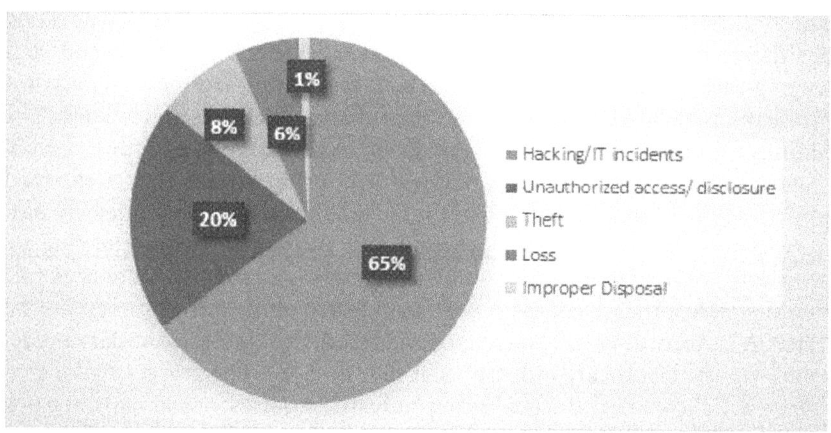

Figure 8.2 Percentage ratio of the healthcare breach types.

any company [2]. According to a survey, 93 percent of healthcare businesses have experienced cyber-attacks [3]. As a result of cybercriminals taking advantage of the COVID 19 pandemic, we've seen an increase in hacking incidents, including ransomware attacks and phishing scams directed toward the healthcare sector. In 2020, there was a 53% increase in the number of confirmed data breaches in the healthcare industry. In total, these incidents exposed almost 12 billion pieces of protected health information (PHI). On their network, 94 percent of healthcare companies experienced cyber-attacks [3]. An annual healthcare industry breach study report [4] reveals that the

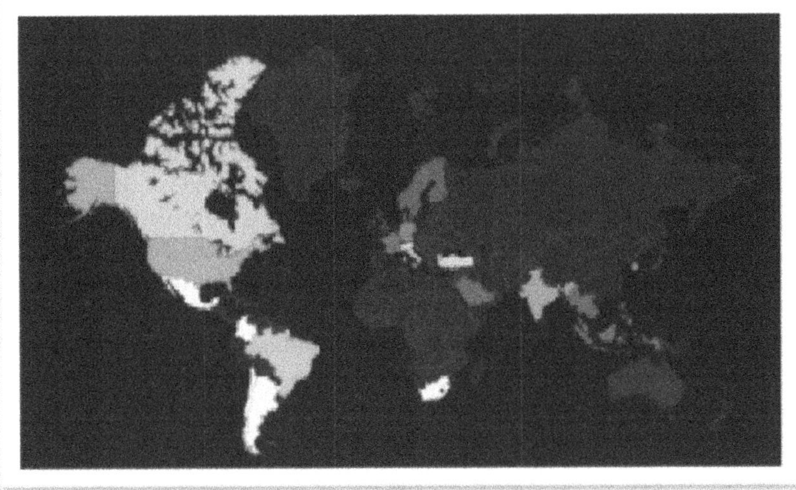

Figure 8.3 Country wise total data record stole (2020–2021).

number of breaches increased almost thrice in 2018 compared to 2017. According to *Internet News*, the average cost of any healthcare document on the dark web is between $1 and $1000 [5]. Cancer Treatment Centers of America (CTCA) [6] identified 20,535, cancer patients by targeting their addresses in 2020. According to an internet report, the American Medical Collection Agency (AMCA) was hacked for seven months in early 2020, resulting in the theft of 27 million patient information. Data considered confidential, such as a patient's credit/debit card record and medication, was stolen during the attack [7]. Figure 8.3 depicts the total amount of stolen records broken down by country.

The most recent data breaches happened in two large healthcare businesses, LabCorp and Quest Diagnostics, with about 19 million patient records being handled through a common service provider [8]. According to a recent research report by Global Market Insights [9], the global healthcare cybersecurity market is estimated to reach 27 billion USD by 2025. Another disturbing instance in 2019 was the data breach of 17.325 patients at the American Baptist Homes of the Midwest, which was caused via compromising emails and Network services [10].

The numbers analysed clearly indicate assault trends, as well as a history of attacks on healthcare in previous years. An in-depth examination of these assaults reveals the current state of data integrity in the healthcare industry. Manipulation of data is a common source of concern. In today's data-driven world, the ramifications of ambiguity are scary. Data integrity violations affect infrastructure, national security, commerce, political institutions, and health. Data tampering is more subtle, compromising not just an industry's capacity to safeguard its data, but also the industry's data's integrity.

8.3 RELEVANT STUDY INITIATIVES

In this study, a literature review was conducted on a number of healthcare-related studies. The majority of these have concentrated on organizational characteristics and criteria, with a few examining various approaches to data privacy and security. The most essential and challenging topic for current security scientists and researchers is data integrity management in the healthcare sector. Researchers have also found that there is not a lot of literature on the topic of healthcare data integrity. Regardless of whatever survey is accessible, it does give some important data. The majority of researchers have focused their research on defining data integrity techniques or healthcare methodologies.

The researchers investigated diverse healthcare data integrity-related publications to perform a literature review. A few studies have looked at administrative assets and needs, while others have looked at various privacy and data security techniques. Researchers also discovered that, even though data integrity management in healthcare sector is the most important and difficult area for today's security specialists and scholars, there is little literature on the area. However, every survey that is available, provides useful information. In their surveys, it is clear that the majority of studies have focused on a healthcare data integrity approach or healthcare method. The majority of research has focused on a particular data integrity or healthcare technique. The following are the sources that were used in this study's research:

Fernández-Alemán et al. [xx] research work is responsible to report the results of a systematic literature review concerning the security and privacy of electronic health record (EHR) systems. In recent years, the research focused on the implementation of legislation and the continuance of directives related to EHR technology protection and privacy. Despite this, more work has to be done to enforce these requirements and implement healthy EHR systems.

Rezaeibagha et al. [xx] The conclusions of a comprehensive analysis of the current literature on the most commonly utilized technological aspects of EHR systems in terms of security and privacy were published. Ansari et al. [33] emphasize the significance of health information integrity problems throughout the first section of the target data. The second section of the study conducts a systematic review of extensive literature and data integrity approaches in the healthcare business.

With the help of literature, the research papers described above give significant healthcare data. The authors saw the necessity for a literature that includes various data integrity strategies and gives guidance for aspiring researchers to use in demonstrating their research efforts. To achieve this goal, the proposed research effort investigates the various data integrity management techniques mentioned in top-quartile research articles.

8.4 RESEARCH PURPOSE/OBJECTIVE

We evaluated previous literature on related themes and retrieved the researcher's proposed approaches and work to perform a literature review on the topic. It is crucial and critical to examine this problem and provide a stand-alone review for research scholars and security specialists to have a better understanding. A review always provides an overview of the current state of the related area. However, in our case, the researchers have supplied all of the necessary facts on the subject that a researcher would need to comprehend the problem. In this part, the scholars have examined the literature's two primary and most significant objectives. These objectives offer a path for the researchers to successfully conduct the literature.

Objective 1: What data integrity methods have been used in the last years around the world to manage the integrity of healthcare information and electronic records?

Motivation: To create effective solutions to prevent data breach occurrences, it's critical to first understand and then gather the existing tactics and approaches. As a result, this literature seeks to incorporate and characterize the material accessible to provide a thorough guide. As a result, the literature will serve as a repository for potential researchers. Furthermore, the major motive for choosing this topic was to draw the scientific community's attention to this critical issue.

Objective 2: In healthcare services, which type of data integrity approach requires the most attention?

Motivation: The purpose of the researchers of this study was to create a prioritized list of data integrity strategies based on their importance to future researchers. Prioritizing earlier research will also aid future scholars in selecting the most successful technique and comprehending the healthcare sector's needs.

8.5 METHODOLOGY

In the manner of conducting a literature review, the researcher's main purpose is containing quality papers relating to data integrity. To achieve this purpose, researchers use different scientific databases like Science Direct, IEEE Xplore, and Google Scholar. However, the following keywords were utilized using a Boolean operator AND to conduct an accurate search: Healthcare Data Integrity, EMR Security, Healthcare IS, MDT, and so on. After conducting all of this research, to find the most relevant publications, we employed inclusion and exclusion criteria. At the preliminary stage, 110 experiments were listed, 89 studies were found in an examination database, and 21 more were found in offline sources such as conference proceedings, symposia reports, and books. For conducting the literature review on healthcare data integrity, the researchers selected 22 related research studies. The following were the chosen inclusion criteria: The paper provided studies addressing data integrity as a security in healthcare concern and proposing some quantitative solutions.

- A literature review is a collection of publications that take a particular approach to the topic of healthcare reputation.
- Only papers from journals Q1 and Q2 are included in the literature review (for reliability and completeness of results).
- A literature study provides research on healthcare credibility concerns with some conclusive data.

The following were used to determine the criterion for exclusion:

- Exclude papers that did not follow the request's terms and the examination's aim.
- Papers that dealt with data integrity but not from a healthcare perspective were excluded.
- To aid the issue of healthcare data integrity, articles that are ineffective and conclusive were excluded.

Based on their study, the researchers removed the publications from the screening and qualifying procedure, as shown in Figure 8.4.

After evaluating the entire report, the researchers removed the review articles from the point. This phase resulted in the omission of 76 articles that were not suitable for the literature evaluation. In addition, after reviewing full-text articles, nine more articles were removed from the eligibility check. Literature Reviews and Meta-Analysis Preferred Reporting Items Researchers utilized 2009 Flow Diagrams to show the criteria for selecting papers, have taken this approach, and established standards for the construction of investigations and meta-analyses [13]. Table 8.1 shows the studies and their percentages in a tabular format.

For a better understanding of the search technique, Table 8.1 summarizes the various search numbers obtained from various digital libraries. The researchers' purpose was to use a review to describe and focus on the relevance of data integrity concerns in the healthcare industry. This was accomplished by compiling information on a variety of security breaches.

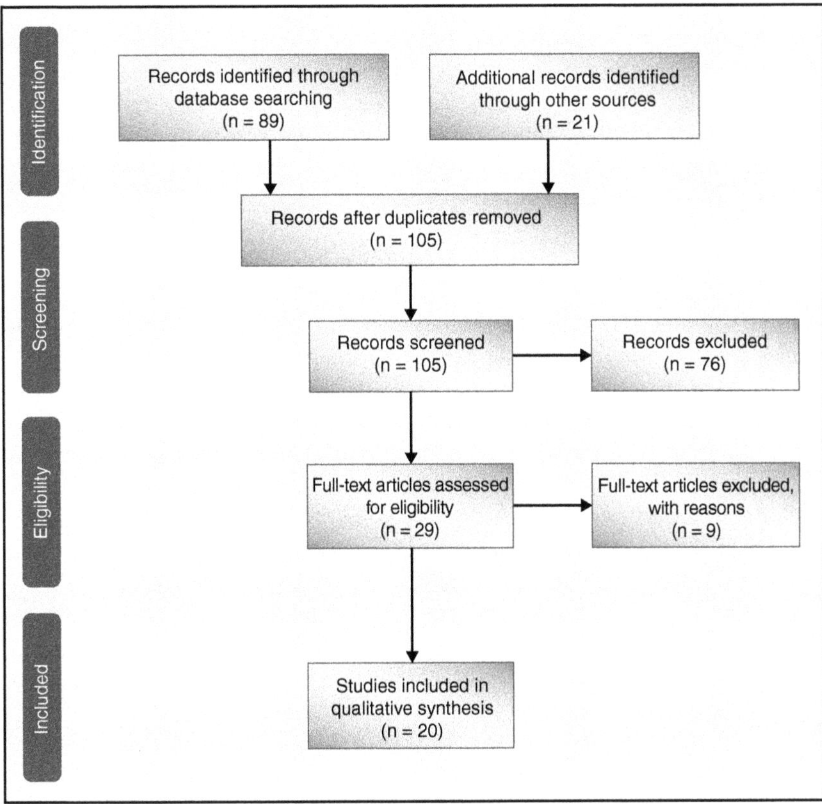

Figure 8.4 PRISMA flow diagram.

Table 8.1 Literature Review Search Figures

Data Repository	Relevant Papers	Not-Relevant Papers	Total	Relevant Percentage (%)
PubMed	4	21	22	18
Science Direct	5	17	22	22.72
Google Scholar	4	17	20	0.2
IEEE Xplorar	4	20	20	0.2
Cite-Seeker	3	14	16	18.75
	20	85	105	19.04

8.5.1 Results of exploratory analysis

A summary of the included research may be found in Table 8.2. The table summarizes the major points of the research findings as well as their related data integrity methods. During the review, it was discovered that some studies referenced ethical and healthcare issues in their material. As a result, for this literature review, the authors incorporated all of the papers for a considerably more extensive study.

8.5.1.1 Blockchain approach technique

Many researchers have used the blockchain approach techniques to manage healthcare data securely as a key aspect in their research, including:

- A study provided by *Ref.* [14] investigates how a blockchain system may be implemented in a healthcare company. The study paper tackles the issues and concerns raised in the article and proposes a new paradigm for boosting blockchain use in healthcare facilities.
- *Ref.* [16] proposed a model that addresses scientific data security and provides an FHIR-Chain architecture based on blockchain
- *Ref.* [26] published a study on blockchain as a health-related storage system.
- *Ref.* [28] devised a unique blockchain system that makes hospital-related data auditable, sharable, and secure.
- *Ref.* [29], used a hybrid fuzzy-based method to evaluate different blockchain technology models' influence and provide future scientists with an original awareness and track.

8.5.1.2 Masked authenticated messaging extension

James Brogan et al. used the masked identity verification message extension component [15] to improve patient data security in connected medical devices. In their article, the authors created a connection between IOTA and disguised authentication message extensions, overcoming wearable technology concerns. The IOTA technique was established with the purpose of acting as the foundation for secure data communication across IoT systems. The IOTA method was created to be lightweight and adaptable. It differs from typical blockchain-based distributed ledger systems in that it takes into account two major issues: latency and expenses. The technique employed in this study is quite useful to prospective researchers.

- Secure-Body Sensor Network (BSN): An article on the BSN strategy in the IoT healthcare scenario was provided by Prosanta Gope et al. In the IoT healthcare context, the BSN method, in which a patient

Table 8.2 Studies on Exploratory Analysis

Sources	Researchers	Years	Study Descriptions	Techniques for Data Integrity
[14]	C. Catalini et al.	2018	The paper describes ways to make the block chain technique more accessible in the healthcare sector. To ensure safe communication, the study also covers the issues that are linked with blockchain.	Blockchain
[15]	James Brogan et al.	2018	The research examines the use of distributed ledger technology in the advancement of electronic health records. For the healthcare organization, the paper offers a cost-effective and unique strategy.	Masked Authenticated Messaging extension
[16]	Peng Zhang et al.	2018	The FHIR-Chain, a blockchain-based architecture for securing Medicare, is proposed in the paper.	Blockchain
[17]	Christian Esposito et al.	2016	A cloud storage system is used in the study to store data from healthcare organizations and patients. The authors employ the blockchain method for secure lab report transactions and communication as well.	Blockchain
[18]	T. Hwang et al.	2015	The Body Sensor Network method is used in the study to provide for a secure and reliable IoT architecture in healthcare.	Secure-BSN
[19]	P. Vimalachandran et al.	2017	For Australian healthcare services, the authors recommended an authorization-based paradigm.	Authentication
[20]	M. ELHOSENY et al.	2018	For safeguarding health records and photos, the paper proposes a stenographic technique using a hybrid encryption mechanism.	Encryption
[21]	EntaoLuo et al.	2018	The article offers a secure sharing-based data transmission in an IoT paradigm for the data security of healthcare organizations.	Slepian-Wolf-coding-based secret sharing (SW-SSS)
[22]	K. Kwak et al.	2010	In this study, the problems and issues associated with wireless sensors in the healthcare industry are explored.	–

(continued)

Table 8.2 (Cont.)

Sources	Researchers	Years	Study Descriptions	Techniques for Data Integrity
[23]	Gunasekaran Manogaran et al.	2017	The authors offer a secure IoT-based organizational paradigm for storing and retrieving information from wearable sensors in medical services.	Secure Cloud
[25]	Jinyuan Sun et al.	2011	Based on cryptographic approaches and the IoT context of the healthcare business, the paper proposes a secure health record system for patient privacy.	Cryptography
[26]	Abdullah Al Omar et al.	2017	The research provides a blockchain-based data management system for healthcare services to help patients.	Blockchain
[27]	Sue Bowmanet et al.	2013	The report addresses the existing issues in healthcare data integrity as well as other mistake reasons in healthcare organizations. The report presents an overview of the current HER healthcare system.	–
[28]	S. Arakliotis et al.	2018	The report outlines a mechanism for using blockchain technology to provide healthcare organizations with auditable and sharable data.	Blockchain
[29]	M. Alenezi et al.	2020	The researchers employed a hybrid fuzzy-based method to assess the impact of various blockchain technology models in the medical sector.	Blockchain
[30]	Karim Abouelmehdi et al.	2018	The authors of this paper explored the problems and conducted a survey of the present status of big data in healthcare.	–
[31]	Anam Sajid et al.	2016	The study examines the security of healthcare medical data to protect patients' privacy. The present procedures and approaches in the healthcare system are also discussed in the paper.	–
[32]	Brihat Sharma et al.	2018	The Merkle tree-based approach is proposed as a paradigm for ensuring the integrity of medical records in the study. The blockchain technology is closely referenced in the software model.	Merkle tree-based approach

Table 8.2 (Cont.)

Sources	Researchers	Years	Study Descriptions	Techniques for Data Integrity
[33]	Katharine Gammon	2018	The paper shows how blockchain can be used in several healthcare areas.	–
[39]	Rateb Jabbar et al.	2020	The purpose of the paper to provide the groundwork for future research targeted at building a decentralised blockchain technology-based EHR management system that ensures confidentiality, authentication, encryption, and data sharing among healthcare services	Blockchain
[37]	Abhishek Kumar Pandey et al.	2020	The study also discusses data integrity strategies, concluding that blockchain is the most important data integrity strategy.	Blockchain

is tracked via tiny compact body sensors, is critical [18]. The study offers a BSN method that is both reliable and controllable in terms of integrity for secure IoT connections with healthcare providers.

- Authentication: The authors advised that Australian healthcare facilities take a step toward identity verification [19]. The identity verification phase allows patients to keep a realistic eye on their data, and this unique technique allows them to keep track of data access.
- Encryption: The authors released an essay on image preservation in healthcare and patient records in image and other formats. To protect healthcare data, the article recommends steganography and a hybrid encryption solution. More study is needed in the future for the method to provide positive results [20].
- Wolf-Coding-Based Secret Sharing: The authors create reliable IoT connection and data sharing between two devices of the IoT [21]. To secure the hospital IoT context, the authors developed a Wolf-based sharing strategy.

8.5.2 Secure cloud

The authors published a security model that addresses the current state of healthcare big data and suggests a constant cloud solution for the handling of large amounts of data in healthcare sector [23]. Furthermore, the safe cloud approach for inter-organizational information transfer was addressed by Benjamin Fabiana et al.

8.5.3 Merkle tree-based approach

The authors released an article that outlines a method for securing data transfer and communication in healthcare facilities [32]. By simulating blockchain technology, the suggested method aims to provide better and more accurate data sharing and statement-making.

8.5.4 Unit analysis

The researchers describe and categorize the research according to their appropriate subfield of healthcare in the analysis, it is a portion of the literature review. Assume that research provides an entire integrity-controlled organization for the entire healthcare sector, the sub-section "Whole Healthcare Systems" is specified in the literature review. Data transfer is a sub-section of that field if a study exclusively includes secure connection between IoT devices. Table 8.3 lists the most recent research papers that have fixated on various facets of the healthcare industry.

Table 8.3 highlights the most recent published research on several aspects of the healthcare industry. The table shows that, in comparison to other aspects of healthcare facilities, fixing healthcare record integrity is important. Multiple data integrity management solutions are also required for the complete healthcare system to have a good integrity-managed process.

8.5.5 Scientometric analysis

In the third stage, the researchers did a scientometric study to determine which data integrity method should be given more study attention. The scientific study that was all about analysing studies qualitatively [33] was the first to describe it in this way. Table 8.4 summarizes the quantitative and qualitative findings of the investigations, including journal indexing, ranking, and quartile. The quartile area comprises all of the sorts of journals given, according to their Indexed category.

The field of Computer Science has the highest number of publications, as shown in Table 8.4. In the Computer Science group, a total of five studies can be found. There are two articles in each of the categories of informatics and health knowledge. There are two publications in the category of medicine (other). For the engineering group, there are another two publications. All of these findings suggest that computer science research aimed at resolving the data integrity problem in the healthcare sector is increasing at a rapid pace.

Only one manuscript was published in each journal, with the exception of the *Journal of Computational and Structural Biotechnology*. Three studies on data integrity were published in the CSB journal. Because there is a dearth of research in this area, paper quartiles explicitly indicate that

Table 8.3 Unit Analysis

Authors	Years	Data Soundness	Data Auditability	Privacy-Preserving	Data Honesty	Data Backup
William J. Gordon et al.	2018		Y	Y		
James Brogan et al.	2018		Y			
Peng Zhang et al.	2018			Y		
Christian Esposito et al.	2016		Y	Y	Y	
Prosanta Gope et al.	2015	Y				
P. Vimalachandran et al.	2017	Y				
M. Elhoseny et al.	2018					Y
Entao Luo et al.	2018			Y		
Moshaddique Al Ameen et al.	2010			Y		Y
Gunasekaran Manogaran et al.				Y	Y	
Benjamin Fabiana et al.	2014		Y			
Jinyuan Sun et al.	2011					Y
Abdullah Al Omar et al.	2017				Y	
Sue Bowman et al.	2013		Y		Y	
Anastasia Theodouli et al.			Y	Y		
Xueping Liang et al.	2017		Y		Y	
Karim Abouelmehdi et al.	2018	Y		Y		
Anam Sajid et al.	2016		Y		Y	
Brihat Sharma et al.	2018					Y
Katharine Gammon	2018					

the standard of research work is extremely beneficial in healthcare data integrity techniques. To accomplish the utilization of free data processing in medical services, it is especially urged to develop and conduct high-quality research on a continuous basis.

8.6 CONCLUSION

By analysing statistical data, this literature review provides an overview of the present state of healthcare data integrity. In addition, this study examines

Table 8.4 Scientometric Analysis

Year	Authors	Journal	Quartile	Category
2018	**William J. Gordon et al.**	*Journal of Computational and Structural Biotechnology (CSB)*	A1	Computer Science & Application
2018	**James Brogan et al.**	*Computational and Structural Biotechnology Journal (CSB)*	A1	Computer Science & Application
2018	**Peng Zhang et al.**	*Computational and Structural Biotechnology Journal(CSB)*	A1	Computer Science & Application
2016	**Christian Esposito et al.**	*IEEE Cloud Computing*	A1	Computer Science (miscellaneous)
2015	**Prosanta Gope et al.**	*IEEE sensors journal*	A1	Electrical & Electronic Engineering
2017	**P.Vimalachandran et al.**	*Orange Technologies:An International Conférence (ICOT)*	–	–
2018	**M. Elhoseny et al.**	*IEEE Access*	A1	Engineering (miscellaneous)
2018	**Entao Luo et al.**	*IEEE Communications Magazine*	A1	Computer Networks and Communications
2010	**Moshaddique Al Ameen et al.**	*Journal of Medical Systems*	A2	Health Informatics
--	**Gunasekaran Manogaran et al.**	*Springer Series in Advanced Manufacturing, Cybersecurity for Industry 4.0.*	–	–
2014	**Benjamin Fabiana et al.**	*Information Systems*	A1	Information System
2011	**Jinyuan Sun et al.**	*Distributed Computing Systems- 31st International Conference*	–	–
2017	**Abdullah Al Omar et al.**	*International Conference on Computation, Communication, and Storage Security, Privacy, and Anonymity*	–	–
2013	**Sue Bowman et al.**	*Perspective Health Information Managing*	A2	Medicine (miscellaneous)
2018	**Anastasia Theodouli et al.**	*Trust, Security and Privacy in Computing and Communications – 17th IEEE International Conference/Big Data Science And Engineering – 12th IEEE International Conference (TrustCom/ BigDataSE)*	–	–

Table 8.4 (Cont.)

Year	Authors	Journal	Quartile	Category
2017	**Xueping Liang et al.**	*Personal, Indoor, and Mobile Radio Communications (IEEE 28th Annual International Symposium) (PIMRC)*	–	–
2018	**Karim Abouelmehdi et al.**	*Big Data Journal*	A1	Information System & Management
2016	**Anam Sajid et al.**	*Journal of Medical Systems*	A2	Health Information Management
2018	**Brihat Sharma et al.**	*9th IEEE Annual Conference on Ubiquitous Computing, Electronics, and Mobile Communication (UEMCON)*	–	–
2018	**Katharine Gammon**	*Nature Medicine Journal*	A1	Medicine (miscellaneous)

past studies on data integrity methods to clarify the working environment in the healthcare industry for data integrity management. The findings of this analysis clearly suggest that the healthcare industry requires a new, more rigorous data integrity strategy. The importance of data integrity issues in healthcare organizations is illustrated in the first section of this literature study. Future researchers are encouraged to embrace and inspire data integrity investigations in the second section (Review portion). Two different objectives led the whole research, as stated in the Discussion section. The initial goal was to offer a concise and detailed overview of prior articles using various techniques of analysis. The second goal was to list all of the approaches' data integrity mechanisms. The amount of time involved in papers evaluated and databases that the authors were able to access is a drawback of this research. Despite the authors' access to several resources, there are certain research and databases that were not able to be included in the highlighted literature study.

REFERENCES

1. Al-Hanawi, M. K., Khan, S. A., &Al-Borie, H. M. Healthcare human resource development in Saudi Arabia: emerging challenges and opportunities – a critical review. *Public Health Reviews*, 40(1), 1, 2019.
2. Healthcare Data Breach Statistics, 2019. Accessed: October 01, 2020 [online]. Available at: www.hipaajournal.com/healthcare-data-breach-statistics/
3. Filkins, B. *New Threats Drive Improved Practices: State of Cybersecurity in Health Care Organizations*. Sans Institute. 2014. chrome-extension://efaid

nbmnnnibpcajpcglclefindmkaj/https://cdn2.qualys.com/docs/mktg/sans-thre ats-drive-improved-practices-state-of-cybersecurity-health-care-organizati ons.pdf (Accessed: March 15, 2024).

4. *Breached Patient Records Tripled in 2018 vs 2017, as Health Data Security Challenges Worsen*, Accessed: October 03, 2020 [online]. Available at:www. protenus.com/press/press-release/breached-patient-records-tripled-in-2018- vs-2017-as-health-data-security-challenges-worsen

5. *Here's How Much Your Personal Information Is Selling for on the Dark Web*, Accessed: October 03, 2020[online]. Available at: www.experian. com/blogs/ask-experian/heres-how-much-your-personal-information-is-sell ing-for-on-the-dark-web/

6. *The 10 Biggest Healthcare Data Breaches of 2019, So Far*, Accessed: October 02, 2020. Available at:https://healthitsecurity.com/news/the-10-biggest-hea lthcare-data-breaches-of-2019-so-far

7. *Healthcare Data Breaches Reach Record High in April*, Accessed: October 01, 2020 [online]. Available at: www.modernhealthcare.com/cybersecurity/ healthcare-data-breaches-reach-record-high-april

8. Alhakmi, W., Baz, A., Alhakami, H., Ahmad, M., & Khan, R.A. "Healthcare device security: insights and implications," *Intelligent Automation & Soft Computing*, 27(2), 409–415, 2020.

9. "Data breach at major healthcare firms," *Computer Fraud & Security*, vol. 2019(6), pp. 3, 19, 2019. https://doi.org/10.1016/S1361-3723(19)30059-4

10. Ansari, M. T. J., Al-Zahrani, F. A., Pandey, D., & Agrawal, A. "A fuzzy TOPSIS based analysis toward selection of effective security requirements engineering approach for trustworthy healthcare software development," *BMC Medical Informatics and Decision Making*, vol. 20(1), pp. 1–13, 2020.

11. Ansari, M. T. J., Baz, A., Alhakami, H., Alhakami, W., Kumar, R., & Khan, R. A. "P-STORE: extension of STORE methodology to elicit privacy requirements," *Arabian Journal for Science and Engineering*, vol. 46, pp. 8287–8310, 2021.

12. Moher, D., Liberati, A., Tetzlaff, J., & Altman, D.G. Preferred reporting items for systematic reviews and meta-analyses: the PRISMA statement," *Journal of Clinical Epidemiology*, vol. 62(10), pp. 1006–1012, 2009. doi:10.1016/j.jclinepi.2009.06.005

13. Gordon, J. W., & Catalini,C."Blockchain technology for healthcare: facili- tating the transition to patient-driven interoperability," *Computational and Structural Biotechnology Journal*, vol. 16, pp. 224–230, 2018. doi:10.1016/ j.csbj.2018.06.003

14. Brogan, J., Baskaran, I., & Ramachandran, N. "Authenticating health activity data using distributed ledger technologies," *Computational and Structural Biotechnology Journal*, vol. 16, pp. 257–266, 2018. doi:10.1016/ j.csbj.2018.06.004

15. Zhang, P., White, J.,Schmidt, C. D., Lenz, G., & Rosenbloom, T. S."FHIR chain: applying blockchain to securely and scalably share clinical data," *Computational and Structural Biotechnology Journal*, vol. 16, pp. 267–278, 2018. doi:10.1016/j.csbj.2018.07.004

16. Esposito, C., De Santis, A., Tortora, G., Chang, H., & Choo, R. K.-K. "Blockchain: a panacea for healthcare cloud-based data security and

privacy," *IEEE Cloud Computing*, vol. 5(1), pp. 31–37, 2018. doi:10.1109/mcc.2018.011791712

17. Gope, P., & Hwang, T. "BSN-care: a secure IoT-based modern healthcare system using body sensor network," *IEEE Sensors Journal*, vol. 16(5), pp. 1368–1376, 2016.

18. Vimalachandran, P., Wang, H., Zhang, Y., Heyward, B., & Zhao, Y. "Preserving patient-centred controls in electronic health record systems: a reliance-based model implication," 2017 International Conference on Orange Technologies (ICOT), 2017. doi:10.1109/icot.2017.8336084

19. Elhoseny, M., Ramirez-Gonzalez, G., Abu-Elnasr, O., Shawkat, S. A., Arunkumar, N., & Farouk, A. "Secure medical data transmission model for IOT-based healthcare systems," *IEEE Access*, vol. 6, pp. 20596–20608, 2018. doi:10.1109/access.2018.2817615

20. Attaallah, A., Ahmad, M., Ansari, M. T. J., Pandey, A. K., Kumar, & Khan, R. A. "Device security assessment of Internet of healthcare things," *Intelligent Automation & Soft Computing*, 27(2), 593–603, 2020.

21. Al Ameen, M., Liu, J., & Kwak, K. "Security and privacy issues in wireless sensor networks for healthcare applications," *Journal of Medical Systems*, vol. 36(1), pp. 93–101, 2010. doi:10.1007/s10916-010-9449-4

22. Manogaran, G., Thota, C., Lopez, D., & Sundarasekar, R. "Big Data security intelligence for healthcare Industry 4.0," *Cybersecurity for Industry*, vol. 4.0, pp. 103–126, 2017. doi:10.1007/978-3-319-50660-9_5

23. Fabian, B., Ermakova, T., & Junghanns, P. "Collaborative and secure sharing of healthcare data in multi-clouds," *Information Systems*, vol. 48, pp. 132–150, 2015. doi:10.1016/j.is.2014.05.004

24. Sun, J., Zhu, X., Zhang, C., & Fang, Y. "HCPP: cryptography based secure EHR system for patient privacy and emergency healthcare," 2011 31st International Conference on Distributed Computing Systems, 2011. doi:10.1109/icdcs.2011.83

25. Bowman, S. "Impact of electronic health record systems on information integrity: quality and safety implications," *Perspectives in Health Information Management*, vol. 10(Fall), 1c, 2013.

26. Theodouli, A., Arakliotis, S., Moschou, K., Votis, K., & Tzovaras, D. "On the design of a Blockchain-based system to facilitate healthcare data sharing," 2018 17th IEEE International Conference on Trust, Security and Privacy in Computing and Communications/12th IEEE International Conference on Big Data Science and Engineering (TrustCom/BigDataSE), 2018. doi:10.1109/trustcom/bigdatase.2018.00

27. Zarour, M., Ansari, M. T. J., Alenezi, M., Sarkar, A. K., Faizan, M., Agrawal, A., & Khan, R. A. (2020). "Evaluating the impact of blockchain models for secure and trustworthy electronic healthcare records," IEEE Access.

28. Abouelmehdi, K., Beni-Hessane, A., & Khaloufi, H. "Big healthcare data: preserving security and privacy," *Journal of BigData*, vol. 5(1), 2018. doi:10.1186/s40537-017-0110-7

29. Sharma, B., Sekharan, N. C., & Zuo, F. "Merkle-tree based approach for ensuring integrity of electronic medical records," 2018 9th IEEE Annual Ubiquitous Computing, Electronics & Mobile Communication Conference (UEMCON).IEEE, 2018. doi:10.1109/uemcon.2018.8796607

30. Price, D. J. *Little Science, Big Science*, New York: Columbia University Press, 1963. doi: 10.7312/pric91844.
31. Zheng, Z., Xie, S., Dai, N. H., Chen, X., & Wang, H. "Blockchain challenges and opportunities: a survey," *International Journal of Web and Grid Services*, vol. 14(4), pp. 352, 2018. doi:10.1504/ijwgs.2018.095647
32. Sobers, R. "The world in data breaches," *Inside Out Security*, October 03, 2020.www.varonis.com/blog/the-world-in-data-breaches/
33. Ansari, M. T. J., Pandey, D., & Alenezi, M. "Store: security threat oriented requirements engineering methodology," *Journal of King Saud University-Computer and Information Sciences*, vol. 34(2), 191–203, 2018.
34. Algarni, A., Attaallah, A., Ahmad, M., Agrawal, A., et al. "A fuzzy multi-objective covering-based security quantification model for mitigating risk of web based medical image processing system," *International Journal of Advanced Computer Science and Applications*, vol.11(1), pp.481–489, 2020.
35. Pashazadeh, A., & Navimipour, N. J. "Big data handling mechanisms in the healthcare applications: a comprehensive and systematic literature review," *Journal of Biomedical Informatics* vol. 82, pp. 47–62, 2018.
36. Kumarpandey, A., Agrawal, A., Kumar, R., & Khan, R. A. "Key issues in healthcare data integrity: analysis and recommendations," *IEEE Access*, 2976687, 2020.
37. Alazeb, A., Panda, B., Almakdi, S., & Alshehri, M. "Data integrity preservation schemes in smart healthcare systems that use fog computing distribution," *Electronics*, vol. 10(1314), 2021. https://doi.org/10.3390/electronics1 0111314
38. Jabbar, R., Fetais, N., Krichen, M., & Barkaoui, K. "Blockchain technology for healthcare: enhancing shared electronic health record interoperability and integrity," *International Conference on Informatics, IoT, and Enabling Technologies*, Doha, Qatar, pp. 310–317. 2020. doi: 10.1109/ICIoT48696.2020.9089570

Secure, low-latency IoT cloud platform enhanced by fog computing for collaboration and safety

Eram Fatima Siddiqui and Sandeep Kumar Nayak

9.1 INTRODUCTION

Cloud computing offers on-demand and on-rent access to various technical resources, such as servers, storage, networking, databases, software platforms, office programs, and natural language processing and analytics. These resources are accessible online, thereby eliminating the need for physical systems. Customers can utilize these services according to their business requirements, paying only for the cloud services they use actually. This lowers operating expenses and improves organizational infrastructure management. Cloud computing is widely used today, reducing the cost of building and maintaining physical servers and networking infrastructure. Software providers now offer their programs online, allowing consumers to subscribe for a set period. These services include Infrastructure as a Service (IaaS), Platform as a Service (PaaS), and Software as a Service (SaaS), which enable users to rent physical servers, virtual machines, and operating systems [1–3].

The Internet of Things (IoT) refers to electronic and digital Things connected via wireless technology, enabling them to gather, share, and exchange data based on their environment and nature. These Things can communicate with biochips and integrated sensors, sensing various factors and providing corresponding information as shown in Figure 9.1. For example, a linked automotive sensor can detect engine, gasoline, tires, and other issues, sending a warning to the driver. This ecosystem enables automatic data exchange and collection, thus enhancing the overall efficiency of the Internet of Things. IoT devices' data can be used for analytics, extracting valuable information, and sharing it with valuable demands and requirements. This technical platform enables management, identification of trends, and automation of operations, enabling efficient management of company demands, identifying trends, and enhancing workflows [4–6].

The technical hour requires combining biochip transponder-based digital equipment with IoT devices to maintain data and provide insights. However, IoT devices have limited power and storage, making it difficult to make decisions. Cloud computing offers a solution to store and handle

DOI: 10.1201/9781003514312-9

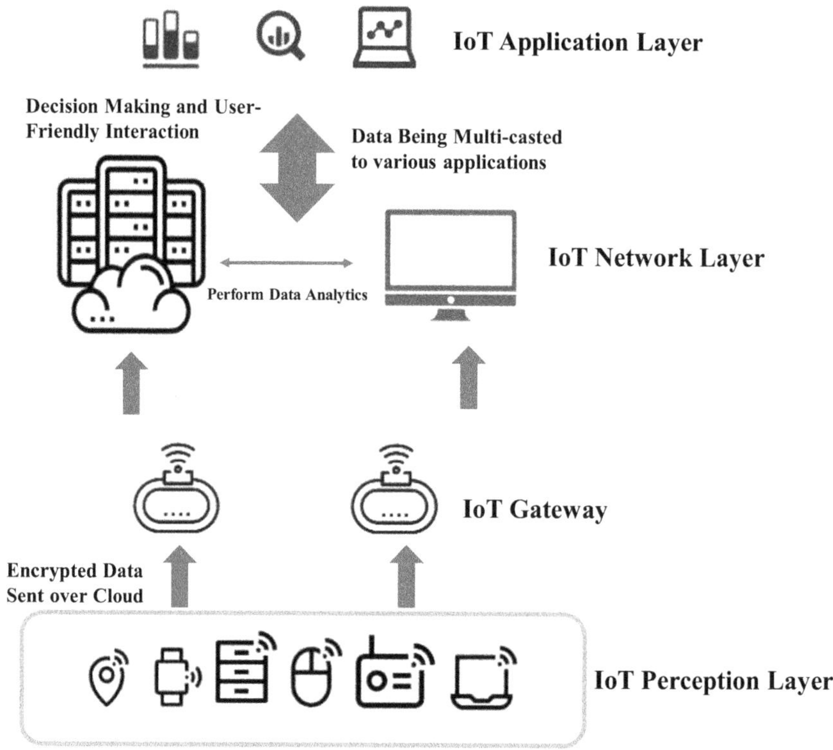

Figure 9.1 Figure Illustrating the Data Communication Model between Sensor-Based Devices and Cloud Layer.

large amounts of data, offering various services to manage these resources effectively [7].

Cloud provides automation, quick calculations, storage space, and flexibility, using artificial intelligence (AI) and machine learning techniques to extract useful insights from massive data collections [8]. Alexa, a cloud-based Internet of Things device, uses human speech to find the best route using Google Maps. It offers high security, encryption for private information, and reduces overhead costs by keeping data in the cloud. This partnership protects against data breaches and threats, ensuring easy access to approved and verified data from anywhere [9].

9.2 POTENTIAL SECURITY THREATS IN CLOUD

Security threats pose a danger to data communication and transmission, causing data loss and issues across different regions. Following are enlisted threats that may cause data loss and other issues in various regions.

9.2.1 Security threats in the cloud region

a. **Cryptojacking:** Hackers exploit cloud users' computer power, slowing down client processes and stealing sensitive data. Latent responsiveness remains, preventing suspicious activity detection and stealing sensitive information.

b. **Data Breaches:** Malicious users unlawfully access, copy, or transfer corporate data, posing significant security risks, including legal obligations, financial loss, and intellectual property loss [10].

c. **Non-Intentional ID Confirmation:** Google Drive's popularity raises concerns about employee privacy, as malicious URLs prompt login information, exposing private information and exposing sensitive data.

d. **Data Loss:** Unauthorized users can lose sensitive data and gain access to systems due to insufficient cloud account encryption.

e. **Use of Open Source Codes:** Most IT programmers create apps using free open source software, which may contain harmful elements. Once used, attackers can access and harm data, targeting a well-defined group [11].

9.2.2 Security threats in IoT region

a. **Botnet Attacks:** A botnet targets IoT devices, including computers and cellphones, and launches disruptive attacks like phishing, DDoS, and Email Bomb. Mirai Botnet, a well-known attack, infected 2.5 million devices, including routers and printers [12].

b. **Denial of Service:** A hostile attacker overloads a system with numerous requests, causing system capacity to decrease, preventing data theft but affecting overall performance.

c. **Ransomware:** Malicious users demand ransoms to restore normalcy in automated commercial spaces or intelligent residences, allowing unauthorized users to unlock critical data or restore functions to smart home gadgets.

d. **DNS Attacks:** Integrating outdated devices with modern technology increase security risks in the IoT ecosystem, making it vulnerable to DNS flaws, and allowing malicious individuals to access data or inject malware.

e. **Tracker Attacks:** IoT gadgets with tracking capabilities are becoming increasingly vulnerable to contamination and DDoS attacks. Shadow IoT devices, such as voice assistants and wireless printers, can be easily accessible to unauthorized users and exposed to botnets, making it difficult to ensure their security in corporate environments [13].

9.2.3 Security threats in cloud–IoT region

a. **Data Convergence through API Gateway in Cloud:** A high-end firewall minimizes external attacks by enabling IoT and cloud integration for data processing through API Gateways. However, if data flows through the API Gateway, potential attackers may discover access opportunities, creating a tunnel for future attacks.

b. **Existence of Loopholes in Communication and Data Flow Model of Cloud-IoT:** Data transmission without proper encryption and security poses a high-security threat, potentially allowing malicious users to access it if security standards and layers are not set at both data exchange and communication points.

c. **Data Privacy and User Authorization:** Business organizations must consider sensitive data storage in public clouds, reevaluating data collection and transmission methods, cloud ecosystem transmission, and access control for Edge and IoT devices.

d. **Improper deployment of IoT devices:** IoT security procedures are not fully implemented, potentially opening a tunnel into a cloud with strong security, posing a threat to important data.

e. **Misconfiguration issues in Cloud:** Improper cloud setup increases data vulnerability to malicious attacks. Therefore, rigorously preventing data loss and manipulation is crucial.

9.3 ARCHITECTURE OF CLOUD-IOT MODEL ENSURING SECURITY

9.3.1 Existing architecture

The existing Cloud-IoT architecture consists of a number of layers which help in data manipulation and data transmission. Figure 9.2 illustrates full traditional Cloud-IoT layered model.

Each layer in Figure 9.2 can be explained as follows:

a. **Device Layer:** The data link layer processes incoming data through a protocol, converting analog impulses into digital signals. It sends data bits via wired or wireless communication media, decoding it and sending it to the device's backend IoT processor. Secure protocols like Ethernet, ZigBee, RFID, LoRaWAN, LTE-A, NB-IOT, and IEEE 802.15.4e are used throughout the process [14].

b. **Data Packet Encapsulation Layer:** The layer addresses data packets through a specific communication channel, either wrapping or decapsulating them to reveal their source and destination IP addresses. It serves as the link between the physical and digital worlds, allowing data to enter IoT networks. Secure protocols like

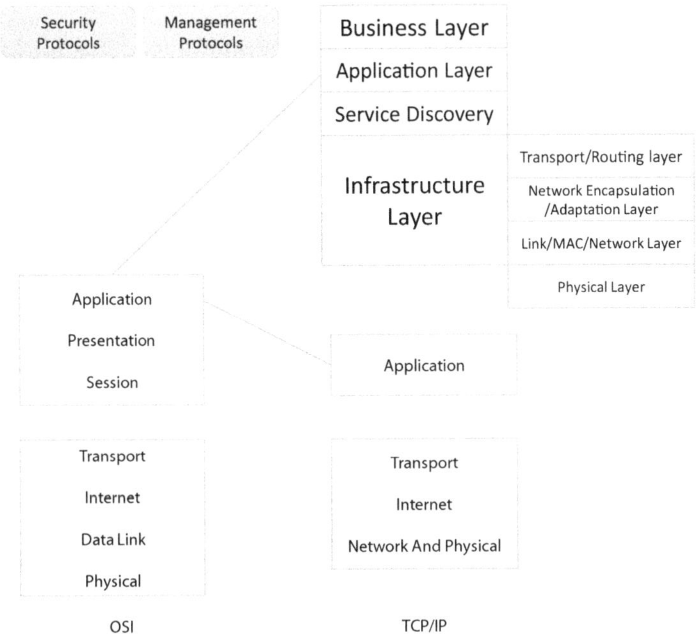

Figure 9.2 Figure Illustrating Traditional Cloud–IoT Data Transmission Architecture.

IPv4, IPv6, 6LoWPAN, 6Lo, AMQP, CoAP, and MQTT are used throughout the process.

c. **Data Processing and Service Discovery:** The edge layer in IoT design routes data packets between devices, ensuring real-time communication and enhanced network performance. It uses secure protocols like Transmission Control Protocol (TCP), User Datagram Protocol (UDP), Datagram Transport Layer Security (DTLS), Transport Layer Security (TLS), and Time Synchronized Mesh Protocol (TSMP). The layer also handles service discovery, resource management, and data processing, collecting, storing, and processing data in two steps: aggregating incoming data and abstracting raw data for business processes. This layer is crucial for IoT devices to locate and connect to the cloud and IoT environment.

d. **Application Layer:** The top layer of the architecture interfaces between applications and IoT devices, using specialized software platforms like web browsers. It supports various protocols like HTTP, MQTT, CoAP, DDS, and SOAP. Intelligence is applied to raw data after processing and analysis, enabling device management and monitoring applications. Raw data can be modified using analytics and machine learning techniques.

e. **Business Logic Layer:** The application layer integrates IoT devices with the cloud for real-time communication and database operations in various fields, such as health, transportation, and education. It is not part of the IoT communication architecture but is crucial for business decisions and long-term plans.

f. **Security Layer:** IoT architecture involves parallel layers, ensuring secure communication across networks. Three areas of security are connection, cloud, and device security. Protocols like X.509 and Open Trust Protocol protect each layer, thereby preventing hostile attacks and data breaches [15,16].

9.4 ROLE OF LATENCY IN DEPRECIATING PERFORMANCE OF CLOUD-IOT MODEL

Latency refers to the time it takes to reply to a user's request, with an increase in latency occurring when a connection is established for the first time. The server manages a queue, and a large number of requests may be waiting to be processed. Some connections may be hoarding resources, causing temporary blockages on both new requests and network resources until the current connections relinquish them. This increases the delay in Cloud-IoT's request-response mode.

There are a number of levels, a network request has to go through before it can actually get the requested data and respond back to the user, as depicted in Figure 9.3.

Level 1: **Domain Lookup:** When a request is generated for a webpage DNS lookup is carried out. If the number of hostnames is more, it may lead to more lookups and hence more latency.

Level 2: **TCP/TLS Handshake:** If the number of connections to be made with the server is more, it will take more time to complete the TCP (SYN-SYN-ACK) handshake. Also, it will take more time to complete a TLS handshake too to ensure the security of each connection.

Level 3: **Transmission of Request:** This defines the number of time instances taken by the user to send the HTTP request from the client location to the targeted server.

Level 4: **Time to Wait:** This defines the total instances of time taken by the server to search and respond to the user with the requested data and totally depends upon the performance of the disk.

Level 5: **Responded Data:** When the requested data arrives at the client side, it has to be downloaded to get the view by the user. The total instances of time also depend upon the size of the data received. If the data has already been cached, it may take a short amount of time [17].

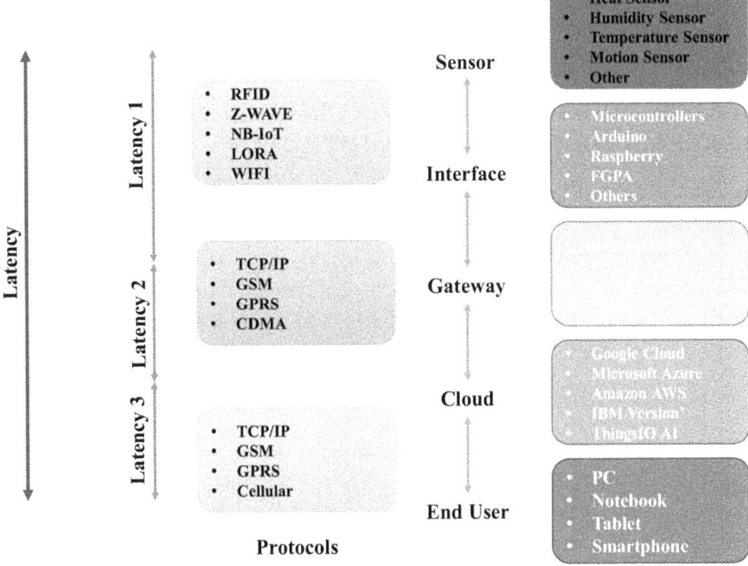

Figure 9.3 Figure Illustrating Occurrence of Latency at Each Layer of Network Cloud–IoT Model.

IoT devices send requests to the cloud from anywhere, but response times may be delayed due to distances and intermediary gateways. This is unacceptable for real-time applications requiring time-critical responses. Factors make it challenging for this model to function in a time-sensitive manner, causing delays. Each cause can be laid down as follows:

- **Bad Configurations:** Improper request specifications and user-based faults can cause query delays, erroneous responses, or no answer, resulting in potential errors.
- **Network Congestion:** In congested networks, packets may take longer time to reach their destination due to resource limitations, limited bandwidth, and poorly configured infrastructure, wasting resources for immediate response.
- **Complex Queries:** Complex queries have longer processing times, increasing the delay factor; cleaning and integration queries often have faster processing and response times.
- **Transmission Mediums and Routing Mechanisms:** Transmission media with higher data transfer rates may cause data loss due to inherent shortcomings. Switches and routers repeatedly open data, causing delays and adding more information.

- **Security Processes:** Security measures like firewalls and antivirus software increase delay in data transmissions, causing mixed and deconstructed transmissions, affecting communication [18,19].

9.5 COMBINED SECURITY AND LATENCY ISSUES IN LAYERED CLOUD-IOT ARCHITECTURE

Cloud-IoT communication tiers face various latency and security issues, with multiple levels of penetration and issue types. This section covers latency development, its causes, intrusion and malware activity, and a simple Cloud-IoT model architecture. Figure 9.4 is responsible to explore the various kinds of security layers provided at each layer.

9.5.1 Physical layer

IoT architecture involves the physical layer processing data from sensors and actuators, transforming it into a usable format, and transferring it for processing. This information may be based on physical factors or smart device identification.

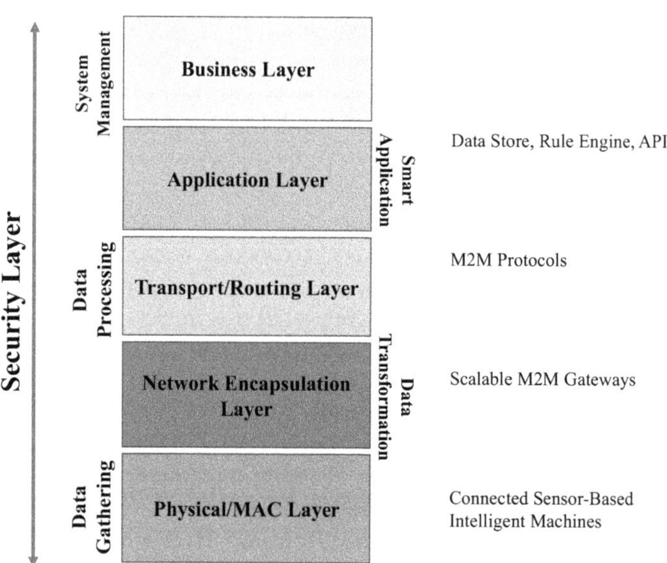

Figure 9.4 Figure Illustrating Various Kinds of Security Layers Provided at Each Layer.

9.5.1.1 Security issues in the perception layer

Malicious users may attack real-time streaming data, alter sensor readings, or access audio and video feeds. Some weaker threats include altering physical parameters or accessing audio and video feeds.

a. **Eavesdropping:** Deliberate internet data theft involves insecure transmission methods, affecting phone conversations, video conferences, and faxes.

b. **Tampering:** A hacker gains control of an IoT device, modifies its operation, or reduces its perceptive power.

c. **Collision:** The attacker waits for a packet to cross a channel, broadcasting its own malicious data packet simultaneously, causing collision and loss, potentially affecting the entire network's functioning.

d. **Node capture:** The perception layer poses significant security risks due to attackers gaining complete control over nodes like IoT gateways, enabling them to view data transmitted.

e. **Jamming:** A jamming device attacks an IoT node, causing it to shut down and interfere with its ability to send and receive data, preventing its functionality and communication.

f. **Malicious Node:** Malicious node connects to the IoT environment, uses resources, and supplies false data, wasting resources and inhibiting data delivery.

g. **Replay attack:** Real-time data stream attacks involve intrusive parties posing as legitimate parties, convincing victims to comply, and attempting to steal information.

h. **Exhaustion:** Resource depletion assault causes frequent collisions, packet losses, and retransmissions, permanently depleting nodes and causing their failure.

i. **Relay Attacks:** An interceptor attack inserts a third party between sender and recipient, allowing them to view and steal data, potentially modifying it.

j. **Timing Attacks:** Harmful assaults aim to learn a system's capabilities, responsiveness, and security methods, exploiting flaws and vulnerabilities to obtain sensitive information [20,21].

9.5.1.2 Latency issues in perception layer

Figure 9.5 can be explained as follows:

- T_{MACLR}: The total time taken to transmit the data packet to move from MAC Layer to the physical layer.

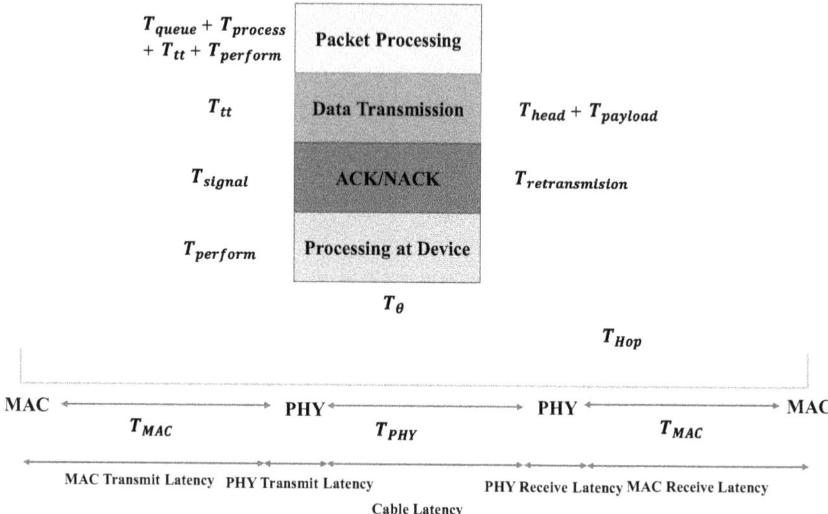

Figure 9.5 Figure Illustrating Different Kinds of Timelines Causing Latency in Data Transmission.

- T_{PHYLR}: The total time taken to transmit the data packet to move from PHY Layer to upper layers of the network stack.
- T_{Queue}: Data packet waits in line before being sent, potentially adding to a lower layer's queue, causing longer processing times as upper layers process.
- T_{Queue}: This is the sum of all the levels the data packet waits in a queue at each layer.
- T_{SIGNAL}: The total time taken for granting the privilege of scheduling of data packet, request the connection from sender to receiver, exchange of signals, and preprocessing of data packet.
- $T_{PERFORM}$: The total time taken in the encoding and decoding procedure of the packet as well as the time taken in the estimation of the channel through which the data packet will arrive in its initial transmission.
- T_{TT}: The total time taken to actually transmit the data from the sender's end over the prospected channel through scheduling the data packet transmission, channel selection and processing beforehand.
- T_{TRANS}: This time is basically a sum of two other times, i.e. T_H and T_P where H stands for Header, and P stands for Payload. When the sender has successfully claimed the transmission medium for its own turn, and it is now its time to send the data packets, the data packets

actually get transmitted through that medium. The data packet is composed of header and payload. So, each time its header is checked for the sender and receiver information, it is symbolized by T_H and its payload is symbolized by T_P. Thus, T_{TRANS} is the total time a data packet takes to get itself transmitted from one end to the other.

- $T_{PROCESS}$: This is the time taken by the medium to get the packet processed as well as the physical and data link layers at which the packet gets processed over its arrival.
- T_θ: When a data packet is arriving over shared resources it may take some time for the initial connection and time to access the medium. If packet loss occurs extra time for the retransmission of the same data packet also has to be done. T_θ is the sum of all these factors.
- T_{HOP}: When the data packet has to traverse through multiple hops before reaching the destination, this is the total time taken at one hop for processing the data packet and forwarding it to the next mediator or the final end point [22–24].

9.5.2 Network layer

The Physical Layer connects to the Application Layer, gathering data from IoT devices and sensors, and transmitting it to the application layer via transfer media.

9.5.2.1 Security issues at network layer

a. **Denial of Service Attack:** The attack involves numerous malicious requests, causing unhandled requests, failures, and communication blocking on the device or user.

b. **Man in the Middle Attack:** The Man in the Middle is an intruder who intercepts and modifies message chunks between sender and recipient, assuming they are speaking directly.

c. **Exploit Attack:** In this kind of attack, vulnerabilities of a system or network are searched and when the control is gained, it tries to steal the information from the system.

d. **Storage Attack:** In this type of attack, the intruder tries to attack the storage devices or the cloud itself where sensitive information of the user is stored and steals them.

e. **RFID Spoofing:** An assault seeks control of an entire IoT system, allowing unauthorized access and introducing harmful material, stealing private information and causing harm.

f. **Sinkhole Attack:** The most risky attack involves a rogue node demonstrating strength entering an ecosystem, attracting other participants, and sending data packets to it [25,26].

9.5.2.2 Latency issues at network layer

a. **Availability of Limited Bandwidth:** Insufficient or restricted bandwidth causes sluggish communication, causing longer data packet journeys and delays for time-critical applications. Packet loss and additional latency can result from out-of-order data packets, affecting time-critical applications.

b. **Presence of Bad Codecs:** Vocal impulse conversion issues can result from noise, disruptions, internal performance issues, or software issues, leading to incorrect data conversion and ineffective data.

c. **Signal Conversion Procedure:** Latency occurs when a signal changes from digital to analog, causing delays in translation and conversion.

d. **Working with Outdated Hardware:** Establishing connections and data exchange is challenging when using outdated technology alongside new gear, causing delays and hindering effective communication.

e. **Distant Communication:** Data communication may experience delay due to distance, RTT, data amount, and page load speed, affecting latency and data transmission [27].

9.5.3 Processing layer

The processing layer efficiently manages external data, utilizing Data Accumulation and Data Abstraction stages. The initial stage retrieves data based on user queries, while the second stage prepares data for consumption by various user apps, merging and converting formats.

9.5.3.1 Security issues at the processing layer

a. **Threat from Application Layer:** The application layer faces various threats, impacting data sent to the processing layer, which may experience problems and complications during data processing and communication operations.

b. **Infrastructure-Based Security Threat:** Unauthenticated port scanning, IP spoofing, and virtualization threats are common attacks. Virtualization uses the host or guest operating system to attack the hypervisor, while IP spoofing involves stealing reliable IP addresses for malicious messages. Unauthorized users scan network ports and services for system weaknesses and data theft.

c. **Data Security:** Cloud service providers' data backups leave plain text without encryption, posing security risks; attention should be paid to prevent malware attacks from modifying or stealing the backed-up data.

d. **Issue with Pool of Available Resources:** Unauthorized intruders can access cloud storage and processing resources, posing a threat to users and subscribers, potentially stealing or harming the entire system.

e. **Exhaustion:** Assault exhausts devices, thus leading to network failure or unresponsiveness.

f. **Presence of Malware:** Malware includes spyware, worms, viruses, trojan horses, and adware [26, 28–31].

9.5.3.2 Latency issues at the processing layer

a. **Storage Latency:** Storage problems like multi-tenancy and disc virtualization cause data loss and delay in retrieval [28–29].

b. **Format Conversion Issues:** Latency problem arises from heterogeneous data conversion, encryption, and decryption during request and response.

9.5.4 Application Layer

The Application Layer is the top layer in the Cloud-IoT ecosystem, offering facilities and services for various IoT-based applications like smart logistics, smart homes, and health. It includes tools like data analytics, mining, and visualization.

9.5.4.1 Security issues at application layer

a. **Corrupted Interfaces:** Cloud service providers offer accessible interfaces and APIs, but require user and organization credentials, increasing security risk.

b. **Buffer Overflow Attack:** System overloaded with data, causing damage, blockages, and unresponsiveness due to assault.

c. **Account Hijack:** Malicious user gains access to authorized user's credentials, monitors account activities, steals sensitive data, communicates using fictitious information, and engages in illegal activities.

d. **Loss of data:** Unauthorized access to account credentials can lead to data loss or modification [32].

9.5.4.2 Latency issues at application layer

a. **Overloaded Server:** Overloading servers can cause slow processing of requests, negatively impacting real-time applications and time-sensitive applications.

b. **Data Query:** Complex user queries or database processing time can cause unexpected delays, affecting network functionality.

c. **Task Distribution and Load Balancing:** Delay in task services occurs when multiple servers work simultaneously, affecting workload balancing.

d. **Slow Response:** Latency caused by various factors, programs, and internal issues.

e. **DNS configuration Issue:** DNS configuration latency caused by delayed website responses, hardware failure, and prolonged waits.

f. **Slow Network Performance:** Slow network performance may cause latency due to factors like data packet size, application faults, and authentication issues.

9.6 COUNTERMEASURES FOR SECURITY AND LATENCY IN LAYERED CLOUD-IOT MODEL

9.6.1 Countermeasures for security issues in perception layer

a. **Hashed-Based Encryption:** This method securely stores data using a key pair value twice as long as the chunk, making it difficult for attackers to break and gain access.

b. **Public Key Infrastructure (PKI) Protocol:** Infrastructure protocol combines authorization, encryption, and intrusion detection using the RSA encryption method.

c. **Lightweight Cryptography:** Limited resource IoT devices often use lightweight encryption methods like public key cryptography, hash functions, and symmetric keys.

d. **Secure Authorization:** This method integrates OAuth, focuses on authorized users, privileges, and security mechanisms using role-based access control and attributes-based access control.

e. **Embedded Security Framework:** A secure architecture protects the entire network, including nodes, interfaces, mediators, and data transmission channels. It requires identifying authorized users, secure communication, reliable storage, and data protection from unauthorized users and hackers [25,26].

9.6.2 Countermeasures for latency issues in perception layer

a. Adaptive Modulation and Coding Schemes: Modulation and coding techniques offer low latency, high transmission rates, and low spectral efficiency.

b. **Waveform Design:** Selecting the right waveform modulation strategy is crucial for efficient data transmission at the physical layer.

c. **Sending Short Packets:** Long data packets have larger payloads, high error rates, and shorter transmission times [33].

9.6.3 Countermeasures for security issues in network layer

a. **Identity Detection Framework:** IoT ecosystem requires authorized nodes; detection framework verifies legitimacy of information exchange nodes.

b. **Using SDN with IoT:** IoT uses software-defined networks for malicious activity detection, and traffic management.

c. **Inter-Nodal Attack Detection:** Ad-Hoc based authentication system is essential to protect against harmful actions in clusters of connecting nodes, preventing data loss and detecting malicious behavior.

d. **Early Detection Adaptive Risk-Based Framework:** This architecture handles environmental intrusion threats by recording changes, taking countermeasures, and taking prompt precautions against unknown attacks.

e. **In-Procedure Watch Framework:** This framework utilizes two-way communication between nodes, ensuring data packets are received and transmitted while monitoring harmful behavior and implementing preventative measures.

f. **Cluster-Based Attack Detection System:** The network is divided into clusters, with a head cluster detecting malicious behavior. Data collected from these clusters is sent to the network level, with the network disabling rogue nodes from exchanging data [25,26].

9.6.4 Countermeasures for latency issues in network layer

a. **Measuring Packet Trip Latency:** Network management must closely monitor packet travel time using tools like Ping to learn and reduce transmission delay.

b. **Routing Optimization:** Routing slowness can be caused by network traffic, hardware, and configuration issues. Network managers can improve routing through various technologies.

c. **Adopting Caching and Compression Mechanisms:** Implement caching and compression techniques for long-distance data packets, reducing load and bandwidth [30].

9.6.5 Countermeasures for security issues in the processing layer

Encryption of Sensitive Data: Encrypted data is crucial for secure communication and cloud storage, using various techniques.

Applying Data Fragmentation: Data fragmentation across the cloud means information is dispersed as individual pieces, preventing hacker access and making it useless due to incompleteness.

Authorized Writing of Data: Authorized and secured data writing prevents customization without pointer indexes, ensuring data safety.

Use of Firewalls: Firewalls detect and respond to harmful activity [31].

9.6.6 Countermeasures for latency issues in processing layer

a. Using Big-Sized Cache Chunks: If larger cache chunks are used by the storage sources over both the cloud and devices, it will not only improve the network performance but also improve the data fetch and response rate.

b. Using Storage Resources Based on Data Performance: Distributing the type of storage, based on the type of data will do a lot. If the data is historic in nature it needs to get stored at very internal sections of the cloud as it will be rarely accessed and vice versa [34,35].

9.6.7 Countermeasures for security issues in application layer

a. Validated and Authenticated Users: An authorized user should be confirmed at the region of data usage. If any unauthorized user gains access to the account, it may steal or alter the confidential data, never mind how much data is secure and encrypted.

b. Use and application of Policies and Permissions: A number of secure policies and permission lists should be made beforehand to prevent any grant of privileges or access control to malicious users.

c. Use of Antivirus and Anti-Spyware: One should definitely use this software to keep the confidentiality and security and privacy of data so that the data remain reliable and consistent.

d. Use of Firewalls: These applications should be used to keep a check on the incoming and outgoing network traffic as well as to detect any kind of malicious activity [31].

9.6.8 Countermeasures for latency issues in application layer

a. **Using Improved TCP Transmission and Retransmission Schemes:** As compared to traditional TCP streams, TCP thin Streams of data can be sent for data transmission which will not only reduce packet loss at interarrival times but also improve the retransmission time of lost packets [37–40].

9.7 FOG COMPUTING AS A SUPPORTIVE INTERMEDIARY FOR CLOUD-IOT MODEL

Fog computing addresses shortcomings in cloud-IoT by storing and processing IoT data near IoT devices, using fog nodes located close to devices. This decentralized technology offers a more efficient and real-time application solution. Fog nodes, such as routers, switches, controllers, or security cameras, are essential for analytical operations and processing tasks in IoT environments. They provide a low latent and safe environment for handling sensitive data, reducing the risk of intrusion and unauthorized user interference. Fog computing offers security for data propagation, ensuring the protection of sensitive information. Fog computing offers low latency due to its close proximity to fog nodes with complete processing and storage capabilities. Only long-term and sophisticated data is transmitted to the cloud, resulting in minimal response time delay for most requests, making it a low latent environment.

Following are characteristics of fog computing within the Cloud-IoT environment which is described in Figure 9.6:

a. **Location-Aware Computing:** The fog nodes have the capability to detect and track the connected IoT devices and are deployed in a decentralized manner near the IoT devices.
b. **Minimum Latency:** As the fog nodes are available at very near regions above the IoT devices, the data request and response cycles as well as low-level data analytics are very fast. Fog helps in supporting time-critical sensitive IoT applications.
c. **Supporting Heterogeneous Platforms:** Fog computing platforms deploy a canvas of differently manufactured network devices which communicate with each other without any major issues while keeping the data consistent and reliable.
d. **Inter-Operable Behavior:** Fog nodes are deployed at different locations and thus are distributed all over a vast region. Their components are capable of inter-operation with various other distributed components.

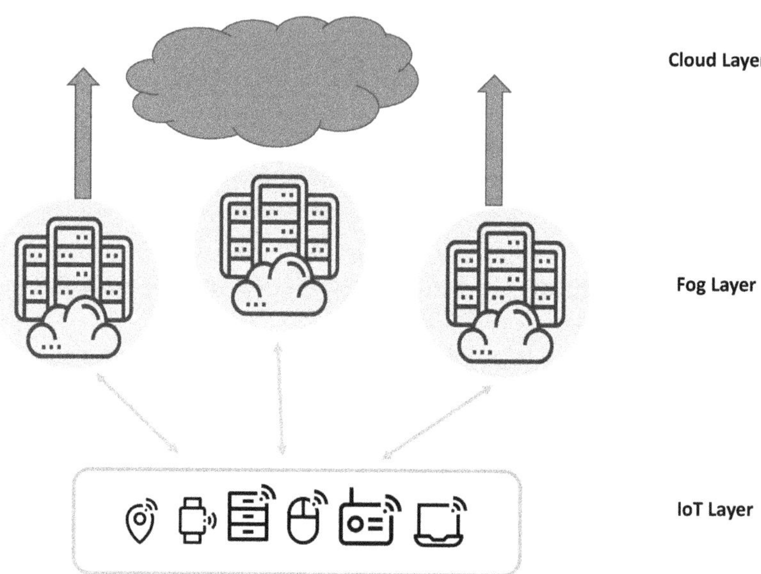

Figure 9.6 Figure Illustrating Cloud–Fog–IoT Layered Model.

    e. **Scalable:** A number of fog nodes, even a cluster of fog nodes, for a targeted group of IoT devices can be added within the environment to enhance the processing and storage capability, thus making the fog technology scalable [41–44].

Following are some advantages of using fog computing within the Cloud-IoT environment:

    a. **Increased Agility in Business:** With the introduction of fog computing the business needs as well as customer needs can be accomplished in a more efficient and accurate manner.

    b. **Enhanced Security:** Fog gives over a number of policies, procedures, and secure mechanisms to make the propagating data more secure.

    c. **Privacy:** Fog computing has the capability to control and keep our data under privacy.

    d. **Lower Operating Overheads:** As the fog nodes are available very near to the IoT devices the transmission and decision making and data processing overhead are very low as compared to that of sending the data over the cloud.

Figure 9.7 explores the latency issue in Cloud-Fog-IoT layered model. Each layer and its various security mechanisms can be explained as follows:

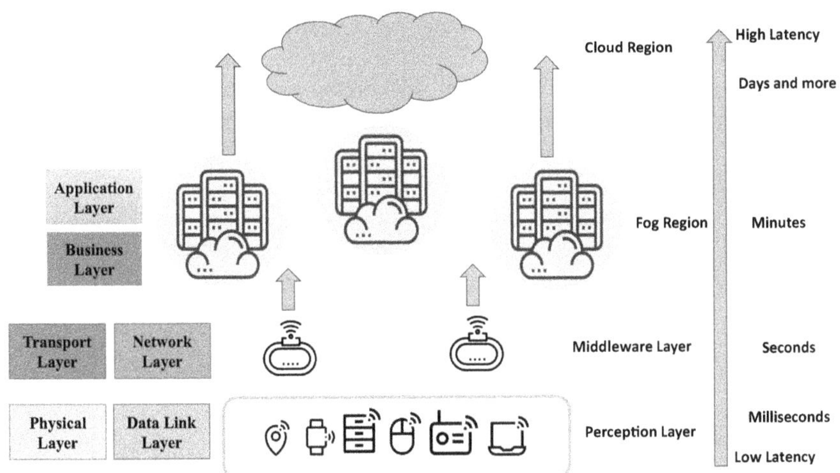

Figure 9.7 Figure Illustrating Latency Issue in Cloud–Fog–IoT Layered Model.

a. **Sensing Layer:** In this layer a number of perception or sensing technologies can be used to sense and collect data, for example, RFID technology. The data they sensed and gathered form a vast number of connected IoT devices and sources. Following are enlisted some security mechanisms that the fog environment provides in order for secure perception:
   i. Use of antivirus application
   ii. Using caching mechanism
   iii. Data verification
   iv. Privacy and data access control
   v. Data encryption
   vi. Session development
b. **Routing and Transport Layer:** When the data is gathered on the perception layer, it needs to get processed and stored at some secure side, which is the main concern. The availability, integrity, confidentiality, and security of the data should be provided at this layer. Following are enlisted some security mechanisms which the fog environment provides in order for secure transmission:
   i. Lower layer encryption
   ii. Authenticated routing
   iii. Identity verification
   iv. Data packet authentication
   v. Password verification
   vi. Secure policies and procedures

   c. **Fog Layer:** At this layer, the processing and storage of user-sensitive data take place when clouds are residing at a very distant point. The fog server provides minimal optimization and data analytic mechanisms near the edge with responsive data availability and fast data access. Following are enlisted some security mechanisms which the fog environment provides in order for secure transmission in the upper layer and cloud [45]:

     i. Use of antivirus

     ii. Data encryption

     iii. Session and boundary inspection and authorization

     iv. Risk assessment.

9.8 CONCLUSION

The need for minimizing security and latency concerns grows as demand for cloud computing rises daily. To maintain the security and safety of data transfer and manipulation, these two factors must be continuously maintained. It is well known that hackers and intrusions take advantage of any openings in latency and security solutions as well as their implementation, such as weak passwords and encryption/decryption techniques. Fog and edge elements in the new cloud architecture will undoubtedly provide new issues that must be resolved as quickly as feasible with several modifications and updates. IoT-based flexible and scalable infrastructure has been suggested for reduced latency. Fog and edge computing will benefit from this, and it will also make it easier to embrace heterogeneous IoT devices in an environment that might change. In this case, fog computing will be crucial in modernizing the Cloud-IoT model's services. Facilities for security, latency, privacy, and storage are included. Fog computing is a decentralized technology that offers safe solutions for data confidentiality and privacy as well as the capacity to minimize overall job computation latency. This article outlines a variety of security and latency concerns that affect the typical Cloud-IoT architecture at each tier. The potential for their countermeasures moving forward has also been considered. The study also explores how fog computing is crucial to the Cloud-IoT architecture and how it may effectively reduce security and latency challenges.

REFERENCES

[1] "What Is Cloud Computing? | Microsoft Azure." https://azure.microsoft.com/en-us/resources/cloud-computing-dictionary/what-is-cloud-computing (accessed: 27 July 2023).

[2] Baron, J. AWS Certified Solutions Architect Official Study Guide. John Wiley & Sons, 2016.

[3] "What Is Cloud Computing? Everything You Need to Know About the Cloud Explained." ZDNET, no date. www.zdnet.com/article/what-is-cloud-computing-everything-you-need-to-know-about-the-cloud/ (accessed: 27 July 2023).

[4] "What Is IoT? Defining the Internet of Things (IoT) | Aeris." Aeris | India, www.aeris.com/in/what-is-iot/.

[5] Violino, Bob. "11 Top Cloud Security Threats." CSO Online, 11 Oct. 2019, www.csoonline.com/article/3043030/top-cloud-security-threats.html.

[6] "The Potential of IoT and Cloud Computing – IoT Tech Expo." IoT Tech Expo, 31 May 2019, www.iottechexpo.com/2019/05/iot/potential-of-iot-and-cloud-computing/.

[7] "The Biggest Cloud Security Challenges in 2021." Check Point Software, www.checkpoint.com/cyber-hub/cloud-security/what-is-cloud-native-security/the-biggest-cloud-security-challenges-in-2021/.

[8] Oracle. "What is the Internet of Things (IoT)?" www.oracle.com, 2022, www.oracle.com/in/internet-of-things/what-is-iot/.

[9] Mamun-Ibn-Abdullah, M., et al. "Convergence platform of Cloud Computing and Internet of Things (IoT) for Smart Healthcare Application." Journal of Computer and Communications, vol. 08, no. 08, 2020, pp. 1–11, https://doi.org/10.4236/jcc.2020.88001 (accessed 6 April 2023.

[10] "Cloud Computing and IoT: 3 Important Things to Understand." Jigsaw Academy, www.jigsawacademy.com/how-cloud-computing-closely-relates-to-iot/.

[11] Clark, Jen. "What Is the Internet of Things, and How Does It Work?" IBM, 17 Nov. 2016, www.ibm.com/blogs/internet-of-things/what-is-the-iot/.

[12] "What Is Network Latency and How to Minimize It? | Versitron." www.versitron.com, www.versitron.com/blog/what-is-network-latency-and-how-to-minimize-it (accessed 27 July 2023).

[13] "Understanding Latency – Web Performance | MDN." Developer.mozilla.org, developer.mozilla.org/en-US/docs/Web/Performance/Understanding_latency.

[14] Priya. "IoT Standards and Protocols: IoT Part 3." Engineers Garage, www.engineersgarage.com/iot-standards-and-protocols-iot-part-3/ (accessed 27 July 2023).

[15] Mayuresh. "7 IOT Layers That You Should Know in 2021." 2021. www.biz4intellia.com/blog/7-layers-of-iot-what-makes-an-iot-solution-comprehensive/

[16] "What Is Low Latency?" Integrate.io, www.xplenty.com/glossary/what-is-low-latency/ (accessed 27 July 2023).

[17] Burhan, Muhammad, et al. "IoT Elements, Layered Architectures and Security Issues: A Comprehensive Survey."Sensors, vol. 18, no. 9, 24 Aug. 2018, p. 2796, www.ncbi.nlm.nih.gov/pmc/articles/PMC6165453/, https://doi.org/10.3390/s18092796

[18] Aarika, K., et al. "Perception Layer Security in the Internet of Things." Procedia Computer Science, vol. 175, 2020, pp. 591–596, https://doi.org/10.1016/j.procs.2020.07.085

[19] "A Security Guide to IoT-Cloud Convergence – Security News." www.tre
      ndmicro.com,   www.trendmicro.com/vinfo/us/security/news/internet-of-thi
      ngs/a-security-guide-to-iot-cloud-convergence.
[20] "8 Types of Security Threats to IoT | IoT Security Threats |." Allerin.com, 22
      May 2019, www.allerin.com/blog/8-types-of-security-threats-to-iot.
[21] "5 IoT Security Threats to Prioritize." IoT Agenda, internetofthingsagenda.
      techtarget.com/tip/5-IoT-security-threats-to-prioritize.
[22] Mdpi.com,   2023,   www.mdpi.com/electronics/electronics-10-01171/art
      icle_deploy/html/images/electronics-10-01171-g001-550.jpg   (accessed   27
      July 2023).
[23] "Security of the Internet of Things: Vulnerabilities, Attacks and
      Countermeasures." 2019. IEEE Communications Surveys & Tutorials, vol.
      22, no. 1, pp. 616–644, First quarter 2020, doi: 10.1109/COMST.2019.
      2953364
[24] Tips, SiteUptime. "6 Ways to Boost Your Web Application Performance for
      Optimal User Experience." SiteUptime Blog, 24 Nov. 2017, www.siteupt
      ime.com/blog/2017/11/24/6-ways-to-boost-your-web-application-performa
      nce-for-optimal-user-experience/.
[25] Siddiqi, M. A. et al. "5G ultra-Reliable Low-Latency Communication
      Implementation Challenges and Operational Issues with IoT Devices."
      Electronics, vol. 8, no. 9, 2 Sept. 2019, p. 981, https://doi.org/10.3390/ele
      ctronics8090981.
[26] "SatMagazine."   www.satmagazine.com,   www.satmagazine.com/story.
      php?number=1342678280 (accessed 27 July 2023).
[27] Puthal, Deepak, et al. "Fog Computing Security Challenges and Future
      Directions [Energy and Security]." IEEE Consumer Electronics Magazine,
      vol. 8, no. 3, May 2019, pp. 92–96,https://doi.org/10.1109/mce.2019.2893
      674 (accessed 4 March 2020).
[28] Bellavista, Paolo et al. "How Fog Computing Can Support Latency/
      Reliability-Sensitive IoT Applications: An Overview and a Taxonomy of
      State-of-the-Art Solutions." 25 Apr. 2020, pp. 139–213,https://doi.org/
      10.1002/9781119551713.ch6. Accessed 27 July 2023.
[29] O'Farrell, Patrick, and Aniruddha Khadye. "Latency in Factory Automation
      All Trademarks Are the Property of Their Respective Owners. (1) Application
      Report Latency in Factory Automation."2015.
[30] Media, IR. "Network Latency – Common Causes and Best Solutions |
      IR."www.ir.com, www.ir.com/guides/what-is-network-latency.
[31] Litoussi,   Mohamed,   et   al.   "IoTsecurity: challenges   and
      countermeasures."Procedia Computer Science, vol. 177, 2020, pp. 503–508,
      https://doi.org/10.1016/j.procs.2020.10.069 (accessed 16 December 2020).
[32] "What Is IoT Security?" Palo Alto Networks, www.paloaltonetworks.com/
      cyberpedia/what-is-iot-security.
[33] Jiang, Xiaolin, et al. "Low-Latency Networking: Where Latency Lurks and
      How to Tame It." Proceedings of the IEEE, vol. 107, no. 2, 1 Feb. 2019, pp.
      280–306,   ieeexplore.ieee.org/abstract/document/8452158,   https://doi.org/
      10.1109/JPROC.2018.2863960.

[34] "Top 4 Causes of Storage I/O Bottlenecks &How to Mitigate Them." www.
solarwinds.com/legacy-assets/solarwinds/swdcv2/licensed-products/storage-
resource-monitor/resources/whitepapers/top_4_causes_of_storage_io_bott
lenecks.ashx?rev=695614ad5aa64580b920990923eadf27?rev=695614ad5
aa64580b920990923eadf27 (accessed 10 March 2024).

[35] "Latency – the Silent Killer of Conversions." ImageKit.io Blog, 22 Jan.
2020, imagekit.io/blog/latency-the-silent-killer-of-conversions/ (accessed 27
July 2023).

[36] Petlund, Andreas, et al. Improving Application Layer Latency for Reliable
Thin-Stream Game Traffic. In Proceedings of the 7th ACM SIGCOMM
Workshop on Network and System Support for Games (NetGames '08).
Association for Computing Machinery, New York, NY, USA, pp. 91–96.
https://doi.org/10.1145/1517494.1517513

[37] Jiang, Xiaolin, et al. "Low-latency Networking: Where Latency Lurks and
How to Tame It." Proceedings of the IEEE, vol. 107, no. 2, 1 Feb. 2019, pp.
280–306, ieeexplore.ieee.org/abstract/document/8452158, https://doi.org/
10.1109/JPROC.2018.2863960 (accessed 14 May 2022).

[38] Jeya, John, and E Baburaj. "Layers Based Security Issues in Cloud
Computing." International Journal of Advance Research in Science and
Engineering IJARSE, no. 4, 2015, www.ijarse.com/images/fullpdf/201.pdf.

[39] "IoT Architecture: The Pathway from Physical Signals to Business Decisions."
AltexSoft, www.altexsoft.com/blog/iot-architecture-layers-components/.

[40] "How to Improve Network Latency in 3 Steps." SearchNetworking,
www.techtarget.com/searchnetworking/tip/How-to-improve-network-late
ncy-in-3-steps.

[41] "Fog Computing along the Cloud-To-Thing Continuum." I-SCOOP, www.i-
scoop.eu/internet-of-things-iot/fog-computing-cloud-internet-things/.

[42] Burhan, Muhammad, et al. "IoTIoTElements, Layered Architectures and
Security Issues: A Comprehensive Survey." Sensors, vol. 18, no. 9, 24 Aug.
2018, p. 2796, www.ncbi.nlm.nih.gov/pmc/articles/PMC6165453/, https://
doi.org/10.3390/s18092796.

[43] Aziz, Tariq, and Ehsan-ulHaq. "Security Challenges Facing IoT Layers and
Its Protective Measures." International Journal of Computer Applications,
vol. 179, no. 27, 20 Mar. 2018, pp. 31–35, https://doi.org/10.5120/ijca201
8916607.

[44] Atlam, Hany F, et al. "Fog Computing and the Internet of Things: A
Review."ResearchGate, MDPI AG, 8 Apr. 2018, www.researchgate.net/publ
ication/324280213_Fog_Computing_and_the_Internet_of_Things_A_
Review.

[45] Media, Open Systems. "How Fog Computing Can Solve the IoT
Challenges."Embedded Computing Design, www.embeddedcomputing.
com/technology/iot/wireless-sensor-networks/how-fog-computing-can-
solve-the-iot-challenges.

Analyzing the impact of security risk for securing healthcare information systems

Kavita Sahu, R. K. Srivastava, Abhishek Kumar Pandey, and Kotaiah Bonthu

10.1 INTRODUCTION

In recent years, the world has witnessed a dramatic shift towards the digital realm, with the increasing popularity of web-based applications taking center stage [1–3]. These applications have not only transformed the way we interact with technology but have also beckoned the attention of a shadowy and sophisticated counterpart—the hacker community [4–7]. In this dynamic landscape, the security risks associated with web-based services, particularly within the critical domain of healthcare, are mounting at an alarming rate [8].

This burgeoning concern has galvanized researchers and security practitioners across the globe into a unified mission: to comprehensively assess and proactively manage the security vulnerabilities inherent to healthcare web applications. The repercussions of neglecting this task have been starkly illustrated by a series of unsettling statistics, which paint a grim picture of the healthcare industry [9–12]. Data breaches in healthcare pose an immediate and tangible threat, endangering not only sensitive patient information but also the very systems that hospitals and healthcare providers rely on. These breaches are not just white-collar offenses; they have the potential to cause immeasurable harm [13–15]. Most notably, the compromise and tampering of highly confidential patient medical records could disrupt treatment operations, leading to grave consequences.

In response to this intensifying challenge, concerted efforts are underway to bolster the security of healthcare web applications [16–19]. These applications have become the linchpin for delivering critical services and disseminating vital information across the vast expanse of the internet. The universality and accessibility of web applications make them indispensable in our daily lives, whether it is conducting online banking, engaging in virtual education, accessing financial data, staying informed through news portals, connecting with friends and family via social media, or even enjoying entertainment through streaming platforms [20–25]. In essence, web applications

DOI: 10.1201/9781003514312-10

have become the lifeblood of our digital existence, an indispensable tool in the contemporary world.

What is particularly intriguing is the relentless proliferation of web-based applications year after year. Although pinpointing the precise number of such applications worldwide is a complex task, the first quarter of 2021 alone witnessed approximately 363.5 million domain name registrations [26–29]. Each of these domains, whether static or dynamic, can be viewed as a potential web application, each with its own unique set of security challenges.

The primary objective of this chapter is to conduct a thorough analysis of the security issues intertwined with healthcare web-based applications and to shed light on the evolving landscape of web applications within the healthcare sector, which has been significantly shaped by the advent of the COVID-19 pandemic [30, 31].

This chapter is organized as follows: Section 10.2 provides an overview of related work in the context of security risks within healthcare web-based applications. Section 10.3 delves deeper into the multitude of security threats that afflict web-based applications in the healthcare industry. Section 10.4 presents a detailed analysis of current data breaches affecting web-based applications in the healthcare sector, underlining the urgency of the situation. Section 10.5 offers invaluable insights into the current state of healthcare web-based applications and the shifts brought about by the pandemic. Finally, in Section 10.6, this chapter concludes by offering guidance for securing healthcare information systems, addressing the pressing need for a more secure future in the realm of healthcare industry.

10.2 RELATED RESEARCH INITIATIVES

In the context of the security risk of healthcare web applications, several notable research initiatives and studies have been undertaken, thus providing valuable insights and approaches to enhance the security of healthcare information systems [2, 9, 22, 30]. These initiatives include:

- Vulnerability Identification Approach by Zech et al. [1]: Zech and his team introduced a novel vulnerability identification approach relevant to the security risk analysis of Web-Based Healthcare Management Systems (WBHMS). Their approach involves knowledge-based security solutions to detect and mitigate vulnerabilities in healthcare web applications.
- Static Analysis Tool for Cross-site Scripting (XSS) Vulnerabilities by Engin Kirda [2]: Engin Kirda's research led to the discovery of Pixy's XSS vulnerability, marking the first instance of a static analysis tool identifying XSS vulnerabilities. Although Pixy may be outdated, it sheds light on the persistence of XSS vulnerabilities in

web applications. The limitations, such as false positives, highlight the need for improved analysis methods.

- Web Application Vulnerabilities (Gate and Liska) [3]: Gate and Liska (3rd) identified web application vulnerabilities that could potentially allow hackers to gain unauthorized access. They emphasized the use of bots, malware, and distributed denial-of-service attacks for exploitation. As online businesses and financial transactions increase, addressing these vulnerabilities becomes critical.
- Fuzzy Set Theory by Dark [15]: Dark applied fuzzy set theory to assess cost and time performance, security risk management, and utilization of healthcare web application development methods. This approach provides a quantitative framework to evaluate security risks in healthcare web applications.
- Security Risk Framework by Shedden et al. [16]: Shedden and colleagues utilized a security risk framework to qualitatively assess the security risk of healthcare web applications. This framework offers a structured approach to evaluating and managing security risks in the healthcare sector.
- Learning-Based Security Assessment by Sunitha et al. [4]: Sunitha et al. introduced a method for learning-based security assessment using logic programming, focusing on WBHMS. This approach effectively detects vulnerabilities, such as Uniform Resource Locator (URL), XSS, and Structured Query Language (SQL) injection attacks, making healthcare web applications more secure over time.
- MITIGATE System by Schauer et al. [5]: Schauer and collaborators proposed the MITIGATE system to analyze the security threats in web application networks. This system provides an updated risk assessment of digital resources, highlighting vulnerabilities and interconnections among different partners in the web application ecosystem.
- Hazard Evaluation Mechanism by Ionita et al. [6]: Ionita et al. explored a hazard evaluation mechanism to address security concerns. Their approach prioritizes security features for product owners and ensures that designers have the knowledge and time required to implement security requirements effectively.
- Model for Internet of Things (IoT) Digital Risk by Radanliev et al. [7]: Radanliev and his team put forth a model that includes a structured procedure and specific hazard evaluation vectors for IoT digital risk. This research identifies gaps in digital risk guidelines and policies, contributing to the development of standards for future digital risk evaluation in the context of IoT.

These research initiatives collectively contribute to the understanding and mitigation of security risks in healthcare information systems, emphasizing

Table 10.1 Comparative Analysis

S. No.	Research Initiative	Vulnerability Detection Approach	Application Focus	Key Findings/Contributions
1	Zech et al.	Knowledge-based security solution	WBHMS	Novel vulnerability identification approach.
2	Engin Kirda	Static analysis tool for XSS vulnerabilities	General web applications	Discovery of Pixy's XSS vulnerability.
3	Gate and Liska	Program design/code vulnerabilities	General web applications	Identification of web application vulnerabilities.
4	Dark	Fuzzy set theory	Healthcare web app development	Quantitative assessment of security risks.
5	Shedden et al.	Security risk framework	Healthcare web applications	Qualitative security risk assessment framework.
6	Sunitha et al.	Learning-based security assessment	WBHMS	Improved security assessment for WBHMS.
7	Schauer et al.	MITIGATE system	Web application networks	Risk assessment for digital resources.
8	Ionita et al.	Hazard evaluation mechanism	General web applications	Prioritization of security features.
9	Radanliev et al.	Model for IoT digital risk	IoT digital risk assessment	Identifying gaps in IoT digital risk policies.

the need for robust security measures in the increasingly digital healthcare landscape. A comparative analysis table (Table 10.1) summarizing the key aspects of the research initiatives mentioned in the context of security risk for healthcare web applications.

10.3 SECURITY RISKS OF WEB-BASED APPLICATION

In recent years, there has been a sudden increase that is both disturbing and concerning in the number of incidents involving security violence, data breaches, loss of patient confidentiality, and loss of integrity of patient data. These occurrences are comparable to cyberattacks launched against patients, employees, and places of business. The digital threat posed to crisis centers is growing at an alarming rate all around the world, and it is not confined to a single nation in particular. The scope of computerized violence is limitless, and it has wreaked havoc on numerous crisis institutions in both high-paying countries like the United States and low-paying ones like Kenya [8].

In a report that was released by the United States Department of Justice, a startling fact regarding ransomware was revealed. Ransomware is only one form of malware that can hinder productivity in the workplace. In 2018, there were around 4000 ransomware ambushes that occurred each day across a variety of regions [9], which is an increase of 300 percent compared to 2017's numbers. The survey also disclosed the top three locations that saw the highest level of malware contamination. In addition to ransomware, the number of malicious computer programming attacks has climbed by a factor of four over the course of the past 2 years, and the wealthy region has become one of the most completely coordinated zones. The completion of information systems for definite, cash-related, and helpful errands by crisis facilities throughout the world, as well as the employment of connected restorative gadgets, distributed capacity organizations, and simultaneously expanding framework systems, are producing anxiety as a result of this development.

The Open Web Application Security Project, also known as OWASP, is an organization that does not seek to profit from its efforts to improve software security. The OWASP is a community focused on the security of web applications that makes its publications, strategies, documentation, tools, and technology available to the public. It is common knowledge that the OWASP Top 10 should be followed as a standard awareness guide for web application security by developers. It is a reflection of the common consensus on the most important security vulnerabilities in web applications. There are three new categories included in the Top 10 for 2021 [10], as well as four categories that have had their naming and scope modified, and there have also been some consolidations, as discussed:

- Broken Access Control: Previously ranked fifth, Broken Access Control has now climbed to the first position. It is worth noting that 94 percent of the assessed applications exhibited some form of flawed access control. This category stands out with 34 Common Weakness Enumerations (CWEs) associated with it.
- Cryptographic Failures: Formerly referred to as "Sensitive Data Exposure," this category has risen to the second position. The emphasis has shifted towards encryption failures, a leading cause of sensitive data leakage and system breaches.
- Injection: Injection attacks involve introducing untrusted data into an application's interpreter, potentially causing harmful commands or queries. This category, now in third place, encompasses the 33 CWEs with the second-highest occurrence in applications. Notably, cross-site scripting has been included in this category.
- Insecure Design: A new addition for 2021, Insecure Design addresses issues associated with design flaws. To address these concerns effectively, the industry must embrace more threat modeling, secure design patterns, principles, and reference architectures.
- Security Misconfiguration: Climbing from its previous ranking of sixth, Security Misconfiguration is now in fifth place. As

more organizations adopt highly flexible software, 90 percent of applications were evaluated for misconfigurations. This category now encompasses the old XML External Entities (XXE) vulnerabilities.

- Vulnerable and Outdated Components: This category, previously known as "Using Components with Known Vulnerabilities," has transitioned to the second position in the Top 10 community survey. It has moved up from the ninth position in 2017, underscoring the challenge of assessing and quantifying the risk associated with this prevalent issue.
- Identification and Authentication Failures: Taking the place of Broken Authentication, Identification and Authentication Failures now ranks as the second most common CWE. It covers issues closely linked to identification failures, and the proliferation of uniform frameworks is aiding in addressing these concerns.
- Software and Data Integrity Failures: A new category for 2021, Software and Data Integrity Failures concentrates on making assumptions about software updates, critical data, and continuous integration and continuous deployment pipelines without validating their integrity. This category includes 10 CWEs with significant impact, with Insecure Deserialization now being part of this broader category.
- Security Logging and Monitoring Failures: Formerly known as Insufficient Logging and Monitoring, this category has climbed from the 10th to the 3rd position in the industry survey. It has been expanded to encompass a wider range of failures, despite being challenging to test. Failures in this area directly affect visibility, incident alerting, and forensics.
- Server-Side Request Forgery: Although this category exhibits a low incidence rate, it enjoys above-average testing coverage and potential exploit and impact ratings. It underscores how the security community recognizes its importance even if it is not highly represented in current data.
- Cross-site Scripting: XSS involves injecting malicious code/scripts into web responses, which are then executed by the web browser. There are three primary types of XSS attacks: Stored XSS, Reflected XSS, and DOM-Based XSS, each with distinct characteristics. Ongoing research provides further insights into XSS attacks [28].

Table 10.2 provides a comparative analysis of the vulnerabilities, their categories, and descriptions to help organizations understand the potential risks and take appropriate security measures.

Table 10.2 Lists of Vulnerabilities

S. No.	Vulnerability	Category	Description
1	Unauthorized access to patient records	Broken Access Control	Lack of control over who can access patient data, leading to unauthorized access.
2	Inadequate user privilege management	Broken Access Control	Poor management of user privileges, resulting in inappropriate access permissions.
3	Lack of proper access controls for critical healthcare data	Broken Access Control	Insufficient controls on sensitive healthcare data, making it vulnerable to unauthorized access.
4	Weak encryption algorithms	Cryptographic Failures	Usage of weak encryption algorithms that can be easily compromised.
5	Poor key management practices	Cryptographic Failures	Inadequate key management leading to vulnerabilities in data protection.
6	Data leakage due to improper encryption	Cryptographic Failures	Data exposure because of encryption failures.
7	SQL injection attacks on healthcare databases	Injection	Exploiting SQL injection vulnerabilities in healthcare databases.
8	Code injection vulnerabilities in healthcare software	Injection	Vulnerabilities that allow attackers to inject malicious code into healthcare software.
9	Injection of malicious scripts into web applications	Injection	Exploiting web application vulnerabilities to insert malicious scripts.
10	Poorly designed user interfaces leading to user errors	Insecure Design	User interface flaws causing user errors that may lead to security issues.
11	Lack of security considerations in system architecture	Insecure Design	Security aspects not considered during system architecture, leaving vulnerabilities.
12	Inadequate threat modeling and risk assessment during the design phase	Insecure Design	Failing to adequately assess threats and risks during system design.
13	Improperly configured servers or databases	Security Misconfiguration	Misconfigured servers or databases that expose vulnerabilities.
14	Default settings not changed	Security Misconfiguration	Using default configurations that may have security weaknesses.
15	Unnecessary services running with excessive permissions	Security Misconfiguration	Unneeded services running with elevated privileges, increasing attack surface.

Table 10.2 (Cont.)

S. No.	Vulnerability	Category	Description
16	Use of outdated or unsupported software libraries	Vulnerable and Outdated Components	Using software libraries that are no longer supported or contain known vulnerabilities.
17	Unpatched vulnerabilities in third-party components	Vulnerable and Outdated Components	Failing to apply patches to known vulnerabilities in third-party components.
18	Lack of a system for tracking and updating components	Vulnerable and Outdated Components	No system in place to monitor and update software components.
19	Weak or easily guessable passwords	Identification and Authentication Failures	Usage of weak or easily guessable passwords, making it easier for attackers to gain access.
20	Inadequate multi-factor authentication	Identification and Authentication Failures	Failing to implement strong multi-factor authentication methods.
21	Authentication bypass or privilege escalation	Identification and Authentication Failures	Weaknesses allowing attackers to bypass authentication or escalate privileges.
22	Unauthorized data modification or deletion	Software and Data Integrity Failures	Unauthorized changes or deletion of healthcare data.
23	Lack of digital signatures to verify data integrity	Software and Data Integrity Failures	Absence of digital signatures for data verification.
24	Data corruption due to system errors	Software and Data Integrity Failures	Data integrity compromised due to system errors.
25	Inadequate or non-existent logging of security events	Security Logging and Monitoring Failures	Lack of proper security event logging.
26	Lack of real-time monitoring for suspicious activities	Security Logging and Monitoring Failures	Failing to monitor for suspicious activities in real time.
27	Failure to alert on security incidents	Security Logging and Monitoring Failures	Inability to detect and alert on security incidents.
28	Unauthorized internal network access via SSRF	Server-Side Request Forgery	Gaining unauthorized access to internal networks through SSRF vulnerabilities.
29	Exploiting SSRF to make requests on behalf of the server	Server-Side Request Forgery	Attackers using SSRF to make requests on behalf of the server to external resources.
30	Access to sensitive information through SSRF vulnerabilities	Server-Side Request Forgery	Exploiting SSRF vulnerabilities to access sensitive information.

10.4 DATA BREACHES ANALYSIS REPORT

Data breaches pose a significant threat to the security of healthcare information systems, with Personal Health Information (PHI), Personally Identifiable Information (PII), trade secrets, and intellectual property being prime targets. When unauthorized individuals gain access to such sensitive data, it constitutes an information breach. While hackers and cybercriminals are often responsible for these breaches, unintentional exposure of sensitive information on the internet is not uncommon. This report delves into the alarming statistics and trends associated with data breaches in healthcare information systems, emphasizing the critical need for enhanced security measures.

10.4.1 Healthcare industry as a prime target

Healthcare online applications are frequent targets for attackers due to the high sensitivity and value of health information. A survey revealed that between 2020 and 2021, 287.7 million individuals were victims of healthcare data breaches, with 157.40 million individuals affected in recent years. Moreover, a total of 2216 data breach attacks were registered worldwide, with 536 breaches originating from the healthcare sector. These statistics underscore the severity of the healthcare industry's vulnerability to cyberattacks.

10.4.2 Financial implications

The financial repercussions of healthcare data breaches are substantial. According to a survey conducted by IBM in 2019, the average revenue loss resulting from data breaches worldwide was approximately $3.92 million, but in the healthcare industry, it was much higher, at $6.45 million. These figures underscore the complexity of healthcare online application security and the urgent need to address these challenges effectively.

10.4.3 Historical data breach analysis

Table 10.3 and Figure 10.1 provide a historical analysis of data breaches in healthcare, which is essential for understanding the evolving landscape of healthcare data breaches. The data reveals a concerning trend with an increase in the number of data breaches over the years, coinciding with the growing use of web applications for healthcare services. The multifaceted nature of healthcare makes it challenging to implement comprehensive security measures, and attackers exploit these vulnerabilities on a large scale.

Table 10.3 HIPAA Data Breach Analysis Report

S. No.	Year	Number of Data Breaches	Exposed Records (Millions)
1	2010	199	5.530
2	2011	200	13.150
3	2012	219	2.800
4	2013	280	6.950
5	2014	314	17.450
6	2015	270	113.270
7	2016	330	16.400
8	2017	360	5.100
9	2018	371	33.200
10	2019	512	41.200
11	2020	642	29.300
12	2021*	674	3.350
13	Total	4371	287.7

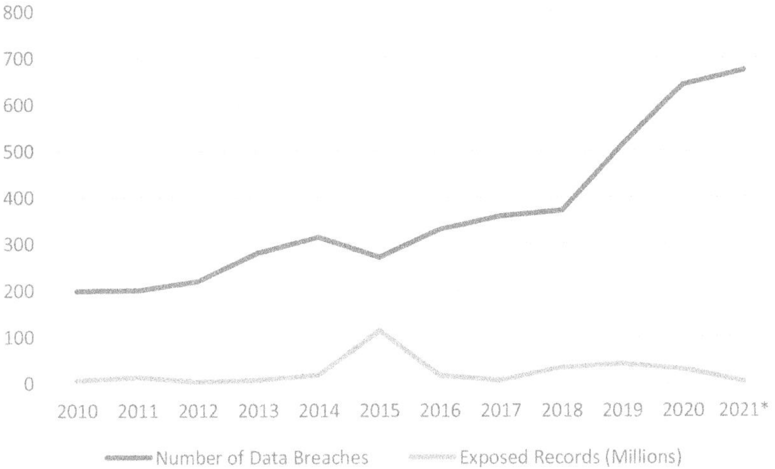

Figure 10.1 Graphical Attack Illustration.

10.4.4 The critical need for security enhancement

Health Insurance Portability and Accountability Act (HIPAA) compliance is essential for healthcare organizations, as depicted in Table 10.3 and Figure 10.1. The statistics illustrate the officially documented cyberattacks on healthcare institutions, revealing a staggering number of breach instances and exposed records. The high number of breach records is a matter of grave concern, as data breaches can result in severe, life-threatening consequences for patients. Ensuring robust online application security is imperative to safeguard healthcare systems and, ultimately, human lives. The healthcare sector faces a growing threat from data breaches, with far-reaching consequences for patients and organizations alike. The complexity of healthcare information systems, coupled with the increasing use of web applications, makes security implementation challenging. Urgent action is required to enhance online application security and protect sensitive health information from malicious actors.

10.4.5 Analysis of OWASP Top-10 for 2021

The OWASP Top-10 is a widely recognized and highly influential document that highlights the most critical security risks facing web applications. OWASP releases periodic updates to this list to reflect the evolving threat landscape. In this comparative analysis, we will examine OWASP Top-10 for 2021 in relation to the lists from previous years to identify changes, trends, and evolving security concerns.

OWASP Top-10 for 2021

- Injection: This category still includes SQL injection, but it now broadens to include NoSQL injection and Command injection. The expansion indicates the growing diversity of injection attack vectors.
- Broken Authentication: Continues to be a top concern, emphasizing the importance of robust user authentication and session management.
- Sensitive Data Exposure: Similar to the previous list, highlighting the critical need to protect sensitive data.
- XML External Entities: XXE was not explicitly mentioned in previous lists, indicating a growing awareness of this vulnerability.
- Broken Access Control: It remains a significant concern, emphasizing the importance of proper access controls.
- Security Misconfiguration: Similar to previous lists, highlighting the prevalence of misconfigured security settings.
- Cross-Site Scripting: Persistent XSS and DOM-based XSS are both covered, indicating the continuous concern about client-side vulnerabilities.
- Insecure Deserialization: A new entry in the list, reflecting the increased focus on this vulnerability.
- Using Components with Known Vulnerabilities: Unchanged from the previous list, underlining the importance of keeping third-party components up to date.
- Insufficient Logging and Monitoring: Remains in the list, emphasizing the importance of real-time detection and response.

Comparative Analysis

- Inclusion of New Threats: The 2021 list demonstrates an evolution in the threat landscape, with the addition of XXE and Insecure Deserialization. This suggests that attackers are exploring new avenues to exploit.
- Continuing Concerns: Several items from previous lists, such as Broken Authentication, Sensitive Data Exposure, and Insufficient Logging and Monitoring, remain top priorities. This indicates that the fundamentals of web application security are still critical.
- Awareness and Emphasis: OWASP's inclusion of specific attack vectors like NoSQL injection and Command injection highlights the need for developers and security professionals to understand a broader range of injection vulnerabilities.
- Client-Side Vulnerabilities: XSS continues to be a significant concern, reflecting the continued focus on protecting against client-side attacks.
- Vulnerability Management: The presence of "Using Components with Known Vulnerabilities" shows that managing third-party components is a persistent challenge.
- Dynamic Threat Landscape: The evolving nature of web application security is evident through the shifting emphasis and the inclusion of new threats in each iteration of the OWASP Top-10.

The comparative analysis of OWASP Top-10 for 2021 with previous years' lists highlights the dynamic and evolving nature of web application security. It emphasizes the need for continuous vigilance, adaptation, and education to address the ever-changing threat landscape and protect against a broader range of vulnerabilities.

10.5 CURRENT SCENARIO OF HEALTHCARE WEB-BASED APPLICATION SECURITY

In the evolving landscape of healthcare, web-based applications have become integral for various aspects of patient care, information management, and administrative processes. However, the security of these applications has come under severe threat, with an alarming surge reported in cyberattacks, particularly after the initiation of COVID-19 vaccine distribution in December. This chapter examines the impact of security risks on healthcare information systems with a focus on the recent trends and challenges faced in the healthcare industry.

10.5.1 Imperva Research Labs findings

According to a report from Imperva Research Labs, there has been a staggering 51 percent surge in web application attacks against healthcare targets during the initial phases of COVID-19 vaccine distribution. This surge marks the culmination of a year of extraordinary cybersecurity activity, highlighting the vulnerability of healthcare web applications in the face of ongoing global health challenges. In 2020, healthcare organizations faced an average of 187 million web application attacks per month, equating to nearly 500 attacks per organization every month. This represents a 10 percent year-over-year increase, indicating a growing concern in the healthcare sector [13].

10.5.2 Types of attacks

Two specific types of attacks have been particularly prominent in the recent wave of cyber threats. XSS attacks increased by 43 percent in December, becoming the most prevalent form of web application attacks. XSS attacks exploit vulnerabilities in web applications, making them infinitely exploitable. SQL injections (SQLi) were the second most common attack type, experiencing a 44 percent increase in volume. Additionally, protocol manipulation assaults, the third largest in terms of volume, saw a substantial increase of 76 percent [13].

10.5.3 Understanding the surge in attacks

The surge in web application attacks can be attributed to several factors. As the healthcare industry rapidly adapted to address the challenges posed by the COVID-19 pandemic, many organizations embraced third-party solutions and JavaScript APIs to provide telehealth services and other essential functionalities. This increased reliance on third-party solutions and the integration of new web applications expanded the attack surface for malicious actors [13, 14]. Organizations often utilize web browsers to access local applications, thereby introducing more potential targets for hackers. Due to the reuse of common code, preventing injection attacks has become a challenging task. For instance, XSS attacks, being infinitely exploitable, pose a persistent threat.

Terry Ray, SVP and Fellow at Imperva, stated that the increase in web application attacks is directly related to the fact that "many of the COVID-19 mitigation efforts are powered by new web applications and services." However, the exact reasons for the surge in attacks around vaccine distribution remain unclear [13]. In conclusion, the healthcare sector's growing reliance on web-based applications, compounded by the pressure of the pandemic, has made it a prime target for cybercriminals. This chapter delves deeper into the evolving landscape of healthcare information system security and strategies to mitigate the increasing security risks faced by healthcare organizations in this digital age.

10.5.4 Issues and challenges

- Rising Cybersecurity Threats: The healthcare industry is experiencing an alarming increase in cyberattacks, with web application attacks becoming more sophisticated and frequent. As demonstrated by Imperva Research Labs, the healthcare sector has seen a 51 percent surge in web application attacks, posing a significant threat to patient data, hospital operations, and public health.
- Vulnerability Due to Rapid Technological Adoption: The rapid adoption of digital health technologies, telehealth services, and IoT devices has expanded the attack surface. This increased reliance on web applications and third-party solutions has created vulnerabilities that malicious actors can exploit.
- Data Privacy and Compliance: Healthcare information systems must adhere to strict data privacy regulations, such as HIPAA in the United States and General Data Protection Regulation in Europe. Ensuring compliance while safeguarding patient data against cyber threats is an ongoing challenge.
- Insider Threats: Insider threats, whether intentional or accidental, continue to be a significant concern. Employees and healthcare professionals with access to sensitive information can inadvertently compromise security.
- Resource Constraints: Many healthcare organizations, particularly smaller ones, face resource constraints when it comes to implementing robust cybersecurity measures. Limited budgets and cybersecurity expertise can hinder their ability to defend against cyber threats effectively.

10.6 FUTURE WORK

- Enhanced Security Protocols: Healthcare organizations should prioritize the implementation of robust security protocols, including intrusion detection and prevention systems, real-time monitoring, and regular security audits. Investing in cutting-edge technologies like artificial intelligence (AI)-based threat detection and blockchain for data integrity will be essential.
- Employee Training and Awareness: Continuous training and awareness programs for healthcare staff are crucial. Employees should be educated about the latest cybersecurity threats and the best practices to prevent security breaches.
- Collaboration and Information Sharing: Healthcare institutions should foster collaboration and information sharing to create a collective defense against cyber threats. Sharing threat intelligence

and experiences can help organizations proactively defend against attacks.

- Data Encryption and Access Control: Encrypting patient data and implementing stringent access control measures can significantly reduce the risk of unauthorized access or data breaches. This includes multifactor authentication and strict role-based access.
- Compliance and Regulation Adherence: Staying updated with changing data privacy regulations and compliance standards is vital. Healthcare organizations should proactively adapt their security policies and procedures to remain in compliance.

10.7 CONCLUSIONS

The security of healthcare information systems is at a critical juncture, with an unprecedented surge in cyberattacks coinciding with the ongoing global health challenges. As digital transformation continues to shape the healthcare industry, it is imperative to address the associated security risks. To secure healthcare information systems effectively, organizations must embrace a multi-faceted approach that combines technology, education, and collaboration. Enhancing security protocols, training employees, and staying compliant with regulations are all crucial elements in safeguarding patient data and maintaining the integrity of healthcare operations. The future of healthcare information system security depends on the commitment of healthcare organizations to adapt, evolve, and prioritize cybersecurity. By doing so, they can ensure that healthcare data remains protected and patient trust is maintained in an increasingly digitized healthcare landscape.

REFERENCES

[1] Zech, P., Felderer, M., & Breu, R. (2012). Towards risk-driven security testing of service-centric systems. In 12th International Conference on Quality Software (pp. 140–143). Xi'an, China.

[2] Jovanovic, N., Kruegel, C., & Kirda, E. (2006). Pixy: A static analysis tool for detecting web application vulnerabilities. IEEE Symposium on Security and Privacy (S&P'06) (pp. 6, 263). Berkeley/Oakland, CA.

[3] Gates, S., & Liska, A. (2018). Securing Web Applications. O'Reilly Media, Inc.

[4] Sunitha, K. V. N., & Sridevi, M. (2009). Automated detection system for SQL injection attack. International Journal of Computer Science and Security, 4(4), 426.

[5] Schauer, S., Stamer, M., Bosse, C., Pavlidis, M., Mouratidis, H., et al. (2017). An adaptive supply chain cyber risk management methodology. In Hamburg International Conference of Logistics (pp. 15–19). Hamburg, Germany.

[6] Ionita, D., Bullee, J., & Wieringa, R. J. (2014). Argumentation-based security requirements elicitation: The next round. In IEEE 1st International

Workshop on Evolving Security and Privacy Requirements Engineering (pp. 7–12). Karlskrona.

[7] Radanliev, P., Roure, D. C. D., Nicolescu, R., Huth, M., Montalvo, R. M., et al. (2018). Future developments in cyber risk assessment for the Internet of Things. Computers in Industry, 102(1), 14–22.

[8] Jalali, M. S., Razak, S., Gordon, W., Perakslis, E., & Madnick, S. (2019). Health care and cybersecurity: Bibliometric analysis of the literature. Journal of Medical Internet Research, 21(2), 52–57.

[9] Argaw, S. T., Pastoriza, J. R. T., Lacey, D., Florin, M. V., Calcavacchia, F., et al. (2020). Cybersecurity of hospitals: Discussing the challenges and working towards mitigating the risks. BMC Medical Information and Decision Making, 20(2), 146.

[10] OWASP. (n.d.). OWASP Top Ten. Retrieved from https://owasp.org/www-project-top-ten

[11] Barona, R., & Anita, E. A. M. (2017). A survey on data breach challenges in cloud computing security: Issues and threats. In International Conference on Circuit, Power and Computing Technologies (pp. 1–8). Kollam.

[12] Wallarm. (n.d.). OWASP Top 10 2021—Proposal based on statistical data. Retrieved from https://lab.wallarm.com/owasp-top-10-2021-proposal-based-on-a-statistical-data/

[13] Cybersecurity Dive. (n.d.). Imperva: Web application attacks surge in healthcare amid coronavirus. Retrieved from www.cybersecuritydive.com/news/imperva-web-application-security-healthcare-coronavirus/593202

[14] WhiteHat Security. (n.d.). FAQ—Top vulnerabilities list. Retrieved from www.whitehatsec.com/faq/content/top-vulnerabilities-list

[15] Dark, M. J. (2004). Assessing student performance outcomes in an information security risk assessment service learning course. In Proceedings of the 5th Conference on Information Technology Education (pp. 1–9). ACM. https://doi.org/10.1145/1029533.1029552.

[16] Shedden, P., Smith, W., & Ahmad, A. (2010). Information security risk assessment: Towards a business practice perspective. In Proceedings of the 8th Australian Information Security Management Conference. Perth, Western Australia. Edith Cowan University. Retrieved from http://ro.ecu.edu.au/ism/98.

[17] OWASP. (n.d.). Category: OWASP CSRFGuard project. Retrieved from www.owasp.org/index.php/Category:OWASP_CSRFGuard_Project

[18] OWASP. (n.d.). Category: OWASP Top Ten project. Retrieved from www.owasp.org/index.php/Category:OWASP_Top_Ten_Project

[19] Excess XSS. (n.d.). A comprehensive tutorial on cross-site scripting. Retrieved from http://excess-xss.com/

[20] OWASP. (n.d.). Information leakage. Retrieved from www.owasp.org/index.php/Information_Leakage

[21] InfoSecPro.com. (n.d.). Computer, network, application, and physical security consultants. Retrieved from www.infosecpro.com/applicationsecurity/a52.htm

[22] The Web Application Security Consortium. (n.d.). Information leakage. Retrieved from http://projects.webappsec.org/w/page/13246936/Information%20Leakage

[23] Thieu, L. (2012). Cross-site request forgery—A small demo. Retrieved from www.synopsys.com/glossary/what-is-csrf.html

[24] Google. (n.d.). Safe browsing. Retrieved from https://developers.google.com/safe-browsing/

[25] Mnif, A., Cheikhrouhou, O., & Jemaa, M. B. (2011). An ID-based user authentication scheme for wireless sensor networks using ECC. In ICM 2011 Proceedings (pp. 1–9). IEEE.

[26] Jemal, I., Cheikhrouhou, O., Hamam, H., & Mahfoudhi, A. (2020). SQL injection attack detection and prevention techniques using machine learning. International Journal of Applied Engineering Research, 15(6), 569–580.

[27] De Ryck, P., Desmet, L., Piessens, F., & Johns, M. (2014). Primer on client-side web security. Springer. https://doi.org/10.1007/978-3-319-12226-7.

[28] Cui, Y., Cui, J., & Hu, J. (2020). A survey on XSS attack detection and prevention in web applications. In Proceedings of the 2020 12th International Conference on Machine Learning and Computing (pp. 443–449). Association for Computing Machinery, New York, NY.

[29] Liu, V., Musen, M. A., & Chou, T. (2015). Big Data breaches of protected health information in the United States. JAMA, 313(14), 1471–1473.

[30] Privacy Rights Clearinghouse. (2020). Big Data breaches. Retrieved from https://privacyrights.org/Big-Data-breaches.

[31] IBM. (2021). Big Data breach. Retrieved from ww.ibm.com/security/Big Data-breach.

Chapter 11

Revisiting supply chain management

Security perspective

Satya Bhushan Verma and Abhay Kumar Yadav

11.1 INTRODUCTION

Supply chain (SC) is a group of interconnected activities through which materials are transported from supplier to end customers. Supply chain management (SCM) is the process of managing the movement of raw materials into a company, the internal processing of materials to transform them into completed products, and the movement of these completed products out of the company to reach the consumers. This process involves the strategic management of various factors to ensure that the appropriate products are available in desired quantity, at the right location and time. It encompasses considerations such as pricing, maintaining product quality, and ensuring that the goods are in optimal condition for the intended customers. It is centered around the oversight and management of the entire SC, which encompasses the journey of raw materials from their initial sourcing to the ultimate delivery of the complete product to end consumer. It plays a pivotal role in the success of contemporary enterprises by facilitating the smooth and effective movement of goods, services, and information between suppliers and end customers. It involves coordinating and controlling various activities, processes, and stakeholders in production, logistics, and distribution of goods to ensure a smooth and efficient flow throughout the entire SC network. Crucial for the success of modern businesses, it ensures the smooth and efficient flow of goods, services, and information from suppliers to end customers. It plays a vital role in improving overall efficiency, enhancing customer service, reducing costs, mitigating risks, promoting collaboration and coordination, fostering innovation and agility, supporting sustainability practices, and providing a competitive advantage in the market. SCM is responsible for the coordination and oversight of the movement of goods and services. It encompasses all the processes involved in transforming raw materials into finished products. This includes activities such as sourcing, procurement, production, logistics, and distribution [1].

DOI: 10.1201/9781003514312-11

Garai et al. [2] proposed eight fundamental sectors of SCM, including client relations management, demand, customer care handling, order completion, product flow management, potential product growth, and reentry management. According to Mohamed Abdel-Basset et al. [3], SCM lowers the associated raw materials price by (1) changing low-cost components, (2) avoiding waste in the business, and (3) removing the product's unnecessary features. Smart SCM is a self-organizing and self-optimizing system. In this system, entire office is integrated and associated with savvy examination zeroed in on diminishing personal time and waste. According to Yuanchun Zhang et al., SCM includes product creation, procurement and production, customer service, logistics, and performance management, among other things [4], as shown in Figure 11.1 (functionality of SCM).

Circular SCM (CSCM) has gained popularity in recent years due to the growing emphasis on circular economy and Industry 4.0. Theofilos D. Mastos et al. [5] highlighted the need for flexibility in SCs, including supply, techniques, and procedures, to enhance productivity, minimize waste, optimize resource utilization, and promote sustainable manufacturing and consumption. CSCM works by creating closed-loop systems, where resources can be reused, recycled, or repurposed, leading to a more sustainable and efficient SC. Various technologies and approaches, such as advanced data analytics, Internet of Things (IoT) devices, forecasting techniques, and blockchain applications, are employed to address these requirements. This concept has gained success within corporate discussions due to increasing demands from all associated components in SC. Companies have recognized the significance of adopting more sustainable practices to meet these demands and promote environmental stewardship [6].

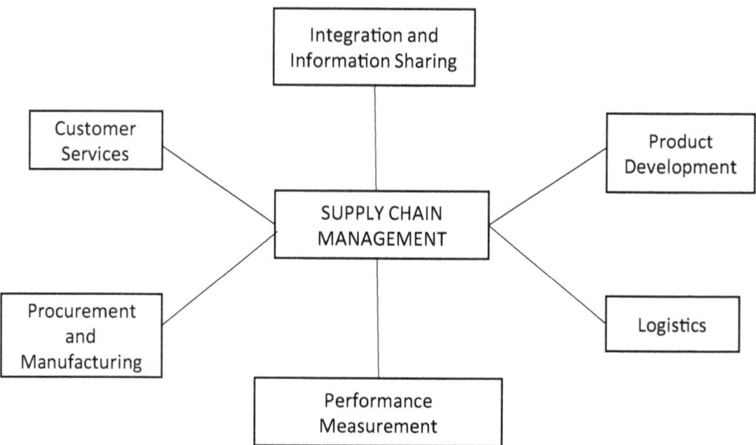

Figure 11.1 Functionality of SCM.

11.1.1 Industrial revolution

Phuyal et al. [7] categorized industrial revolution into four different revolutions.

11.1.1.1 First industrial revolution

In early 18th century, worker-based cottage industry gave way to an initial mechanical economy. The first European and American industrial revolution occurred between around 1760 and 1820, with the birth of a new manufacturing technique. During that era, the textile industry emerged as a dominant sector in terms of capital investment, output, and employment opportunities. It was at the forefront of adopting modern manufacturing methods. The expansion of railway networks and increased trade activities facilitated economic growth, enhanced human interactions, and facilitated the movement of materials. Additionally, the development of coal mining and the introduction of steam engines ushered in a new era of energy, driving progress across various industrial processes.

11.1.1.2 Second industrial revolution

This technical revolution occurred between 1830 and 1914. New technological breakthroughs ushered in a newer energy source including electricity, gas, and oil spawning new innovations. Innovations such as the telephone and telegraph enabled faster and more efficient long-distance communication. Simultaneously, the development of automobiles and airplanes expanded the possibilities of transportation, allowing for quicker and more convenient travel. These technological breakthroughs had a profound impact on society, transforming the way people communicated and connected with one another. All of these breakthroughs were made feasible by concentrating research and resources into a new economic and industrial paradigm centered on big factories.

11.1.1.3 Third industrial revolution

After World War I, industrial progress continued till 1969 creating this next industrial revolution. The working electromechanical systems were changed by computer-based control systems, with the biggest ideas and implementations in the industrial automation system being Programmable Logic Controllers and robots. It introduced nuclear energy as the newer energy provider. The transistor and microprocessor were born during this revolution. It brought about transformative changes across industries, improving overall efficiency, precision, and productivity in manufacturing. It also paved the way for innovations in products, services, and SCM, revolutionizing how industries operate.

11.1.1.4 Fourth industrial revolution

The newer industrial revolution described integration of internet with industrial applications. The German Government originally initiated this revolution with its Industry 4.0 program; afterward it was revisited and supported in various nations and territories by other policymakers [8]. In 2011, Industry 4.0 integrated the IoT and industrial internet to the production system, allowing machines to be able to communicate with each other and apply intelligent decisions depending on system's requirement. Artificial intelligence, autonomous robotics, flexible industrial automation systems, additive manufacturing, and augmented reality (AR) are all part of Industry 4.0.

11.1.2 Industry 4.0

Industry 4.0 represents a new idea where industrial processes are automated and digitized, leading to self-optimization and self-control within businesses. This innovative business model aims to enhance service delivery by leveraging advanced technologies [9]. The key benefits of Industry 4.0 are the ability to achieve a high degree of mass customization, enabling businesses to offer customized products similar to the craftsmanship era. The application of Industry 4.0 is detailed in the following text.

11.1.2.1 Smart factories

Smart factory is a notion for the ultimate objective of production digitalization. The commonly accepted definition of a smart factory is a technologically advanced manufacturing facility that relies on interconnected devices, gadgets, and production systems to continuously gather and exchange data. This digital shop floor is equipped with various smart technologies to enable seamless communication and data sharing throughout the manufacturing process. Then the data may be used through self-optimization devices or throughout the company to proactively tackle challenges, optimize production processes, and handle new requirements. In the other words, we can say the concept of smart factories is described as a facility that combines physical manufacturing processes with digital technology, smart computation, and big data to make enterprises focus on manufacturing and SCM more opportunistic. The intelligent factory is a new wave of the industrial revolution, which relies significantly on real-time data, built-in sensors, networking, machine learning, and automation.

11.1.2.2 Smart product

An intelligent or connected product is a gadget, which is connected to internet to communicate information on itself, its atmosphere and users.

These take input from the location and condition of the things they cover and include motor vehicles, medical devices, industrial equipment, and enhanced packaging process. Data sharing intelligent goods allows them to perform efficiently, to make life easier and safer for their owners and to enhance operations. It can boost assistance by monitoring and optimization of the use, intelligence to see the state or location of health, and by predicting service demands breakdowns or downtimes.

11.1.2.3 Smart cities

In smart cities, environmental issues like climate change and air pollution are also monitored and addressed. In addition, smart technology can be used for refining the sanitation and waste management, whether or not using IOT-enabled waste cans and waste collection based on IoT and disposal fleet management mechanism.

11.1.3 Evolution of operator

According to Tommaso Gallo et al. [10], for handling the increasing unpredictability in productivity, Industry 4.0 allows newer connections among machines and operators, interactions to revolutionize industrial workforce [10]. According to Romero et al., humans operating "manual and dexterous work" with the assistance of mechanical tools are referred to as the Operator 1.0 generation. The Operator 2.0 generation is a human being who conducts aided work with computer tools. The Operator 3.0 age epitomizes a human substance associated with "agreeable work" with robots and different machines and PC devices, otherwise called human–robot joint effort. The Operator 4.0 portrays the "future operator," a skilled worker who does "work assisted by machines" as needed. The evolution of operator is divided into four stages. This is detailed in Table 11.1.

Wittenberg et al. [11] state that the main purpose of Operator 4.0 is to administer automated production utilizing upgraded supervision systems.

Table 11.1 Evolution of Operator

S. No.	Operator 1.0	Operator 2.0	Operator 3.0	Operator 4.0
1.	Operator 1.0 manually operated machine with the help of mechanical tools.	Operator 2.0 operated with computer tools.	Operator 3.0 operated with robots and computer.	Operator 4.0 works on operator based on "work aided" by machine
2.	Manual and dexterous work	Assisted work	Cooperative work	Automated work

11.2 LITERATURE REVIEW

11.2.1 Smart manufacturing

Xue-Feng Shao et al. [12] enhance the processing productivity and performance, smart manufacturing leverages data obtained from various business operations. The initial step involves collecting data from the production environment, which encompasses information on inputs as well as output data. Subsequently, the collected data is analyzed in cloud-based data centers, serving as a central hub for additional activities. The monitoring stage serves in this respect as a quality checker, resulting in process readjustment with any changes in process parameters. During the last step, i.e., the processing of problems, the information is used to forecast new issues and viable solutions. Tao et al. [13] stated that smart manufacturing can be implemented in achieving tasks including tasks such as complete monitoring, optimization of production and data simulation.

Büchi et al. have shown enhanced flexibility in production, greater performance, decreased mistakes, enhanced efficiency, and decreased installation times owing to intelligent manufacturing efforts [14]. In addition, as a result of intelligent acceptance, Büchi et al. pointed to an enhanced efficiency factor and higher production capacity [15]. In addition, the data that is driven by intelligent production, which comprises product creation, self-organization, intelligent execution (transportation of material and processing), and self-regulatory and inside a system, were also underlined by Tao et al. [13]. Further improved production planning may be done through the use of the data for the best allocation of resources and the network optimization. In addition to diagnostics, which ultimately cause preventative maintenance, the data on the machine may also be utilized to forecast possible failure of equipment.

11.2.2 Enabling technology

Different researchers have tried to elaborate the idea of smart manufacturing, the 4.0 technologies in terms of virtual reality (VR), additive manufacturing, IoTs, and cyber systems. There are different terms denoting integration: (1) data analytics and artificial intelligence, (2) adaptive robotics, (3) sensors and actuators, (4) cloud, (5) virtualization technologies, (6) cyber security, (7) embedded systems, (8) additive manufacturing, simulation, (9) communication and network, (10) mobile technology, and (11) radio-frequency identification and real-time locating systems technologies. Additive manufacturing can employ 3D computer-aided design (CAD) as a digital source.

According to Lee et al. [16], cyber physical system (CPS) is a combination of techniques that constitute a network that integrates and maintains assets in computing. There are two fundamental processes in a CPS. The first one is the gathering of data via feedback from cyberspace and the actual world, which comes with improved connection; and the second is the cyber environment, which is built and nurtured through smart data management, data analytics, and computing capabilities.

11.2.3 Supply chain

Supply chain is the arrangement of firms that bring goods or services to market. According to Carter et al. [17], conventional SC is a mesh of data and information, raw materials, processes, services that define supply, transformation, and demand. -

According to Mohamed Abdel-Basset et al. [3], traditional SCM systems face several challenges, including problems like overstocking, delivery delays, and stockouts. These issues arise due to the complexity and unpredictability inherent in real SCs. Smart SC (SSC) enters the globe to address these shortcomings.

11.2.4 Smart supply chain and digital transformation

According to Tsai-Chi Kuo et al. [8], many items are integrated with sensor devices to provide improved visibility, automation, and intelligent decision-making capabilities. This SSC development offers tremendous cost-cutting and efficiency-improving potential. SSC has the features of connectivity, cooperation, and customization, as well as the features of functional and structural flexibility. Raab et al. [18] stated that traditional SC is built on a combination of digital processes and on-paper documentation. Because the organizational structures are generally unfamiliar with exchanging information, they function poorly. On the other side, SSC uses Internet of Things (IoT) technologies, such as autonomous decisions, to digitalize logistics and SCs.

11.2.5 Collaboration in the supply chain predicts information sharing

According to Fahimnia et al., an SC planner's job is to reduce the entire cost of the SC as well as the SC's environmental effect. Collaboration is widely recognized as one of the most effective methods for achieving these objectives. The supply partnerships identify four major contents: (1) Information sharing, (2) strategic planning, (3) sharing of resources, and (4) development and usage of knowledge synergies. Information sharing across SC stakeholders has been identified as a vital facilitator for achieving long-term cooperation or collaboration. One of the most important strategies for reducing the Bullwhip effect is information-sharing.

11.2.6 Industry 4.0: Circular economy

According to Tseng et al. [19], studies have emphasized the need of innovations for improving CSCM by utilizing tools and technology such as big data and IoT. The implementation of CSCM is driven by operational and material management efficiency, which results in cost savings and enhanced resource utilization. Critical components for CSCM have been recognized as information interchange among SC partners and forecasting systems that deliver trustworthy outcomes and prevent waste creation. According to Hassini et al. [20], stakeholder pressures from regulatory bodies, non-governmental organizations (NGOs), SC partners, competition, and delayed market entry, in addition to the aforementioned critical factors, have shown that companies must pay equal attention to effect on environment. Mastos et al. [21] discovered that implementing Industry 4.0 technologies leads to long-term gains at the business and SC levels. More precisely, their findings revealed a 20% timesaving through checking scrap and a 4% reduction in scrap metal creation due to maintenance processes.

11.2.7 Smart cities in the transition

According to Luisa Franchina et al. [22], an SC is an important system for getting products and services to customers. It consists of transition of raw resources into finished goods, and hence occupies a prominent position in the current debate over social, political, economic, and environmental factors. Traditional supply networks are fundamentally harmful to the environment, which poses a problem. This element should be included in a genuine conceptualization of green SCM, which would advocate for a reorientation of the industrial network. A greener SC is now more of a notion than a reality due to various existing impediments to its implementation. As previously stated, the top-down approach to sustainability is only used in a limited way. Between 2005 and 2017, these seem to be successful in member countries of the SAARC. "Three or more companies (or organisations) are actively involved in upstream and downstream flows of products, services, funds and/ or informative transfers from one source to the client" [23]. They classify supply networks as "direct," "extended," or "ultimate." To ensure their complexity, they might begin with the fundamental relationship between supplier agency and customers (a direct SC) to include first suppliers of suppliers and consumers (expanded SCs and ultimate SCs).

11.2.8 Industry 4.0 for small- and medium-sized enterprises

According to Ali Turkyilmaz et al. [24], it enables digitalization and the integration of technology, industrial, and business activities across the firm for product creation and procurement through manufacturing, distribution, and after-sale services. Industry 4.0 consists of following:

1. **Big Data:** The amounts of data that exceed the capacity of current technology to successfully store, manage, and analyze it efficiently.
2. **Horizontal and vertical integration of system:** This network integration provides a comprehensive linkage of all sections of complete SC, via highly dynamic system.
3. **Simulation:** It is supposed to be a digitized implementation of scheming manufacturing systems and be used for obtaining CPSs real-time data.
4. **Cloud technology:** This is also known as cloud computing enabling "on-demand" digital information interchange between clouds and related smart devices.
5. **Augmented reality:** AR encrusts digitized data on actual world, allowing clients and CPSs to interact.
6. **Additive manufacturing:** The 3D printing of actual products is additive manufacturing. It can also employ 3D CAD as a digital source.
7. **Cyber security:** Cyber security is the goal of preventing cyber-attacks on digitized data and smart devices.

Small enterprises encounter more obstacles in embracing Industry 4.0 compared to multinational enterprises due to limited resources and skills. However, small- and medium-sized enterprises (SMEs) possess significant method of achieving digital transformation. In SMEs, the leader's influence and decision-making power are substantial and influential due to close relationships with employees [25]. Martin Prause et al. [26], on the other hand, believed that the flat organizational structure of SMEs aids easy communication and tight control inside the firm, lowering the chance of failure. The high product specialization of SMEs is another major argument in their favor when it comes to digital transformation. According to J.M. Muller, SMEs may make better use of limited resources to produce more value, which is a smart place to start for Industry 4.0.

According to Ali Turkyilmaz et al., for SMEs, Industry 4.0 has created whole new needs for production systems and equipment [24]. Furthermore, SMEs encounter several challenges arising from the Industry 4.0's execution. Alongside the lacking expertise and a limited strategic roadmap, SMEs must navigate substantial investments and employee apprehension toward uncertainty. Therefore, it is crucial to initially consider key influencing factors, such as challenges and opportunities.

11.2.9 Industry 4.0

Continuous system and software engineering: Because of the unique nature of Industry 4.0 systems, Elisa Yumi Nakagawa et al. suggest a more exact and relevant word to describe their evolution, development, and operation processes: "Software Engineering and Continuous Systems for Industry 4.0" [27].

1. **Continuous Twinning:** According to Elisa Yumi Nakagawa et al., macro activity should include a variety of continuous activities to especially construct and evolve digital twins (e.g., continuous architecture, testing, integrating and deployed) [27]. As these twins have to reflect real-world elements precisely, such ongoing processes demand great accuracy and rigor. Digital twins may be utilized during the design times of such entities, for example through simulation, to enable their continual check and tests. The macroactivities (continuous twinning) and those of Industrial 4.0 systems development must thus be perfectly interacting.

2. **Twin Operation:** According to Elisa Yumi Nakagawa et al., this is another macro action in which the most significant activity is the continuous runtime monitoring [27]. In fact, in real-world entities digital twins are particularly vital. Digital twins are continually monitoring the conduct of their twins and collecting real-time information with their real-world twins. Consequently, it becomes imperative to observe real-world entities in real time, whether through human intervention or automated processes, and simulate the behavior to enable early identification of issues such as performance deterioration, unexpected failures, and predictive maintenance, thereby facilitating continuous improvement.

11.2.10 Pertinent reviews of Industry 4.0

As per Changhun Lee et al. [28], Industry 4.0 is not a single issue but a wide agenda. Many scholars examined its technology and uses, discussing obstacles and research difficulties. A systemic literature review was carried out by several researchers through the article identification, screening, evaluation of eligibilities and quantitative and qualitative analysis of selected articles. The results of their investigation include categorizing Industry 4.0, identifying the common terms associated with Industry 4.0, identifying major study topics, and implementation domains of Industry 4.0 and suggesting agendas. In the study described, smart industries, intelligent goods, and intelligent cities were the three key applications of Industry 4.0 [29].

Although there are a number of major facts and insights in current reviews on Industry 4.0, the majority of studies are selective in the literature used and can be constrained by the amount and range of sources. In particular, existing assessments focus primarily on manufacturing or engineering in general. Despite the advantages of knowing the whole range of Industry 4.0

as the basis of learning and encouraging them from technology, there still needs to be research that offers many application questions of Industry 4.0. Moreover, given the diversity and number of documents about Industry 4.0, it is vital to use a quantitative strategy to resist Industry 4.0, as these papers can only be reviewed or integrated by a person researching them.

11.2.11 Internet of Things and supply chain management

Internet of Things is defined in a variety of ways. According to Christopher Mims [30], IoT is a network of different entities that connect and cooperate for humankind. IoT allows for communications to take place at any time, in any location, using any media. The IoT can be used in any point of our lives. Industrial Internet of Things (IIoT)-A CPS and the IoT are capable of transforming a traditional manufacturing into a smart factory. According to Mohamed Abdel-Basset et al., SCM is the practice of managing SC operations for maximizing customer loyalty and achieve a long-term strategic advantage [3].

11.3 THEORETICAL MODEL: REDUCING THE SUPPLY CHAIN CHANNEL COST

11.3.1 Manufacturer–retailer supply chain model

According to Xiang Li [31], a producer is someone who trades any product to market via a store. A channel cost c0 including operational cost is split between the producer and the retailer in alpha and 1-alpha ratios, respectively. To put it another way, the manufacturer pays alphac0 and the retailer (1-alpha) c0. As an exogenous parameter, we choose alpha belongs [0, 1]. Furthermore, by using a certain type of SSC technology, this channel cost may be decreased. For example, by planning an intelligent logistics dispatching technologies, transportation costs from the producer to the store may be decreased, and SC trading costs may be decreased by developing an enhanced information sharing system. $K(C)$ for reducing the cost from c0 to c corresponds to the investment necessary to implement the SSC technology (0, c0), which fulfills $K'(\cdot) < 0$, $K'(\cdot) > 0$, $K(0) = +$infinite $K(c0)$. In particular, an assumption that the constants $K(c) = a - b \ln c$ is supplied for b and $a = b \ln c0$. Cost formulation is implemented in production line investment. Here b states the cost criterion for the economics of investment in SSC technology. The evaluation of the product by the end user is distributed equally in (0,1), on the other hand. In economics, commercialization and operational management, uniformly distributed customer evaluations were broadly embraced. As a result, demand may be represented as $1-p$, where p is the retailer's market price. We assume that the manufacturing and sales expenses are zero without a loss of generality.

According to Xiang Li et al., the participants engage in Stackelberg depending on price contract: a price w is offered by the fabricator and a market price p is set by the retailer [31]. The firm, which is the game leader in Stackelberg, also initiates the investment in SSC technology. This assumption is compatible with most findings by our business and conforms to the modeling of the SC in most current literatures.

There are therefore two instances that we have:

1. The C case: producer and retailer operate together to optimize the benefits of whole SC.
2. The MR case: producer provides in SSC system and calculates the channel cost c for channel cost c belongs to $(0, c0)$. The maker is called "she" to eliminate any misunderstanding, and the merchant is called "he."

11.3.2 Platform supply chain model

Analyze an SC of a platform consisting of a platform and a manufacturer. This approach relates to prominent e-business structure known as agency selling, in which a platform firm provides marketplace for creators to get access to clients. There are two main differences between the platform SC sales model of the agency and the conventional manufacturer/detailer SC reselling style. First of all, market price is determined by the producer under the selling format of the agency; then, in the reselling format, by the retailer. The second contrast is that interim companies make a margin from the wholesale price to the sale of the retailer, while a company that sells the platform receives a specific portion of the manufacturer's income. In the SC of the platform, channel costs are incurred through the execution of transactions or the delivery of merchandise between the manufacturer and the end client. Again, reinvesting in SSC can lower a channel cost to $K(c) = a - b \ln c$ from c0 to c. In our modeling environment, two instances in actual industry are taken into account:

1. **The scenario MP:** the producer starts investing in SSC technologies and sets cost of channel c to $(0, c0)$.
2. **The instance P:** the platform commences investing in SSC technology and assesses the cost of channel c belongs to $(0, c0)$.

According to Xiang Li et al. [31], for denoting variables for MP and P situations, we utilize subscripts "MP" and "P." We call the maker "she" and the platform "it," in line with the prior narration. In both scenarios, the market price P is a changeable variable for the fabricator and the percentage charge theta $(0, 1-c0)$ is an exogenous parameter based on negotiating strength of the fabricator with the platform. It can be envisaged that the percentage fee is controlled by a more powerful SC member.

11.4 DISCUSSION AND ANALYSIS

11.4.1 Supply chain

An SC is a combination of a series from inception to final customer activity with materials. Many authors studied regarding SC and gives some results, so we discussed recent studies regarding SC in Table 11.2.

Table 11.2 Studies Regarding Supply Chain from First Inception to Final Customer

S. No.	Author and Year	Area	Technique Used	Remark
1	Booneet et al. 2019 [1].	Sales forecasting in the SCM analytics in the big data era.	Sales forecasting and prediction analysis	Purpose of this paper is to examine the influence and improvement of that data explosion on product prediction.
2	Carteret et al. 2019 [17].	Sustainable SCM: continuing evolution and future directions.	With modifications of the focus, theoretical lenses, the analysis unit, methods and kind of analysis, the field of sustainable SCM continues to evolve.	They conduct a thorough evaluation of the literature on sustainable SCM in primary logistics.
3	Hobbset et al. 2020 [32].	Food SCs in COVID-19.	Evaluation of potential food SC disruptions	They discuss the longer-term impacts of COVID-19 pandemic on character of the food supply networks, particularly the expansion in the on-line food supply industry, and the excessive priority given to "local" food supply networks by consumers.
4	Nasiri et al. 2020 [33].	Digital SC management: by smarttechnologies	Confirmation analysis was implemented for checking the reliability and validity.	They conclude how firms' digital transformations can drive smart technologies, resulting in increased relationship performance.

(continued)

Table 11.2 (Cont.)

S. No.	Author and Year	Area	Technique Used	Remark
5	Mastos et al. 2021 [5].	Use case of Industry 4.0 for circular SCM.	Re-SOLVE model was used.	They develop a waste-to energy generation solution for providing sustainability and evaluating the circular SCM.
6	Franchina et al. 2021 [22].	Thinking green smart innovations in making urban communities' waste and SC's flow.	Circular economy, green SC, circular SC.	They analyze smart cities' green potential in developing sustainability at the level of the person and supplier chain.
7	Xue-Feng Shao 2021 [12].	Implementation of multistage framework for smart SCM under Industry 4.0.	It proposed cross-functional approach.	They studied to document how Industry 4.0 principles are being implemented across several SC layers
8	Tsai-ChiKuo et al. 2021 [8].	A cooperative data-driven investigation of material asset the executives in shrewd production network by utilizing a cross-breed Industry 3.5 procedure.	To improve the satisfaction rate of its clients, use the 3.5 hybrid industry approach.	They study an information exchange-based methodology for material resources management, and allocation among SC participants is proposed.
9	Zhang et al. 2021 [4].	Holistic cognitive conflict chain management framework.	Design of the consumer behavior prediction statistical model.	They offer a complete framework for improving customer service quality and the efficiency of SCM.

11.4.2 Industry 4.0

Industry 4.0 refers to an automatic and self-optimizing process paradigm, which enables companies to give a higher quality of service through the automation and making digital of industrial procedures with an innovative business model. Many authors studied regarding Industry 4.0 and gave some results; so we have discussed the recent studies regarding Industry 4.0 in Table 11.3.

Table 11.3 Studies with Implementation of Industry 4.0

S. No.	Author and Year	Area	Technique Used	Remark
1.	Tseng et al. 2018 [19].	Circular economy meets Industry 4.0.	Framework for a circular economy with industrial and urban synergy approaches.	Enhance the benefit of big data-driven symbiosis presents advantages; corporations have recognized that in cross-industry networks social, environmental, and economic aspects.
2	Moeuf 2020 [25].	Documentation of serious factors affecting success, risks factors, and occasions in SMEs.	Delphi study by Regnier's abacuses.	Paper aimed at identifying opportunities, risks, and critical success values for established Industry 4.0
3.	Büchi et al. 2020 [14].	Smart factory in Industry 4.0.	Using regression models, it is shown that: Industry 4.0's width and depth allow for more possibilities and local micro-level systems function well.	Verifying the prospects for firms with Industry 4.0 in costs and skills is crucial in Industry 4.0.
4.	Paola Fantini 2020 [8].	Introduction of operators at Industry 4.0 design	This study addressed difficulty by presenting a methodology for the creation and evaluation of diverse working configurations.	Two industrial scenarios show the applicability of the concept and advise training of employees and the improvement of the cyber social system as a whole.
5.	Mastos et al. 2020 [21].	Industry 4.0 sustainable journal pre-proof SCs	The approach provided is advantageous for reduction of CO_2emissions, availability of resources, and optimization of reaction time.	The aim of this article is to demonstrate how an IoT solution might affect the performance of sustainable SCM

(continued)

Table 11.3 (Cont.)

S. No.	Author and Year	Area	Technique Used	Remark
6.	Tommaso Gallo et al. 2021 [10].	Industry 4.0 and human factor integration	Big data analytics implementation for data management	The objective is to examine how the position of maintenance operators has altered in a digitalized world.
7.	Elisa Yumi Nakagawa et al. 2021 [27].	Continuous systems	Fundamental aspects, including twinning, which is also initially described.	This paper presents the view of Industry 4.0 continuous systems and software engineering.
8.	Ali Turkyilmaz 2021 [24].	Kazakhstan's challenges and opportunities in using Industry 4.0.	SWOT (strengths, weakness, opportunities, and threats) analysis.	This study's objective is analyzing the current concept of Industry4.0 and the impact and applicability on SMEs.
9.	Changhun Lee et al. 2021 [28].	From technology creation to social advancement	Connection infrastructure development, data-driven intelligence development, system and processing optimization, innovation in industry and societal progress.	This study attempts to build a common understanding of Industry 4.0 across a variety of disciplines, allowing research and development for industrial innovation and societal advancement.

11.4.3 Supply chain with Internet of Things

When SC works with IoT in a collaboration and gives improved production foresight and enhances automation and decision-making capabilities, then it known as smart SC. Now we analyze the following paper as shown in Table 11.4.

11.4.4 Smart manufacturing

Smart manufacturing specifies the networked devices inside the CPS for providing an automated evolving environment capable of managing deviations and suggesting the best alternative and direct ways. Many writers studied smart manufacturing and presented their findings. As a result, we have discussed some papers on smart manufacturing in Table 11.5.

Table 11.4 Study on Supply Chain Implementation in IoT

S.No.	Author and Year	Area	Technique Used	Remark
1.	Mohamed et al. 2018 [3].	IoT and SC: architecture for creating secure and efficient systems.	NDEMATEL and analytic hierarchy process (AHP).	The goal of this paper to solve all conventional SCM obstacles and create a safe environment for SCM procedures.

Table 11.5 Studies on Smart Manufacturing in Supply Chain

S. No.	Author and Year	Area	Technique Used	Remark
1	Büchi et al. 2018 [15].	Scaling of economies and networks in Industry 4.0.	Enabling technologies.	They work on theoretical character and reflect on the significance of Industry 4.0 in management by offering fascinating lines which will be extended in future research areas.
2.	Tao et al. 2018 [13].	Data-driven smart manufacturing.	Big data.	They studied the function of big data to promote smart manufacturing.
3.	Phuyal et al. 2020 [7].	Opportunities, challenges, and future of smart manufacturing	Cyber Physical System, Internet of Things	They examine the disparity between today's production system and expected smart manufacturing system of the future.

11.5 CONCLUSION

The objective is to explore that the phenomenon of SCM has been successfully achieved. We found that SC is the combination of series where the raw material moves to semi-finished product, then to finished product, then to the distributor, and lastly to the end customers. This movement of material is conventional as well as digitalized. The conventional is known as traditional SC and digitalized is known as digital SC. When the management of SC is self-organizing and self-optimizing, then it is known as *smart supply chain management*. In the management of SC, product creation, procurement and production, customer service, logistics, and performance management are key for better outcomes. Circular SCs are established in response to the need for flexible supply, techniques, and procedures within SCs to enhance productivity, minimize or eliminate waste generation, optimize resource utilization, and promote sustainable practices in manufacturing and consumption. The circular economy concept is gaining significant importance in corporate agendas due to growing demands from diverse stakeholders, along with environmental and social concerns. Enterprises have recognized the importance of adopting sustainable practices to address these challenges effectively.

During coronavirus disease 2019 (COVID-19) pandemic situation where the people remained at homes with completely no physical involvement in any task, food SC played a very important role. Food is the most prioritized issue after medication in COVID-19 pandemic. Food SC is the combination of activities, which involves getting food from farms to our plates. The food SC is separated into two parts: first is supply and second one is demand. Farming, harvesting, processing, and distribution are all part of the SC, and they all require time-bound activities to ensure high-quality commodities and customers are on the demand side.

REFERENCES

[1] Boone, Tonya, Ram Ganeshan, Aditya Jain, and Nada R. Sanders. "Forecasting sales in the supply chain: Consumer analytics in the big data era." *International Journal of Forecasting* 35.1 (2019): 170–180. https://doi.org/10.1016/j.ijforecast.2018.09.003

[2] Garai, Arindam, and Tapan K. Roy. "Multi-objective optimization of cost-effective and customer-centric closed-loop supply chain management model in T-environment." *Soft Computing* 24.1 (2020): 155–178.

[3] Abdel-Basset, Mohamed, Gunasekaran Manogaran, and Mai Mohamed. "Internet of Things (IoT) and its impact on supply chain: A framework for building smart, secure and efficient systems." *Future Generation Computer Systems* 86.9 (2018): 614–628.

[4] Zhang, Yuanchun, Carlos Enrique Montenegro-Marin, and Vicente García Díaz. "Holistic cognitive conflict chain management framework in supply chain management." *Environmental Impact Assessment Review* 88 (2021): 106564.

[5] Mastos, Theofilos D., et al. "Introducing an application of an industry 4.0 solution for circular supply chain management." *Journal of Cleaner Production* 300 (2021): 126886.

[6] Rossini, Matteo, Daryl John Powell, and Kaustav Kundu. "Lean supply chain management and Industry 4.0: A systematic literature review." *International Journal of Lean Six Sigma* 14.2 (2023): 253–276.

[7] Phuyal, Sudip, Diwakar Bista, and Rabindra Bista. "Challenges, opportunities and future directions of smart manufacturing: A state of art review." *Sustainable Futures* 2 (2020): 100023.

[8] Paola Fantini, Marta Pinzone, and Marco Taisch. "Placing the operator at the centre of Industry 4.0 design: Modelling and assessing human activities within cyber-physical systems." *Computers & Industrial Engineering* 139 (2020): 105058. https://doi.org/10.1016/j.cie.2018.01.025

[9] Khanuja, Anurodhsingh, and Rajesh Kumar Jain. "The conceptual framework on integrated flexibility: An evolution to data-driven supply chain management." *The TQM Journal* 35.1 (2023): 131–152.

[10] Gallo, Tommaso, and Annalisa Santolamazza. "Industry 4.0 and human factor: How is technology changing the role of the maintenance operator?" *Procedia Computer Science* 180 (2021): 388–393.

[11] Wittenberg, Carsten. "Human-CPS interaction-requirements and human-machine interaction methods for the Industry 4.0." *IFAC-PapersOnLine* 49.19 (2016): 420–425.

[12] Shao, Xue-Feng, et al. "Multistage implementation framework for smart supply chain management under industry 4.0." *Technological Forecasting and Social Change* 162 (2021): 120354.

[13] Tao, Fei, et al. "Data-driven smart manufacturing." *Journal of Manufacturing Systems* 48 (2018): 157–169.

[14] Büchi, Giacomo, Monica Cugno, and Rebecca Castagnoli. "Smart factory performance and Industry 4.0." *Technological Forecasting and Social Change* 150 (2020): 119790.

[15] Büchi, Giacomo, Monica Cugno, and Rebecca Castagnoli. "Economies of scale and network economies in Industry 4.0." *Symphonya. Emerging Issues in Management* 2 (2018): 66–76.

[16] Lee, Jay, Behrad Bagheri, and Hung-An Kao. "A cyber-physical systems architecture for industry 4.0-based manufacturing systems." *Manufacturing Letters* 3 (2015): 18–23.

[17] Carter, Craig R., et al. "Sustainable supply chain management: Continuing evolution and future directions." *International Journal of Physical Distribution & Logistics Management* 50.1 (2020): 122–146.

[18] Raab, Martin, and Belinda Griffin-Cryan. "Digital transformation of supply chains." *Creating Value–When Digital Meets Physical, Capgemini Consulting* (2011). efaidnbmnnnibpcajpcglclefindmkaj/www.capgemini.com/wp-content/uploads/2017/07/Digital_Transformation_of_Supply_Chains.pdf

[19] Tseng, Ming-Lang, et al. "Circular economy meets industry 4.0: Can big data drive industrial symbiosis?." *Resources, Conservation and Recycling* 131 (2018): 146–147.

[20] Hassini, Elkafi, Chirag Surti, and Cory Searcy. "A literature review and a case study of sustainable supply chains with a focus on metrics." *International Journal of Production Economics* 140.1 (2012): 69–82.

[21] Mastos, Theofilos D., et al. "Industry 4.0 sustainable supply chains: An application of an IoT enabled scrap metal management solution." *Journal of Cleaner Production* 269 (2020): 122377.

[22] Franchina, Luisa, et al. "Thinking green: The role of smart technologies in transforming cities' waste and supply Chain's flow." *Cleaner Engineering and Technology* 2 (2021): 100077.

[23] Nasiri, Mina, et al. "Managing the digital supply chain: The role of smart technologies." *Technovation* 96 (2020): 102121.

[24] Turkyilmaz, Ali, et al. "Industry 4.0: Challenges and opportunities for Kazakhstan SMEs." *Procedia CIRP* 96 (2021): 213–218.

[25] Moeuf, Alexandre, et al. "Identification of critical success factors, risks and opportunities of Industry 4.0 in SMEs." *International Journal of Production Research* 58.5 (2020): 1384–1400.

[26] Prause, Martin. "Challenges of industry 4.0 technology adoption for SMEs: The case of Japan." *Sustainability* 11.20 (2019): 5807.

[27] Elisa Yumi Nakagawa, Pablo Oliveira Antonino, Frank Schnicke, Rafael Capilla, Thomas Kuhn, and Peter Liggesmeyer. "Industry 4.0 reference architectures: State of the art and future trends." *Computers & Industrial Engineering*, 156 (2021), 107241. https://doi.org/10.1016/j.cie.2021.107241

[28] Lee, Changhun, and Chiehyeon Lim. "From technological development to social advance: A review of Industry 4.0 through machine learning." *Technological Forecasting and Social Change* 167 (2021): 120653.

[29] Romero, David, et al. "The operator 4.0: Human cyber-physical systems & adaptive automation towards human-automation symbiosis work systems." *Advances in Production Management Systems. Initiatives for a Sustainable World: IFIP WG 5.7 International Conference, APMS 2016, Iguassu Falls, Brazil, September 3–7, 2016, Revised Selected Papers.* Springer International Publishing, 2016.

[30] Mims, Christopher. "Here's the one thing someone needs to invent before the internet of things can take off." *Backbone Magazine* February–March (2013): 12–13.

[31] Li, Xiang. "Reducing channel costs by investing in smart supply chain technologies." *Transportation Research Part E: Logistics and Transportation Review* 137 (2020): 101927.

[32] Hobbs, Jill E. "Food supply chains during the COVID-19 pandemic." *Canadian Journal of Agricultural Economics/Revue canadienne d'agroeconomie* 68.2 (2020): 171–176.

[33] Nasiri, Mina, Juhani Ukko, Minna Saunila, and Tero Rantala. "Managing the digital supply chain: The role of smart technologies." *Technovation*, 96–97, (2020). https://doi.org/10.1016/j.technovation.

Chapter 12

Software vulnerability prediction using deep learning on the unbalanced datasets

Wasiur Rhmann, Ishrat Amaan, Jalaluddin Khan, and Jamal Akhtar Khan

12.1 INTRODUCTION

Software vulnerabilities can pose significant risks to IoT, blockchain and web applications. These may be due to coding error, design flaw, or insufficient testing, and when executed, they result in a security breach. Any error introduced at early cycle of software development can come as security issue during operational time of software. Software flaws are difficult to find unless they interfere with the software's normal operation. A viable strategy is to predict software vulnerability early in the life cycle. Execution of software vulnerabilities may cause software failures and companies may lose their financial assets. If a security breach comes from the execution of software, then it has a possibility of a flaw in any artifacts used during software development. If a software flaw leads to breach of security policy, it may be due to weakness in the any artifact produced during development and this flaw is known as software vulnerability (McGraw & Potter, 2004). The software system may be unable to carry out the necessary or expected function if a bug is unintentionally activated (Shin & Williams, 2008). Software vulnerabilities are those software faults that, if exploited, could compromise security. Software metrics are utilized to predict various software quality attributes. During investigation of vulnerability life cycle, it was found that half of the vulnerabilities discovered remain unfixed for more than one year and they are mostly fixed by developers during maintenance phase (Iannone, 2023). The quality of software architecture (QSA) can be measured effectively using different software metrics. To predict software maintainability, change, and defect prediction, software metrics are frequently employed in literature (Bansal, 2017) Many research findings have shown the connection between structural software metrics and software design (Gimothy et al., 2005). Even though these indicators are often utilized, there is no established protocol for using them to predict software vulnerabilities with deep learning. Structural metrics may help organizations in developing dependable and safe software by early detection of software vulnerabilities. Jabeen et al. (2022) have applied machine learning and statistical techniques for

DOI: 10.1201/9781003514312-12

software vulnerabilities prediction on different datasets. Neural network, multilayer perceptron, support vector machine, M5Rrule, bagging, feed-forward back propagation neural network and adaptive neuro fuzzy inference system, from machine learning and logistic regression, linear regression, and Alhazmi–Malaiya model from statistical techniques are used for software vulnerability prediction. It was found that machine learning techniques are producing better results compared to statistical techniques for software vulnerability prediction. Rhmann (2022) used Grey Wolf-Optimized Random for software vulnerability prediction. However, in this work, unbalanced nature of datasets is not considered. In this study, balanced techniques with deep neural network (DNN) and 1D convolution neural network (1D CNN) are used to forecast software vulnerabilities. The remainder of this chapter is organized as follows: Section 12.2 is used to discuss the associated research work, and Section 12.3 contains the description of datasets. Section 12.4 discusses our software vulnerability prediction framework. Measures used to evaluate performance are described in Section 12.5. Section 12.6 provides experimental setup and prediction model findings. Comparative examination of the work is offered in Section 12.7, and Section 12.8 describes how the work is applied. Section 12.10 is used to wrap up the work after Section 12.9 addresses the dangers to the validity.

12.2 RELATED WORK

To forecast software quality, various machine learning and statistical techniques are applied. The prediction of software quality attributes is greatly influenced by several structural measurements. A framework that predicted the vulnerabilities by using the cyclomatic complexity metric was put out by Chowdhury et al. (2011). Different data mining approaches are applied software for vulnerability prediction models. Abunadi and Alenezi (2016) have shown the use of classification algorithm for software vulnerability prediction. In an empirical study, software metrics are utilized to investigate insecure code in three sizable open-source web projects. Professionals in the software industry might prioritize verification efforts once insecure code has been found. Software metrics are successfully applied for software vulnerability prediction and security audit of projects. In their 2014 study, Walden et al. contrasted software metrics-based models versus text mining-based models for predicting vulnerabilities. Text mining models outperformed other models in terms of accuracy. Software failure prediction is done using soft computing and search-based techniques (Mafarja M. et al., 2023; Vandecruys et al., 2008; Carvalho et al., 2010). The use of publicly accessible databases to attack vulnerabilities was foreseen by Bozorgi et al. (2010). They used data elements from the vulnerability report, including text fields, timestamps, and other entries, to train their model. Their methodology outperformed the industry norm for predicting vulnerability based on heuristics and expert knowledge. Models for predicting vulnerability based on supervised and semi-supervised learning have been developed by Shar et al. (2015). It was discovered that semi-supervised learning had better performances with the limited amount of labeled vulnerabilities such as SQL injection, etc. (Filus and Domańska 2023). Threat in TensorFlow implementation is identified using Common Weakness Enumeration. It was found that majority of vulnerabilities are due to missing or incorrect statements. They also provide recommendation to improve code quality. Wang et al. (2023) have proposed a software vulnerability prediction approach based on weighted word vectors and fusion neural network. Proposed vulnerability model TCNN-BiGRU is made of Text CNN and bidirection gated recurrent neural network models of deep learning. They have used CNN with gated recurrent unit (GRU) neural network. Experiments are performed on National Vulnerability Database dataset. Lomio et al. (2022) identified that basic machine learning algorithm rarely performs well for software vulnerability prediction; however, use of ensemble techniques based on boosting improved the performances and there was no direct relation in performances improvement when more metrics are added. Kekül et al. (2021) have used natural language processing techniques for multi-class classification. Kelul et al. (2022) have presented an approach for word embedding technique based on natural language processing technique for software vulnerability prediction. Bag of words, n-gram and term frequency–inverse document frequency are used for software vulnerability prediction. Georg et al. (2018) have proposed token-based and neural network based on framework for detecting software vulnerabilities with injection attacks. Jabeen et al. (2021) have used Morkov chain model for prediction of seriousness of vulnerabilities. Windows 10, Firefox, and Adobe Flash Player are used for experimental purposes. Sharma et al. (2023) have proposed a vulnerability discovery model for prediction of the number and severity of different vulnerabilities. Zhang et al. (2022) have proposed an approach for software vulnerability prediction based on n-gram feature extraction and ensemble learning. Datasets related with buffer overflow and resource management are used for experiment purpose. A novel approach, VULPREDICTOR, predicts vulnerable files, analyzes software metrics, and uses text mining to build a composite prediction model. Zhang et al. (2015) Chen et al. (2023) have presented a temporal CNN for software vulnerability prediction. Their model is based on Deep Residual Shrinkage Network (DRSN) and temporal CNN named BiTCN_DRSN. Three datasets BE-ALL, HY-ALL, and RM-ALL are used for applying BiTCN_DRSN and obtained results improved accuracy compared to traditional temporal CNN. Chakraborty et al. (2022) have investigated the drop of performance of deep learning models and identified main causes as data duplication and unrealistic distribution of vulnerable classes. Due to these shortcomings, these approaches do not learn correct features to identify vulnerabilities. Bassi and Singh (2023) have done studies to assess the impact of dual hyperparameter on metrics based on software vulnerability prediction models. Optimal hyperparameters for machine learning as well as data balancing techniques are assessed. Hyperparameter on both machine learning technique and data balancing techniques performed better compared to only hyperparameter on machine learning and data balancing technique.

12.3 DATASETS USED IN THE STUDY

Due to the widespread use of PHP for open-source web development and it has a bad reputation for security, we have used PHP datasets in this study Chowdhury et al. (2011). Due to the engagement of numerous developers in identifying and fixing software code flaws, open-source software is typically regarded as secure (Meneely and Williams, 2010). Along with software measurements, the dataset includes information about PHP files' vulnerabilities. The dataset includes a variety of programs including Drupal 6, Moodle 2, and phpMyAdmin 3.3. Drupal6, Moodle, and phpMyAdmin3.3 are three PHP-based software applications. Different metrics taken in datasets are based on lines of code, functions, complexity metrics, and datasets with total features are mentioned in Table 12.1.

12.4 PROPOSED FRAMEWORK FOR SOFTWARE VULNERABILITY PREDICTION

This section describes class imbalance handling techniques like SMOTE, SMOTETomek, and deep learning techniques (DLT): DNN and 1D CNN for software vulnerability prediction models (Khalid et al. 2018).

12.4.1 SMOTE

SMOTE is a popular technique used in machine leaning for handling the class imbalance issue. Class imbalance refers to the cases where number of instances of one class are high compared to number of instances of other classes. To balance the datasets, minority samples are artificially created by adding new data to the original datasets (Chawla, 2002). Utilizing k-nearest neighbor, new instances of the minority class are generated and

Table 12.1 Datasets with Total Features

Dataset	Number of Features	Number Vulnerable Files	Total Files
Drupal 6	13	62	202
Moodle 2	13	24	2942
phpMyAdmin3.3	13	27	322

added to balance the datasets. Synthetically generated minority instances help balance the datasets.

12.4.2 SMOTETomek

SMOTETomek method oversampled the original datasets before removing the Tomek linkages, creating a balanced dataset with a distinct class cluster (Batista et al., 2003).

12.4.3 Deep learning

The DNN comes from the family of deep learning. In recent research, deep learning has shown very strong accuracy in wider categories of applications, such as image-based classification, audio, and video classification. Deep learning is applied in software engineering for software fault prediction, code clone detection, software requirement engineering, software testing, etc. (Yang, 2022). Neural networks are inspired and simulate the functioning of the human brain. Each deep learning model contains a number of layers, and each layer contains connected neurons. Figure 12.1 shows simple DNN architecture for software vulnerability prediction.

$$f = \varsigma W_3 \varsigma(W_2 \varsigma(WX_i + b_1) + b_2) + b_3$$

Here W_i and b_i are connection's weights and biases, respectively
$\quad f = 1$ for vulnerable class
$\quad f = 0$ for nonvulnerable class

Cost function for weight optimization is given by eq. (12.1)

$$C(x) = \frac{1}{n}\sum_{i=1}^{n}\frac{1}{2}\big(y(Xi) - F(Xi)\big) \tag{12.1}$$

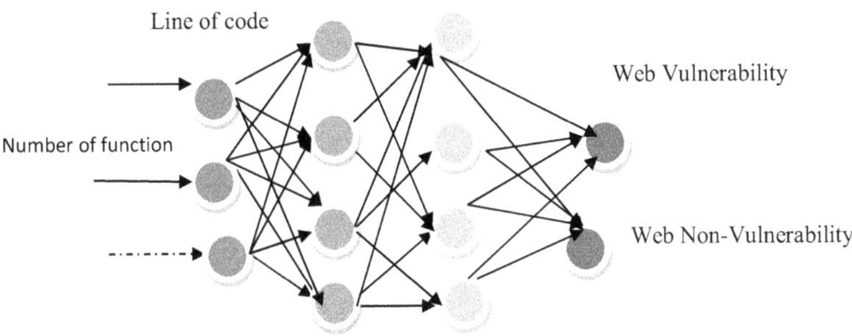

Figure 12.1 Architecture of Deep Neural Network for Software Vulnerability Prediction.

Various optimization algorithms like Adam and Stochastic Gradient Descent (SGD) are used for obtaining minimized cost function. SGD is similar to gradient descent algorithm; however, it uses subset of data. Adam is an adaptive learning rate algorithm. This algorithm finds the learning rate for each parameter.

12.4.4 1D convolution neural network

In the present study, 1D CNN is used for identifying software vulnerabilities. 1D data is the input to the input layer, and the output layer is dense layer and having same number of neurons as classes. When processing 1D data, different CNN layers are used, and they produce characteristics that are helpful for the classification task. In comparison to 2D CNNs, 1D CNNs are less difficult as convolution operations in 1D CNNs are just weighted sums of two 1D arrays. Raw data is processed by CNN layer to extract features, and extracted features are used for classification task by MLP-layer. In 1D CNN, 1D convolution are simply weighted sum of 1D arrays. Kiranyaz et al. (2021) show the 1D CNN that CNN consists of mainly two types of layers:

1. Convolution layer: Convolutional procedures are applied to the input to extract features from the data. Data input for 1D CNN is one-dimensional.
2. Multilayer feed forward layers (fully connected layer): Based on characteristics extracted by convolution layers, it determines the class of the input or forecasts the value of the output.

In Figure 12.2, there are three convolution layers and the last layer is fully connected layer.

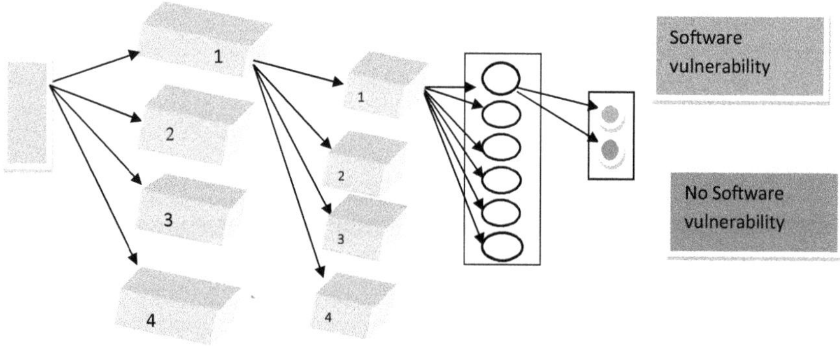

Figure 12.2 1D CNN for Software Vulnerability Prediction Framework.

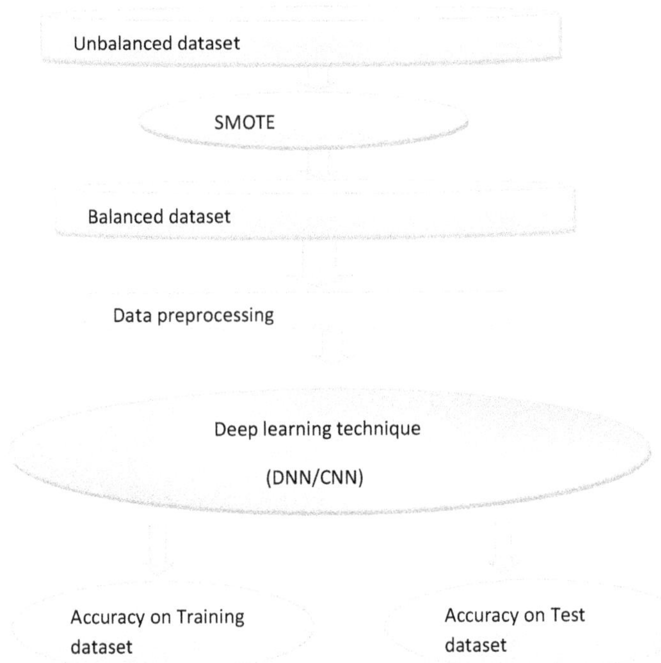

Figure 12.3 Proposed Software Vulnerability Prediction Framework.

Deep learning-based software vulnerability prediction model is given in Figure 12.3. Three freely available unbalanced datasets are used in the study, which are first balanced by SMOTE (Synthetic Minority Oversampling Technique). Then each feature is standardized and preprocessed and datasets are divided into 80:20 in a stratified manner, which is trained with DNN and 1D CNN and 10-cross validation is used to assess performances.

12.5 MEASURES FOR EVALUATING PERFORMANCE

To determine the deep learning method that most effectively forecasts software vulnerabilities, we conducted trials. Positive cases in the current study represent vulnerable classes, whereas negative cases represent nonvulnerable classes. Because each data collection has an unbalanced number of susceptible and nonvulnerable groups, performances are measured using precision value, recall value, and the F-measure values. The following is a description of these measures.

12.5.1 Precision

Precision is defined as the ratio of vulnerable instances accurately predicted to all instances identified as vulnerable by the model. It can be obtained by the formula given below:

$$\text{Precision} = \frac{TP}{TP + FP}$$

12.5.2 Recall

The ratio of accurately predicted vulnerable instances to all vulnerable instances in the data is known as recall. It can be obtained by the formula given below:

$$\text{Recall} = \frac{TP}{TP + FN}$$

12.5.3 F-measure

The harmonic mean of recall value and precision value is all that the F-measure value for binary classification is. Values for all these metrics lie between 0 and 1.

12.6 TEST ENVIRONMENT AND RESULTS

Python packages Keras, TensorFlow, and Scikit-learn (Pedregosa et al., 2011) are used for implementing DNN and 1D CNN. Each dataset is divided into 80:20 ratio, and then 10 cross-validations are used for measuring the performances.

Different hyperparameters of deep learning techniques are given in Table 12.2 and various performance measures for different techniques obtained are given in Table 12.3.

Figures 12.4, 12.5, and 12.6 present the graph of performance measures of different techniques on three different datasets phpMyAdmin, Moodle, Drupal. It is clear that for SMOTETomek-1D CNN precision, recall, and f-measure values are best on all three datasets.

From Figures 12.7, 12.8, and 12.9, it is clear that performance in terms of different metrics is improved when SMOTE and SMOTETomek techniques are applied. Performance of SMOTE-DNN is better than DNN on all datasets and performance SMOTETomek-1D CNN is better than CNN on all datasets.

Table 12.2 Different Hyperparameters of Deep Learning Techniques

Technique	Hyperparameters
DNN	DenseLayer (32, activation='relu', input_dim=13)) DenseLayer (layers.Dense(16)) DenseLayer(layers.Dense(2, activation='softmax')) network.compile(optimizer='rmsprop', loss='categorical_crossentropy', metrics=['accuracy']) network.fit(training_data, training_labels, epochs=50,batch_size=64) test_loss, test_acc = network.evaluate(testing_data, testing_labels)
1D-CNN	layers.Conv1D(130, 2, activation='relu', input_shape=(13,1))) layers.MaxPooling1D((1))) layers.Conv1D(100,1, activation='relu') layers.MaxPooling1D((1))) layers.Conv1D(100, 1, activation='relu') layers.Flatten() layers.Dense(2, activation='softmax')) model.compile(optimizer='Adam', loss='categorical_crossentropy', metrics=['accuracy']) model.fit(training_data, training_label, epochs=20,batch_size=32)

Table 12.3 Various Performance Measures of Different Techniques

Dataset	Technique	Precision	Recall	F-measure
phpmyadmin3.3	SMOTE-DNN	0.76	0.76	0.76
	DNN	0.97	0.60	0.65
	1D CNN	0.58	0.96	0.62
	SMOTETomek-1D CNN	0.78	0.79	0.78
moodle2	SMOTE-DNN	0.93	0.93	0.93
	DNN	0.50	0.50	0.50
	1D CNN	0.50	0.49	0.50
	SMOTETomek-1D CNN	0.95	0.95	0.95
Drupal 6	SMOTE-DNN	0.73	0.73	0.73
	DNN	0.69	0.66	0.67
	1D CNN	0.65	0.73	0.66
	SMOTETomek 1D CNN	0.75	0.75	0.75

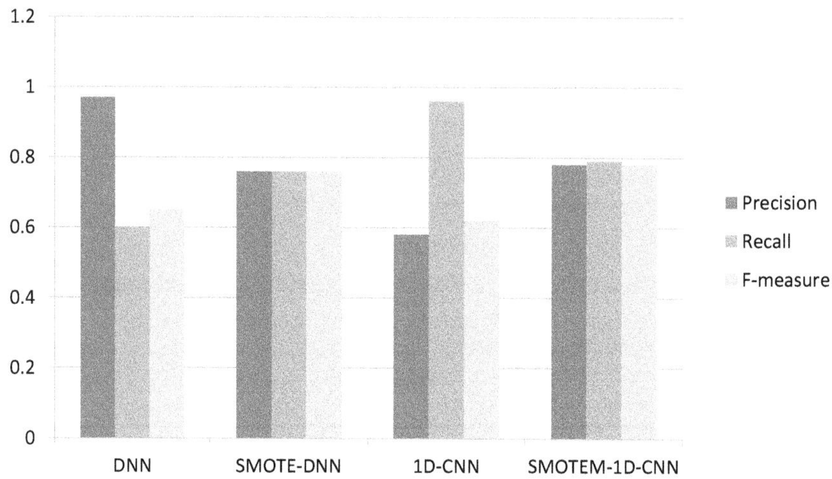

Figure 12.4 Performance of Different Techniques on phpMyAdmin Dataset.

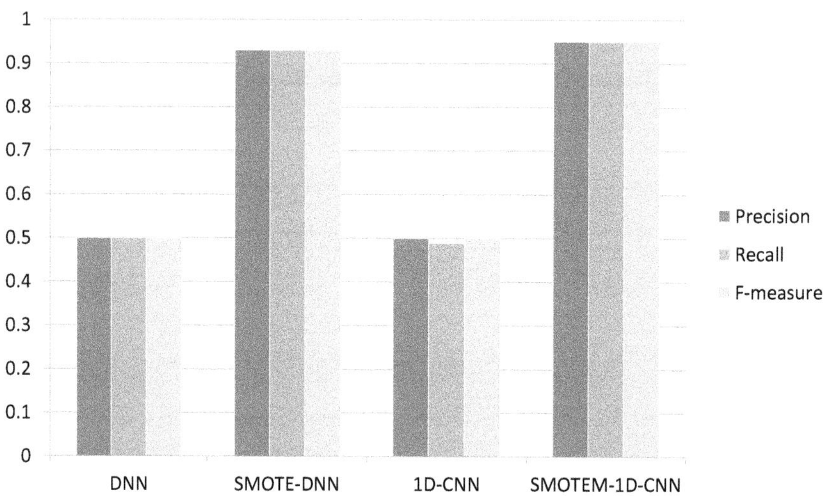

Figure 12.5 Performance of Different Techniques on Moodle Dataset.

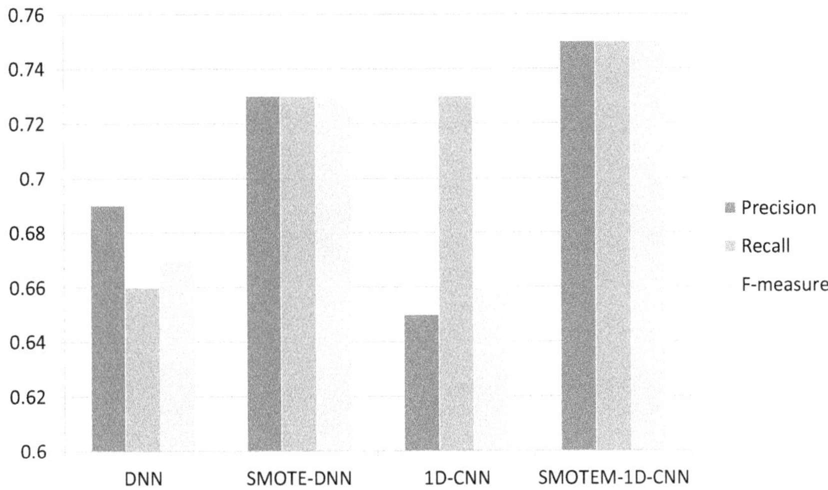

Figure 12.6 Performance of Different Techniques on Drupal Dataset.

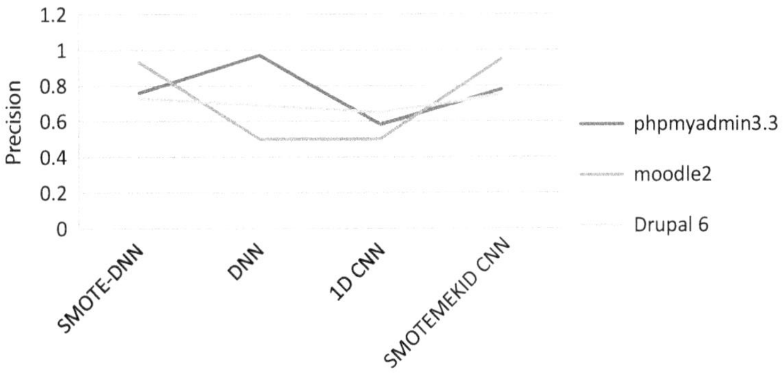

Figure 12.7 Precision of Different Techniques.

1. The study's findings can be helpful for researchers, software professionals, and security experts in the following ways: Software testers can use DLT to predict web vulnerabilities and can identify the classes of web applications that are most vulnerable early on. The effectiveness of deep learning approaches for identifying vulnerable classes is suggested by this study.

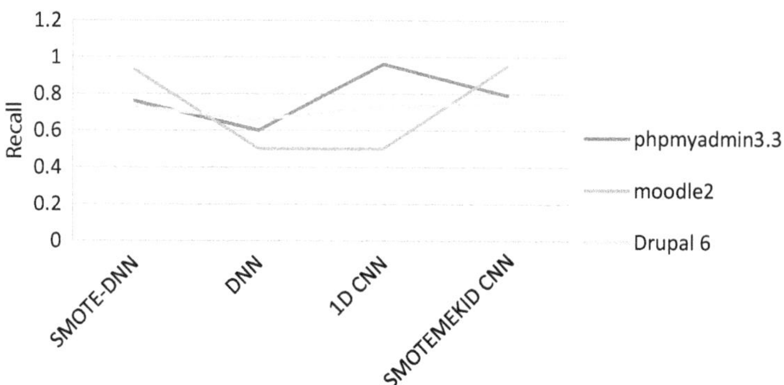

Figure 12.8 Recall of Different Techniques.

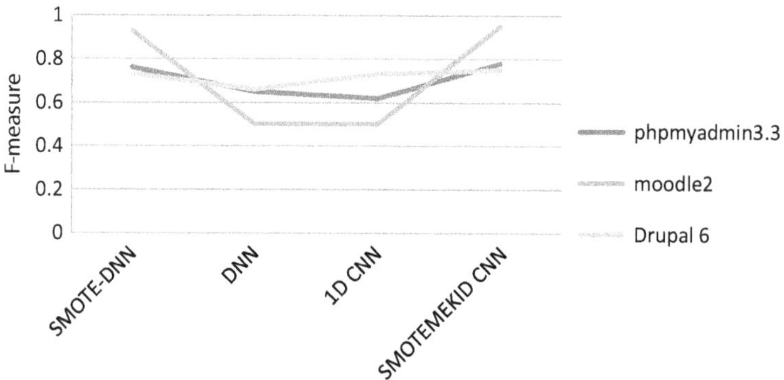

Figure 12.9 F-measure of Different Techniques.

2. Techniques for controlling data imbalance have demonstrated better performance in identifying software vulnerabilities in online applications.
3. The metrics used in this study can be used by security professionals, web developers, and the software industry to establish benchmarks for secure web applications.

12.7 THREATS TO VALIDITY OF STUDY

If the variables used in work are not precisely measured, its validity is threatened (Zhou et al., 2009). The theoretical meaning of variables and their practical measurement differ. Although the severity of the vulnerability is not taken into consideration in our analysis because it may be arbitrary, we have used datasets provided by the data repository publicly available. Three datasets from the literature are available for our analysis as external validity is dependent on how broadly generalizable a study's findings are. Datasets based on PHP are being examined for vulnerability checks; hence it is important to investigate different programming languages, such as (Java), python, ASP.net, and JavaScript also. Therefore, to duplicate the study and enable generalization of the research's findings, it is necessary to investigate various datasets from other application domains.

12.8 CONCLUSION AND SCOPE OF FUTURE WORK

Accurate software vulnerability prediction can reduce the risk of unauthorized system access and it can also prevent system from failures. Software failure risk or unauthorized access can be reduced by accurate and early vulnerability prediction. In the early stages of software development, significant efforts can be made to fix the weak software classes. This study's major goal is to compare the effectiveness of DLT and class imbalance handling techniques for the identification of different classes of vulnerable software and to look into how well each performs when predicting which software would be vulnerable.

The following are the main conclusions of our research:

- Precision, recall, and f-measure of SMOTEMK-1D CNN are better compared to all other techniques.
- Performance of DNN and 1D CNN are improved when they are combined with class imbalance handling techniques.
- F-measure of DNN is slightly better than 1D CNN for some datasets.

In the future, we may extend our research to vast datasets across multiple fields, and we may increase the performance DLT by metaheuristics techniques for better findings and unique architecture can also be proposed for software vulnerability prediction.

REFERENCES

Abunadi, I., Alenezi, M. (2016). An empirical investigation of security vulnerabilities within web applications, Journal of Universal Computer Science, vol. 22, no. 4, pp. 537–551.

Alenezi, M., Abunadi, I. (2015). Evaluating software metrics as predictors of software vulnerabilities, International Journal of Security and Its Applications, vol. 9, no. 10, pp. 231–240.

Bansal, A. (2017). Empirical analysis of search based algorithms to identify change prone classes of open source software, Computer Languages, Systems & Structures, vol. 47, pp. 211–231.

Bassi, D., Singh, H. (2023). The effect of dual hyper parameter optimization on software vulnerability prediction models, e-Informatica Software Engineering Journal, vol. 17, no. 1, 230102. doi: 10.37190/e-Inf230102.

Batista, G., Bazzan, B., Monard, M. (2003). Balancing training data for automated annotation of keywords: a case study, in: II Brazilian Workshop on Bioinformatics, December 3–5, Macaé, RJ, Brazil, pp. 10–18.

Bozorgi, M. et al. (2010). Beyond heuristics: learning to classify vulnerabilities and predict exploits, in: Proceeding KDD 16th ACM Conference on Knowledge Discovery and Data Mining, July 25–28, Washington, DC, USA, pp. 105–114.

Carvalho, A. B. (2010). A symbolic fault-prediction model based on multi-objective particle swarm optimization, Journal of System and Software, vol. 83, pp. 868–882.

Chakraborty, S., Krishna, R., Ding, Y., Ray, B. (2022, Sept. 1). Deep learning based vulnerability detection: are we there yet?, IEEE Transactions on Software Engineering, vol. 48, no. 9, pp. 3280–3296. doi: 10.1109/TSE.2021.3087402.

Chawla, N.V., Bowyer, K.W., Hall, L.O., Kegelmeyer, W.P. (2002). SMOTE: synthetic minority over-sampling technique, Journal of Artificial Intelligence Research, vol.16, pp.321–357.

Chen, J. et al. (2023). BiTCN_DRSN: an effective software vulnerability detection model based on an improved temporal convolutional network, Journal of Systems and Software, vol. 204, 111772.

Chowdhury, I., et al. (2011). Using complexity, coupling, and cohesion metrics as early indicators of vulnerabilities, Journal of Systems Architecture, vol. 57, pp. 294–313.

Filus, K., Domańska, J. (2023). Software vulnerabilities in TensorFlow-based deep learning applications, Computers & Security, vol. 124, 102948.

Georg, T. K. et al. (2018). Token based detection and neural network based reconstruction framework against code injection vulnerabilities, Journal of Information Security and Applications, vol. 41, pp. 75–91.

Gyimothy, T. et al. (2005). Empirical validation of object-oriented metrics on open source software for fault prediction, IEEE Transactions on Software Engineering, vol. 31, pp. 897–910.

Iannone, E., Guadagni, R., Ferrucci, F., De Lucia, A., Palomba, F. (2023, Jan.). The secret life of software vulnerabilities: a large-scale empirical study, IEEE Transactions on Software Engineering, vol. 49, no. 1, pp. 44–63. doi: 10.1109/TSE.2022.3140868.

Jabeen, G. et al. (2021). Vulnerability severity prediction model for software based on Markov chain, International Journal of Information and Computer Security, vol. 15, no. 2–3.

Jabeen, G. et al. (2022). Machine learning techniques for software vulnerability prediction: a comparative study, Applied Intelligence, vol. 52, pp. 17614–17635.

Kekül, H. et al. (2021). A multiclass hybrid approach to estimating software vulnerability vectors and severity score, Journal of Information Security and Applications, vol. 63, 103028.

Kekül, H. et al. (2022). A multiclass approach to estimating software vulnerability severity rating with statistical and word embedding methods, Journal of Computer Network and Information Security, vol. 4, pp. 27–42.

Khalid, M. N. et al. (2018). Predicting web vulnerability in web applications based on machine learning, intelligent technologies and applications. INTAP 2018. in: Communications in Computer and Information Science, vol. 932, Springer, Singapore.

Kiranyaz, S., Avci, O., Abdeljaber, O., Ince, T. Gabbouj, M., Inman, D. J. (2021). 1D convolutional neural networks and applications: a survey, Mechanical Systems and Signal Processing, vol. 151, 107398.

Lomio, F., Iannone, E., DeLucia, A., Palomba, F., Lenarduzzi, V. (2022). Just-in-time software vulnerability detection: are we there yet?, Journal of Systems and Software, vol. 188, 111283.

Mafarja, M. et al. (2023). Classification framework for faulty-software using enhanced exploratory whale optimizer-based feature selection scheme and random forest ensemble learning, Applied Intelligence, https://doi.org/10.1007/s10489-022-04427-x.

McGraw, G., Potter, B. (2004). Software security testing, IEEE Security & Privacy, vol. 2, no. 5, pp. 81–85.

Meneely, A., Williams. L. (2010). Strengthening the empirical analysis of the relationship between Linus' law and software security, in: Proceedings of the 2010 ACM-IEEE International Symposium on Empirical Software Engineering and Measurement, ACM, pp. 1–10.

Pedregosa, F. et al. (2011). Scikit-learn: machine learning in python, Journal of Machine Learning Research, vol. 12, pp. 2825–2830.

Rhmann, W. (2022). Software vulnerability prediction using grey wolf-optimized random forest on the unbalanced data sets, International Journal of Applied Metaheuristic Computing (IJAMC), vol. 13, no. 1, pp. 1–15.

Shar, L. K. et al. (2015). Web application vulnerability prediction using hybrid program analysis and machine learning, IEEE Transactions on Dependable and Secure Computing, vol. 12, no. 6, pp. 688–707.

Sharma, R., Shrivastava, A. K., Pham, H. (2023) Software security evaluation using multilevel vulnerability discovery modeling, Quality Engineering, vol. 35, no. 2, pp. 341–352. doi: 10.1080/08982112.2022.2132404.

Shin, Y., Williams, L. (2008). Is complexity really the enemy of software security? in: Proceedings of the Fourth ACM Workshop on Quality of Protection, Alexandria, Virginia, USA, pp. 47–50.

Vandecruys, O. et al. (2008). Mining software repositories for comprehensible software fault prediction models, Journal of Systems and Software, vol. 81, pp. 823–839.

Wang, Q., Jiadong Y.G., Zhang, R.B. (2023). An automatic classification algorithm for software vulnerability based on weighted word vector and fusion neural network, Computers & Security, vol. 126, 103070.

Yang, Y., Xia, X., Lo, D., Grundy, J. (2022). A survey on deep learning for software engineering, ACM Computing Surveys, vol. 54, no. 10, article no.: 206, pp. 1–73. https://doi.org/10.1145/3505243.

Zhang, Y., Lo, D., Xia, X., Xu, B., Sun, J., Li, S. (2015). Combining software metrics and text features for vulnerable file prediction, in: 2015 20th International Conference on Engineering of Complex Computer Systems (ICECCS), IEEE, pp. 40–49.

Zhang, B. et al. (2022). Approach to predict software vulnerability based on multiple-level *N*-gram feature extraction and heterogeneous ensemble learning, International Journal of Software Engineering and Knowledge Engineering, vol. 32, no. 10, pp. 1559–1582.

Zhou, Y. et al. (2009). Examining the potentially confounding effect of class size on the associations between object-oriented metrics and change-proneness, IEEE Transactions on Software Engineering, vol. 35, pp. 607–623.

Chapter 13

Implementation challenges faced by artificial intelligence for national security

Prabhat Kumar Srivastava, Basudeo Singh Roohani, Rohit Aggarwal, and Nitin Sharma

13.1 INTRODUCTION

Making complex algorithms that mimic the functioning of the human brain is at the heart of artificial intelligence (AI). Although AI is often discussed as if it were its own technology, in reality, it facilitates the development of several other technologies. The versatility of AI is what sets it apart from other technologies. Voice assistants like Alexa and Siri, social networking platforms, and e-commerce websites are used in numerous fields, including the service industry, healthcare, agriculture, climate change, and the financial sector, to name a few. Defence-related uses for AI range from ISR (intelligence, surveillance, and reconnaissance) to cyber security to military logistics to autonomous vehicles to fully autonomous weapons systems.

Improvements in computer power, the explosion of organized and digitalized data with low-cost storage options, and developments in machine learning (ML) algorithms have all contributed to the meteoric rise of AI use in both the public and private sectors over the past few decades. Machine intelligence has surpassed human intelligence in some functional domains because of advancements in ML models. For instance, in 1996, the renowned IBM Deep Blue programme defeated Garry Kasparov, the current world chess champion. In the GO game, Deep Minds Alpha GO beat Lee Seldol, the global champion, in 2016. However, the development, testing, and validation of these systems required a significant amount of human input. It will be some time before computing power is sufficient to run a general AI.

The United States and China, two global powerhouses, are using AI to strengthen their national security frameworks and preserve an edge in the global economy. Additionally, it highlights India's initiatives in the realm of military AI as well as the legal and moral issues raised by AI's dual-use applications.

DOI: 10.1201/9781003514312-13

13.2 LITERATURE REVIEW

AI has emerged as a critical technology for enhancing national security across various domains. However, the implementation of AI for national security is not without its challenges. This literature review aims to explore and summarize the existing research on the implementation challenges faced by AI in the context of national security. The review covers key areas such as lack of clear strategy for AI, inadequate talent with the necessary skill set for AI work, limited availability of key infrastructure in terms of computing power, development of ethical standards, enhanced susceptibility to cyber attacks, theft vulnerability, limited private sector role in defence, technology cannot be completely controlled, the geopolitical environment around India, economic constraints, privacy risks, robustness and bias intelligence sharing and security cooperation and regulatory frameworks. By examining the literature, this review provides insights into the current state of AI implementation for national security and identifies areas for future research and improvement [1].

This review provides a thorough comprehension of the difficulties encountered while implementing AI for national security by synthesizing the available literature on the topic. The findings can inform policymakers, researchers, and practitioners in developing effective strategies to overcome these challenges and harness the potential of AI for enhanced national security.

13.3 ARTIFICIAL INTELLIGENCE AND NATIONAL SECURITY

Defence analysts in the defence sector are interested in AI. Recent breakthroughs in AI, for example, have changed the realm of hybrid warfare. AI "will be a source of great power for the industries and countries that harness them," according to the US National Security Commission on Artificial Intelligence (NSCAI).

AI is becoming more prevalent, and this is leading to developments in the information, economic, and military sectors. AI has greatly improved capabilities for information operations, including advanced data production, analysis, and gathering. Common uses of AI-enabled systems include deep fake/counterfeit identification, geospatial data analysis, audio/video analysis, and image categorization from drone footage.

Private expenditures in AI reached US $93.5 billion in 2021, which was more than double the amount spent in 2020; as a result, it is expected that sales will approach $300 billion by 2024. McKinsey predicts that by the year 2030, AI will have contributed an additional $13 trillion to global economic activity. The commercial sector has been so successfully penetrated by AI that practically every company today either already uses AI in their systems and products or plans to do so in the future. It is generally agreed that AI has the potential to spark a new economic revolution in which the size of the people will no longer be as important for national strength [2].

New autonomous capabilities are becoming available to a range of entities in the military space because of AI. The dual use of AI has improved visibility and opportunities for poor governments and non-state entities to strengthen their capacity.

The military potential of AI has changed the character of battlegrounds by introducing more autonomous systems into the security environment. The defence systems' interaction with this technology has improved the alternatives for asymmetric warfare. AI is utilized by the military in a wide variety of settings, some of which include intelligence gathering, surveillance, reconnaissance, cyberspace operations, military logistics, information operations, and deep fakes. The military makes use of both autonomous and semi-autonomous systems, as well as integrated command and control systems. AI has been given a central role in the development of the country's national defence due to its successful incorporation into applications for missiles, aircraft carriers, and other naval assets, as well as its inclusion in Command, Control, Communications, Computer (C4) and intelligence Surveillance and Reconnaissance [3–5].

Use of AI in security:

1. Accessing knowledge about the elusive adversary and deciding how to deal with it are the two halves of security. The nature of "reaction" in the area of security is often strengthened by AI. AI can assist in honing the "action" tools. The goal of AI is to make technology "smarter," and technology makes a process "smarter." The collection and analysis of information already available are where AI is most useful in that situation. To maintain national security, it is necessary to have access to information regarding the enemy's well-kept secrets, about which little can be learned by looking at publicly available data.

2. No one can deny that in the Internet age, there is always a chance that an enemy would leave behind "footprints" in that medium. Because of this, the security configuration places a high importance on conducting a scan of the various communication networks. The fundamental goal of intelligence agencies is still the tough work of gathering human intelligence through covert methods. It is also true that, to accomplish that goal, a technology-based endeavour like AI can never fully replace the human component in intelligence gathering. Examining publicly available information would reveal a lot.

3. Due to recent breakthroughs in computing power, a substantial decrease in the cost of data storage, and the explosion of digitized

and structured data, the inclusion of AI into our systems has recently seen an uptick.

Here are a few crucial areas where AI can be used:

1. Analysing a large amount of material and finding patterns: AI is excellent at spotting patterns and insights in enormous datasets that are just invisible to humans. AI can combine data from various sources, and its algorithms are built to deliver results instantly.
2. Unlike passive robots, which can only generate predetermined readings, AI-assisted systems may immediately evaluate data from digital information, satellite photos, visual information, text, or unstructured data to generate a signal. Nevertheless, the success of AI depends, in the end, on how well all human input into algorithms is carried out.
3. AI has a tremendous impact on analytical and assessment activities across a variety of disciplines, but "human monitoring and control" must still be prioritized when it comes to national security. Using the massive amounts of data and video that surveillance has gathered, AI can be used to inform "human analysts" about "patterns" or anomalous and suspect behaviour.
4. The moral considerations of warfare had been sidelined since the advent of "proxy battles," which added to the enemy's manoeuvrability and made gathering intelligence and formulating an action plan there a challenging task.
5. The attacker had the upper hand in a proxy or virtual war, but AI boosts the efficiency of the counteroffensive by pinpointing the attack's sources in the fog of war and allowing for a "punitive" response that generates deterrence.
6. The covert dumping of weapons and explosives in Kashmir, as well as the deployment of drones and quadcopters for monitoring, show how the proxy war's new dimensions have made security so much more difficult to manage.
7. When it comes to nuclear weapons, the "deterrence" theory prevents a country from starting a conflict. Secret attacks, on the other hand, seem to be the norm, which has increased the significance of AI-assisted intelligence collecting and collation systems as well as of countermeasures.

Relevance of AI to National Security:

1. **Security's evolving nature:** As a result of technological advancements, older security measures are being replaced by newer ones like cyber security and hybrid warfare.

2. **Massive AI-based tools** owing to the duality of their uses and the absence of international regulations.
3. **Artificial intelligence cannot be ignored:** AI will always be present when utilized in conjunction with other technologies. Consider the application of AI to social networking.

Principal motivators for the rapid advancement of AI technology:

1. increased accessibility to massive datasets for ML systems training
2. the development of better ML methods
3. substantial and accelerating commercial investment
4. accelerating improvement in computing performance

13.4 INDIA'S ARTIFICIAL INTELLIGENCE ROADMAP

After the National Institution for Transforming India Aayog released the National Strategy on AI in 2018, India began its AI journey. Unfortunately, it did not include defence-related industries but rather focused on commercial and private ones (such as agribusiness, healthcare, education, intelligent cities and infrastructure, intelligent mobility, and intelligent transportation). Despite entering the defence market later than its main rivals, India has been swiftly catching up through increasing investment, domestic development, and bilateral and international alliances focused on the application of AI [4–5].

Both the Defence Artificial Intelligence Council (DAIC) and the Defence Artificial Intelligence Project Agency (DAIPA) have been established, and each has a yearly budget of Rs. 1,000 crore dedicated to operations that are enabled by AI. An AI-based Signal intelligence system is currently being developed by 9 CAIR, a laboratory of the Defence Research and Development Organization (DRDO), with the goal of improving the capabilities of the armed services in terms of intelligence collection and analysis [4,5].

The Ministry of Defence released 75 Def (Symp) systems with AI in July 2022. In addition, 140 sensor systems with AI capabilities have been installed along the Pakistani and Chinese borders. The Indian Army plans to test unmanned all-terrain vehicles equipped with AI for use in logistics and surveillance in the region of Ladakh.

The Government of India's Ministry of External Affairs established the New Emerging and Strategic Technologies Division in 2020 with the goal of leading the world in technological diplomacy and fostering cross-national collaboration in the field of AI. Considering India's rising geopolitical status, the US National Security Commission on AI in 2020 recommended the formation of a US–India Strategic Tech-Alliance (USISTA) to establish an Indo-Pacific Strategy on emerging technologies. The 2+2 meeting between India and the United States emphasized the need to deepen bilateral cooperation on new technologies. Cooperation in the field of AI was highlighted during the QUAD Summit 2022. In June of 2022, India and Japan met to examine the state of cyber security, information and communication technology (ICT), and fifth-generation (5G) networks, among other areas of bilateral cyber cooperation. India and Finland have also committed to collaborate in new technology sectors such as AI and quantum computing [6–9].

13.5 ARTIFICIAL INTELLIGENCE IN INDIAN ARMED FORCES

1. An increased focus on border security has been placed in response to current national security concerns due to a sharp increase in bloodshed by terrorist groups across the globe. The Indian armed forces are responsible for maintaining India's 15,000 km of land borders, 7,516 km of maritime borders, and a large amount of airspace. The capabilities and efficacy of all forces would increase substantially with the intelligent and appropriate employment of AI in the armed forces.

2. The core of the force structure is logistic management. The AI can be used for supply chain management forces' logistical maintenance. The most effective delivery platforms now are unmanned aerial vehicles (UAVs).

3. It might potentially decrease the amount of supplies kept at the front lines and speed up resupply. The forward posts in the border region can receive supplies like food, clothing, medicine, and ammunition on a regular basis from drones and unmanned helicopters. It can be applied to India's situation to provide a high-altitude landscape in the Siachen glaciers.

4. With the advent of UAVs, often known as UAVs, nations may be able to carry out military operations with greater efficiency and less risk than they were able to in the past when planes were piloted by people. Surveillance and reconnaissance can be accomplished with UAVs and drones.

5. The speed of aid missions can be increased by using modern, autonomously navigating air and land vehicles. Future military operations will be supported by technology as well, saving lives by lowering risk and speeding up work. Forces have access to small drones for tactical operations. To assist commanders in making tactical decisions, real-time intelligence would be helpful. UAVs can engage enemy artillery and tanks with adequate fatal firepower.

6. In situations where there are standoffs and force engagements, situation awareness is essential for border security. ISR missions are crucial for situational awareness.

7. Improve border management architecture: Radars and sensors might eventually improve border management architecture if they are used to monitor India's land borders. Monitoring and preventing infiltration in border regions would be beneficial. To identify the enemy when being monitored by drones and UAVs, the forces can also deploy "facial recognition" system platforms.

8. Close-quarters combat would incorporate the use of machines; they would have the ability to penetrate the target area and launch attacks against enemies. It has more deadly capabilities than people do. Armed forces and paramilitary forces can use this technology efficiently in counterinsurgency and anti-terror operations.

9. Grey zone warfare combines psychological tactics with media and information warfare, among other things. The intensity of grey zone warfare is high enough to disrupt the internal operations of the enemy. Its primary application is in urban conflicts.

10. In regard to Jammu and Kashmir, Pakistan routinely engages in "grey zone" warfare against India. As a result, the Government of India ought to seriously contemplate conducting research into the military and safety implications of questions about social media, information, and the mainstream media.

11. New AI applications for cyber defence, or immune computer systems, with the capacity for self-adaptation, are appearing in International Bank Note Society. These applications help to prevent sophisticated cyber threats, reverse-engineer malware, and predict future cyber scenarios, all of which contribute to a heightened awareness of the threat landscape. Software like Crowdstrike and DarkTrace, which efficiently used AI to detect cyber threats, led to enormous value for the parent firms. Among its clients were the National Security Agency (NSA), Government Communications Headquarters (GCHQ), Military Intelligence (MIS), and the Central Intelligence Agency (CIA) [10].

12. AI in battle tanks and two-way communication between tanks and drones would give the Armoured Corps a decisive advantage. Another consideration for the Indian Army is the use of drones in combat against traditional military assets like tanks, artillery, and troops. Pakistan may therefore be able to utilize drones and UAVs to fight India's concept of cold start.

13. AI technology integration in the naval platform: Naval combat would rely heavily on the employment of autonomous submarines. The PLA Navy could make use of China's 56 tested "swarm sharks" by 2020. Introducing AI into naval equipment could provide the Indian Navy a strategic advantage in the Indian Ocean.

14. Drones, UAVs, unmanned helicopters, and other AI-enabled devices can be used for surveillance and reconnaissance. Armed with ammunition and short-range torpedoes, unmanned submarines or unmanned

undersea vehicles (UUVs) would be an effective deterrent against the enemies. It would provide the Indian Navy with a strategic advantage in the seas.

15. Improve the Air Force's target acquisition capacity: By integrating AI, the Air Force will be better able to acquire the target and evade air defence radars from the opposition, among other things. The AI technology also advises fighter pilots to utilize weaponry. It gives the pilot the advantage in a bad scenario or dogfight. The US employs the Automatic Logistics Information System (ALIS) on its F-35 to estimate the next maintenance date and the problems that need to be fixed by using live data from the plane's engines and other onboard equipment [11].

16. India also has a nuclear command and control system, an air defence system, and a ballistic missile defence system. Many missiles and missile defence systems are used in the services. India can enhance its capabilities and make precise decisions based on data analysis by integrating AI technology.

17. ML and other forms of AI can be embedded in physical devices or software to grant them autonomy. AI can be implemented on a wide variety of military hardware, from rockets and missiles to radars and fighter planes, to frigates, destroyers, aircraft carriers, and submersibles.

18. Computer attacks have increased in frequency recently. Consider the 2016 disclosure of confidential information from 3.2 million debit cards, the 2017 data breach at Zomato, etc. as examples. In cyber security, AI is used to rapidly evaluate massive amounts of data to detect threats like zero-day exploiting malware and risky user habits that could lead to phishing or the download of malicious software [12].

13.6 INDIA'S INITIATIVES TO USE ARTIFICIAL INTELLIGENCE FOR MILITARY APPLICATIONS

1. The DAIC, which is chaired by the defence minister and is made up of the leaders of the three services, the secretary of defence, and the secretary of defence production, has been added to the Indian defence structure. The creation of "25 defence-specific AI products" by 2024, according to Defence Minister Rajnath Singh, had previously been announced [13,14].

2. Under the secretary of defence production, a new organization called the DAIPA has also been established, with an annual budget of Rs. 100 crore set out for AI-enabled projects.

3. The Army is involved in a number of projects utilizing cutting-edge and developing AI technologies. The Military College for Telecommunication Engineering in Mhow has advanced AI knowledge together with cooperation with civil enterprises.

4. The Indian Naval Ship Valsura at Jamnagar, which already has a state-of-the-art facility for AI and Big Data research, is being converted into an AI centre of excellence by the Navy.

5. A Memorandum of Understanding (MoU) has been signed by the Army Training Command and Rashtriya Raksha University (RRU) to create the Wargame Research and Development Center (WARDEC) in New Delhi.

6. RRU is a Gandhinagar-based Institute of National Importance that was founded by the government of Gujarat. Tech Mahindra is in charge of setting up the lab, and it will be used to develop new military ideas, test out tactics, and evaluate fighting formations to advance military expertise.

7. Defence India Startup Challenge: The sixth Defence India Startup Challenge has been launched by Defence Minister Rajnath Singh as part of the Innovations for Defence Excellence (iDEX) programme.

8. The goal is to give enterprises that specialize in creating technology for the armed forces financial support. Examples of such technologies include AI, advanced imaging, sensor systems, Big Data analytics, autonomous drone systems, and secure communication systems. Projects with budgets ranging from 1.5 crore to 10 crore rupees may be eligible for assistance through this programme. Our scientific and technological specialists now have the chance to understand cutting-edge technologies like AI, augmented reality, blockchains, and space technologies because of the iDEX programme [15–18].

13.7 REGULATORY AND LEGAL METHODS FOR MANAGING ARTIFICIAL INTELLIGENCE SYSTEMS

1. India presently lacks a thorough framework for utilizing AI systems. The development of such a framework will be crucial for advising numerous stakeholders on how to responsibly handle AI in India [19].

2. There is not yet a comprehensive body of law in India pertaining to AI. The Personal Data Protection Bill (2019), which is now just a draught but is intended to be comprehensive legislation that describes many different areas of privacy safeguards that AI solutions need to comply with, is the piece of legislation that comes closest to this. On December 12, the "Bill" was given to a Joint Committee of both Houses of Parliament, sometimes known as the JPC, for the purpose of review and suggestions.

3. Important suggestions made by the committee include expanding the scope of the proposed legislation to cover non-personal data, requiring earlier notification of data breaches, imposing new rules

on social media platforms and intermediaries, clarifying the role of data protection officers (DPOs), and establishing fiduciary obligations for hardware manufacturers.

4. The data localization recommendations are the most noteworthy and decisive ones. The local storage and processing of data mandated by India's data privacy law has social media businesses worried that this will drive up the price and complexity of providing our services.

5. National Mission for Interdisciplinary Cyber-Physical Systems (NM-ICPS): NM-ICPS is a Pan India Mission that encompasses all aspects of Indian society, such as the central ministries, state governments, businesses, and educational institutions.

6. It was first proposed in 2018 and, according to the Department of Science and Technology, the completion of it would require a total expenditure of Rs. 3,660 crores spread out over a period of 5 years. The New Mexico Institute for Computational and Predictive Sciences (NM-ICPS) will provide funding for cutting-edge research in modern domains such as AI, the Internet of Things (IoT), ML, deep learning (DL), and quantum information sciences.

7. AI and ML in cyber security will benefit from the maturation of the research and development ecosystem fostered by the AI Strategy.

13.8 CHALLENGES FACED BY ARTIFICIAL INTELLIGENCE FOR NATIONAL SECURITY

AI developments will gradually increase the threats, difficulties, and opportunities in terms of security provided to the nation. Major powers like China and the United States are making significant investments in AI-capable devices to help them maintain their military superiority. To be better prepared for the battlefield of the future, India is likewise making advancements in the field of military AI. The revolutionary effects that AI will have on national security present a one-of-a-kind set of challenges for those working in the field of national security, challenges that span both the ethical and regulatory worlds.

1. **Lack of Clear Strategy for AI:** When working on a technology with a national security reputation, such as AI, it is important to have a complete awareness of all the implications for its deployment in cutting-edge technological advancement. For instance, trained data is necessary for an AI-based system to function. The majority of individuals might not be aware that labelling the training data takes a considerable amount of human labour to generate a decent dataset. Additionally, to attain our security-based technological development for the country,

there must be amply huge and complete datasets that may be used for training.

2. **Inadequate Talent with the Necessary Skill Set for AI Work:** One of the biggest roadblocks to the country's ambitions of becoming a global power is dearth of expertise in the field of AI, which prevents it from setting its sights on short-term gains. In the fiercely competitive globe, there is a never-ending battle for AI talent. If the talents and growing investments in AI and ML are not put to good use, the country runs the risk of experiencing financial losses and missing out on possibilities [19–21].

   In terms of national security, all decisions on the response are made by people. It is possible that Command and Control will one day be forced by competition to hand off some choices to AI platforms in specific situations, but it is most likely that autonomous deadly systems, for example, will be used without "placing people in the loop."

3. **Limited Availability of Key Infrastructure Regarding Computing Power:** The ability of a machine to execute cognitive tasks akin to those performed by human minds (such as perceiving, thinking, learning, and problem-solving) is the standard definition of AI. However, AI also has a number of other skills that make it useful for solving business challenges. AI requires a lot of data and complex algorithms, so it is crucial to have reliable technology and open data banks domestically. To process and transport this time-sensitive data from the sensors and data banks to the command centre's servers, India requires a network that is both safe and fast. The overall computing power is explained in Figure 13.1.

4. **Development of Ethical Standards:** AI ethics are a set of guidelines that provide direction for the development and use of AI. There are numerous cognitive biases that affect people and may be noticed in our actions and, consequently, in our data. Recency and confirmation bias are just two examples. The use of AI in defence would raise many ethical concerns, such as who would be held accountable if the AI did not perform as planned. How might the existing military procedure be altered to include AI? Prior to using AI for national security, these moral standards must be established.

5. **Enhanced Susceptibility to Cyber Attacks:** As AI systems proliferate, there will be more "hackable things," kinetic energy systems like moving cars, which could lead to tragic repercussions for exploitative activities. Threat actors acquired the personal information, banking information, and Social Security numbers of about 1.5 million people. The healthcare industry is particularly vulnerable to cyber assaults because its systems store so much Personal Identifiable Information.

6. **Theft Vulnerability:** Because they are virtually totally software-based, AI systems are particularly susceptible to theft. Data value has

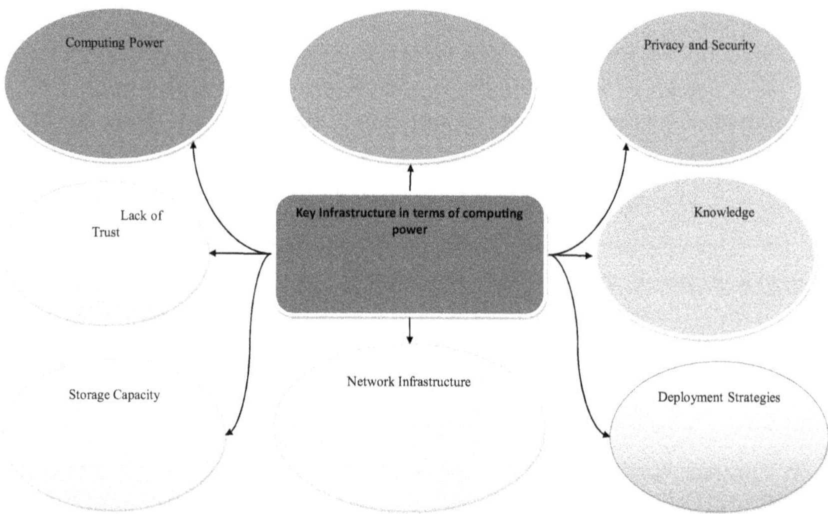

Figure 13.1 Key Infrastructure in Terms of Computing Power.

expanded with time, and more and more cyber security experts, such as analysts and engineers, are being hired to maintain business operations for the benefit of the country. The effectiveness of security measures is increased, and the damage caused by successful attacks is decreased with effective vulnerability management. As a result, the country needs a well-established vulnerability management system. When AI is used to do a detailed analysis, it can sometimes give a real-time picture of a threat to national security. However, deciding what to do about it will require a human to use his or her judgement and take into account things that a "machine" was not designed to handle.

7. **Limited Private Sector Role in Defence:** First and foremost, there is a strong public sector, including DRDO, Defence Public Sector Undertakings, and Ordnance Factories, which are established in terms of infrastructure, skilled labour, and manufacturing capabilities and tailored to suit the operational needs of defence services. Therefore, it is not unexpected that these businesses receive the majority of the large contracts for defence equipment. Any shortages in the public sector are filled by the private sector through outsourcing and subcontracting. The private sector serves only as a backup option in this scenario, with the potential only for spillover or leftover participation in defence programmes. In addition to accelerating the production process, the private sector is expected to introduce new technologies to make up for serious shortcomings. As a result, there

is a need to purchase technology from foreign original equipment manufacturers (OEMS) as the sector is not yet prepared to produce home-grown defence technologies. This is not a straightforward issue because there are broad trends in military-industrial technologies that the major countries view as crucial for the development of their strategic military capabilities and as significant from the standpoint of export controls, accelerating the production process to make up for serious flaws.

8. **Technology Cannot Be Completely Controlled:** Technology is a method of comprehending the world, not an instrument. Technology is not a human activity, yet it evolves in ways that are uncontrollable by humans to give the country security. The ability of the opponent to spread "misinformation" with the aid of AI would have risen; therefore it must be kept in mind when we assess the intelligence obtained from numerous sources is made easier and qualitatively superior as mentioned in Figure 13.2.

9. **The Geopolitical Environment Around India:** The Sino-Russian, Sino-Pakistani, and Russia-Pakistani relationships, together known as the "Trio," have grown in importance in the current geopolitical climate. To develop an AI strategy, India will need to evaluate these relationships.

   A system's deployment environment, design environment, and data reliance are all security weaknesses. Several of these dangers are specific to ML systems and may influence various phases of ML product development. Adversarial ML attacks aim to take advantage of security holes in an ML system, which could have serious practical consequences. By adding "perturbations" to the incoming data, adversarial attacks modify an AI system's output.

10. **Economic Constraints:** India's policies are constrained by economic factors because it is a developing country. As a result, it is unable

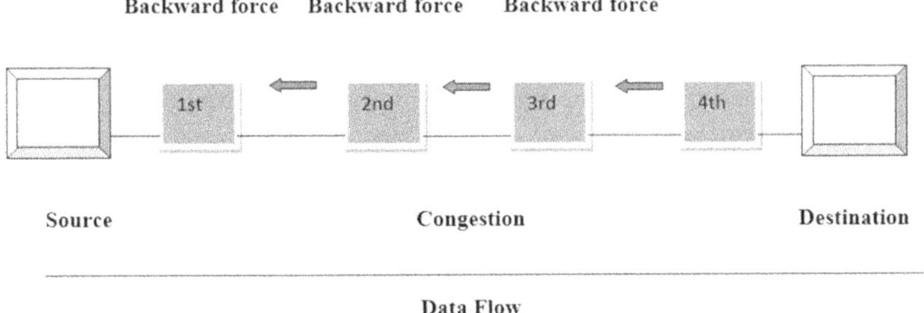

Figure 13.2 Data Flow.

to dedicate money to the necessary forces to provide security to the nation. However, AI is able to work quickly and efficiently in place of manpower. As a result, funding for AI platform research and development is required. It offers the chance to create a defence budget at the lowest possible cost.

11. **Privacy Risks:** When privacy protections are insufficient, technology may be used to completely capture and analyse a person's private life without that person's awareness or consent, seriously hurting that person's interests by disregarding their wishes for the use of data. One such issue is the use of facial recognition technologies for surveillance. Such injury may be emotional, such as when a person's private information is discussed in public, or it may be financial, such as when credit card information is stolen.

12. **Robustness and Bias:** Ensuring the robustness and reliability of AI systems is essential for national security. AI algorithms should be tested rigorously to identify vulnerabilities, biases, or unintended consequences. Bias in training data or algorithmic decision-making can have detrimental effects on national security efforts, leading to unfair targeting, discrimination, or unintended escalation of conflicts.

13. **Intelligence Sharing and Security Cooperation:** The readiness of each side to share the pertinent technologies is now being put to the test in relation to intelligence sharing and security cooperation among friendly nations. Because of the substantial convergence on national security problems that had already been reached with democracies like the United States, Japan, and Israel, India is fortunately in a good position to command the trust of its friends overseas.

However, there are limited areas where improvement can be made. They are as follows:

- To demonstrate the increased Indian patrol presence in the area, which is crucial for demonstrating supremacy over the area. To enhance patrol frequency, it is crucial to acquire remotely operated and autonomous ground and aerial vehicles. Working on making tools like logistics, predictive analysis, and image and video processing that can be used to help secure the borders remotely.

- Creating a strategy to investigate and address border security issues with the incorporation of cutting-edge technology like AI.

- The force structure of India is in a state of transformation. It has made the decision to create five theatre commands: a fully functional unified air defence command, a northern command, a western command, a peninsular command, and a maritime command. The logistical management must be adaptable enough to satisfy the needs of a unified command structure for the three armed forces wing commanders. AI will thereby improve logistical management synergy in such situations [22].

- Operation Parakram, which was started following the terrorist attack on Parliament in December 2001, served as inspiration for the Cold Start. It indicates that Indian armies are advancing quickly and forcefully into Pakistan. Such attacks will not be too extensive so as to provide Pakistan with no justification for starting a massive counterattack. The Cold Start concept emphasizes the importance of surprise. The mobilization of troops is a time-consuming process for traditional offences. At that point, the opponent nation not only has time to prepare for retaliation but also has time to set up diplomatic channels to prevent an assault from India.
- Strong AI algorithms should be developed with an eye towards the future needs of border and national security.

13.9 CONCLUSION

The development of AI will gradually grow, from the point of view of national security, the threats, the problems, and the opportunities. The military application of AI has the potential to be revolutionary as it can be used as a weaponization tool to automate weapon systems and improve cyber warfare. Many AI-enabled systems are now in the hands of state actors as well as non-state players as a consequence of the fact that they can be put to dual uses. This is a cause for concern for maintaining strategic stability and deterrence as a result of the fact that it can be used in both these applications. There will be major obstacles in the way of building a healthy AI ecosystem, including AI governance, ethics, data bias concerns, and legislation.

For India to be able to develop an environment that is supportive of AI, investments would need to be made in key infrastructure, the private sector innovation ecosystem would need to be harnessed, and India would need to capitalize on the advancements made by leading AI nations. The use of education, legislation, and regulation, as well as the development of human resources, is crucial for fostering trust in AI while simultaneously acknowledging the difficulties and risks connected with its use. In-house research and development, as well as international and regional collaboration centred on the use of AI, will be essential to the process of adding value to our military systems. These could include collaborative technological development, technology exchange, and participation in global policymaking and standardization.

REFERENCES

1. Christopher Gorman, "Recent Developments in AI and National Security: What You Need to Know," Lawfare, 3 March 2022.
2. "Deploying AI in Defense Organizations: The Value, Trends, and Opportunities," Research Brief, IBM Centre for Business of Government, 2021.
3. Ryan Fedasiuk, Jennifer Melot, and Ben Murphy, "How the Chinese Military is Adopting Artificial Intelligence," Centre for Security and Emerging Technology, October 2021.
4. "National Strategy for Artificial Intelligence #AIFORALL," NITI Aayog, 2018.
5. "Raksha Mantri Launches 75 Artificial Intelligence Products/Technologies During First-ever 'AI in Defence' Symposium & Exhibition in New Delhi," Press Information Bureau, Ministry of Defence, Government of India, 11 July 2022.
6. Greg Allen and Taniel Chan, "Artificial Intelligence and National Security," Belfer Center for Science and International Affairs, Harvard Kennedy School, 2017.
7. Tim Dutton, "An Overview of National AI Strategies," Politics+AI, MEDIUM, 29 January 2018.
8. "Task Force for Implementation of AI," Press Information Bureau, Ministry of Defence, Government of India, 28 March 2022.
9. Ajay Banerjee, "140 Artificial Intelligence-based Systems along Border to Keep Watch on China, Pak," The Tribune, 7 August 2022.
10. Snehesh Alex Philip, "Army to Conduct Trials of AI Enabled Unmanned All-terrain Vehicles in Ladakh Next Month," The Print, 11 July 2022.
11. "US Body on AI Calls for Creating India-US Strategic Tech Alliance," Financial Express, 14 October 2020.
12. Saroj Bishoyi, "India-US Forging Tech Alliance Since Long. Now Use 2+2 Dialogue to Push it Further," The Print, 11 April 2022.
13. "India, Japan Discuss Cooperation in 5G Technology," The Hindu, 30 June 2022.
14. T. Radhakrishna, "Quantum Computing: India and Finland Agree to Set Up Virtual Centre of Excellence for Technical Cooperation," The Economic Times, 21 April 2022.
15. AI and National Security: Major Power Perspectives and Challenges. www.idsa.in/issuebrief/ai-and-national-security-ssharma-120922 (Accessed 10 March 2024).
16. The Defenders: Artificial Intelligence. www.drishtiias.com/loksabha-rajyasabha-discussions/the-defenders-artificial-intelligence (Accessed 10 February 2024).
17. Artificial Intelligence Policy in India: A Framework for Engaging the Limits of Data-Driven Decision-Making, Vidushi Marda. https://doi.org/10.1098/rsta.2018.0087 (Accessed 15 October 2018).
18. Shivaram Kalyanakrishnan, Department of Computer Science and Engineering Indian Institute of Technology Bombay Opportunities and Challenges for Artificial Intelligence in India.

19. "AI Strategy of India: Policy Framework, Adoption Challenges and Actions for Government." www.researchgate.net/scientific-contributions/Sheshadri-Chatterjee-2175911236.
20. Shashi Shekhar Vempati, India and the Artificial Intelligence Revolution, Carnegie India.
21. Kelley M. Sayler, Artificial Intelligence and National Security, Congressional Research Service.
22. Internal Security. https://pwonlyias.com/upsc-notes/internal-security/ (Accessed 01 January 2024).

Chapter 14

Quantum computing and new dimensions in network security

Mohit Joshi, Alka, and Manoj Kumar Mishra

14.1 INTRODUCTION

Quantum computing is the branch of computation that harnesses the ability of the quantum mechanical world which is essentially probabilistic and reversible in nature [1]. It is the intersection of properties of quantum mechanics with the application in computational science. The groundwork of the subject was laid by Paul Benioff in his work on the Quantum Mechanical Model of Turing Machine [2]. Richard Feynman in his keynote presentation titled "Simulating Physics Using Computers" [3] presented the idea of doing physical computation on a quantum device instead of a classical one. Peter Shor, following the ideas of Bernstein-Vazirani [4] and Simon's algorithm [5], provided a breakthrough to the field by solving the problem of factorization [6] in polynomial time and, in turn, challenged the security of modern classical cryptography. Grover's search algorithm [7] was another such example that showed the power of a fault-tolerant quantum computer with large quantum bits.

Quantum mechanics, in a nutshell, is the description of elementary particles in a fundamentally statistical way. This statistical picture presents to us a reality where an object can be present at more than one place (superposition), can pass through barriers without breaking them (quantum tunneling) and communication between particles can happen despite wide separation (entanglement) [1]. Such particles essentially behave as both wave and particle-like in nature. In a wave-like nature, objects act as a continuous entity and hence can be used for analog computing. In particle-like nature, objects act as discrete entities and we can do digital computation on such entities. The particle-like nature is exhibited only while interacting with matter. This helps us to base a computation device on a quantum system that can process information in the exponential space of continuous variables and then collapse the result into a discrete basis of measurement [8]. Hence, in the simplest terms, quantum computation tries to achieve the computation capability of analog devices with the classical techniques of error correction by mapping the unit of computation on a quantum system of choice [9]. It

DOI: 10.1201/9781003514312-14

is now proven that there are many approaches to doing universal quantum computation. Some of these approaches include an adiabatic model of computation [10], topological modeling, measurement-based model of computation [11], and a circuit model of computation [12]. Among all the models of computation, circuit-based models enjoy heavy support from both the hardware and software side. This chapter primarily discusses the elements in the circuit model of computation. The structure of the chapter is as follows: Section 14.2 of this chapter discusses the basics of quantum computing. Within its subsections, the concept of a qubit, the principle of superposition, the measurement postulate, the entanglement theorem, and the impossibility theorems of quantum computing have been examined. The significance of unitary transformation in evolving the quantum algorithm has also been examined. Section 14.3 then introduces the gate transformation in the circuit model of computation. Section 14.4 establishes the need for and importance of quantum computation in the cryptographic domain and tabulates some of the famous cryptographic algorithms.

14.2 QUANTUM COMPUTING

Quantum computing uses the quantum mechanical principles of a physical system to perform computation. Although the quantum transformation of classical solutions is most suited for scenarios with low input/output space and high computational space, any classical algorithm can be transformed in the quantum world. Hence, it can be regarded as the superset of classical computation [8]. Some examples of physical systems with quantum mechanical properties are nuclear magnetic resonance, nitrogen-vacancy center in a diamond atom, semiconductor quantum transistor, super-conducting qubits, electron spin, photon polarization, and trapped ions with ytterbium atoms [13].

These quantum systems exhibit a wave function that is then evolved into an output state using sets of unitary operations on them. The collection of these sets of unitary operations being applied to a wave function creates quantum algorithms. After the evolution, the result is then measured which collapses the wave function leaving us with probabilities of output states.

The subsections within define the elements of quantum computing [14].

14.2.1 Qubits

The smallest unit of information in a quantum system is called a quantum bit, or simply qubit. These qubits are mathematically represented in n-dimensional Hilbert space where the state of a qubit is expressed as a vector in this Hilbert space. The space here can be understood as all the possible outcomes in a given system of qubits and the state is the amplitude of the possibility of finding a qubit in a particular state. Table 14.1 gives the space

Table 14.1 Space and State Vectors in One-Qubit System

Space	State of 0	State of 1
$\begin{pmatrix} a_0 \\ a_1 \end{pmatrix}$	$\begin{pmatrix} 1 \\ 0 \end{pmatrix}$	$\begin{pmatrix} 0 \\ 1 \end{pmatrix}$

Table 14.2 Space and State Vectors in Two-Qubit System

Space	State of 00	State of 01	State of 10	State of 11
$\begin{pmatrix} a_{00} \\ a_{01} \\ a_{10} \\ a_{11} \end{pmatrix}$	$\begin{pmatrix} 1 \\ 0 \\ 0 \\ 0 \end{pmatrix}$	$\begin{pmatrix} 0 \\ 1 \\ 0 \\ 0 \end{pmatrix}$	$\begin{pmatrix} 0 \\ 0 \\ 1 \\ 0 \end{pmatrix}$	$\begin{pmatrix} 0 \\ 0 \\ 0 \\ 1 \end{pmatrix}$

and state representation of a one-qubit system. Table 14.2 represents the space and state representation of a two-qubit system.

14.2.2 Principle of superpositions

The principle of superposition states that any linear combination of two or more state vectors is another valid state in Hilbert space. It distinguishes quantum computation from its counterpart by validating any intermediate state between zero and one and hence opens an avenue for exponential computation space. Equation (14.1) gives the mathematical expression of the superposition where $\alpha, \beta \in \mathbb{C}$ represent the amplitude of the respective basis component and should follow Born's rule as $(\alpha^2 + \beta^2) = 1$:

$$|\psi\rangle = \begin{pmatrix} \alpha \\ \beta \end{pmatrix} = \alpha|0\rangle + \beta|1\rangle \tag{14.1}$$

Table 14.3 details some of the famous superposition states with their state vector and value representation in terms of zero and one qubit. For visual understanding, the superposition states are represented as vectors in the Bloch sphere, which is a three-dimensional method of qubit representation, parameterized by γ, δ, θ. The wave function is then represented in equation

2 where $\alpha = e^{i\gamma}\cos\left(\dfrac{\theta}{2}\right)$, $\beta = e^{i\delta}\sin\left(\dfrac{\theta}{2}\right)$, $\varphi = \delta - \gamma$.

Table 14.3 Some of the Famous Superposition States

State Symbol	State Vector	State Value
$\|+\rangle$	$\begin{pmatrix} \dfrac{1}{\sqrt{2}} \\ \dfrac{1}{\sqrt{2}} \end{pmatrix}$	$\dfrac{\|0\rangle + \|1\rangle}{\sqrt{2}}$
$\|-\rangle$	$\begin{pmatrix} \dfrac{1}{\sqrt{2}} \\ -\dfrac{1}{\sqrt{2}} \end{pmatrix}$	$\dfrac{\|0\rangle - \|1\rangle}{\sqrt{2}}$
$i\|+\rangle$	$\begin{pmatrix} \dfrac{1}{\sqrt{2}} \\ i\dfrac{1}{\sqrt{2}} \end{pmatrix}$	$\dfrac{\|0\rangle + i\|1\rangle}{\sqrt{2}}$
$i\|-\rangle$	$\begin{pmatrix} \dfrac{1}{\sqrt{2}} \\ -i\dfrac{1}{\sqrt{2}} \end{pmatrix}$	$\dfrac{\|0\rangle - i\|1\rangle}{\sqrt{2}}$

$$|\psi\rangle = \cos\left(\frac{\theta}{2}\right) + e^{i\varphi}\sin\left(\frac{\theta}{2}\right)|1\rangle \tag{14.2}$$

Figure14.1 depicts the diagrammatic representation of the Bloch sphere with the z-axis chosen as the computational basis. Table 14.4 shows the symbolic form for the coordinate on each axis.

14.2.3 Measurement postulate

The act of measurement in quantum computing is essentially a projection operator which collapses the wave function along the orthogonal subspace of the state. It is the only irreversible operator present in quantum computing. The probability of finding the qubit $|0\rangle$ in a certain state $|\psi\rangle = \alpha|0\rangle + \beta|1\rangle$ is given below:

$$prob(|0\rangle) = |\langle 0|\psi\rangle|^2 = \left(\alpha\langle 0|0\rangle + \beta\langle 0|0\rangle\right)^2 = |\alpha|^2 \tag{14.3}$$

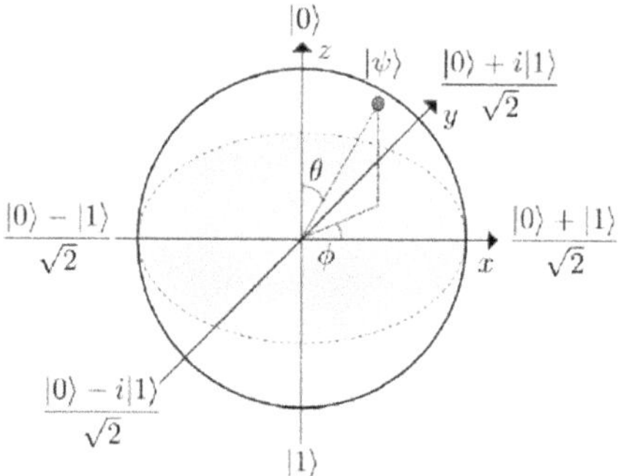

Figure 14.1 Bloch Sphere with Labeled Coordinate Axes [15].

Table 14.4 Coordinates on the Corresponding Axes with Their State Representatives and State Values in Terms of Zero and One Qubit

Axis	State Symbol	State Vector	State Value
Z axis	$|0⟩$	$\begin{pmatrix} 1 \\ 0 \end{pmatrix}$	$|0⟩$
	$|1⟩$	$\begin{pmatrix} 0 \\ 1 \end{pmatrix}$	$|0⟩$
X axis	$|+⟩$	$\begin{pmatrix} \frac{1}{\sqrt{2}} \\ \frac{1}{\sqrt{2}} \end{pmatrix}$	$\frac{|0⟩+|1⟩}{\sqrt{2}}$
	$|-⟩$	$\begin{pmatrix} \frac{1}{\sqrt{2}} \\ -\frac{1}{\sqrt{2}} \end{pmatrix}$	$\frac{|0⟩-|1⟩}{\sqrt{2}}$
Y axis	$i\,|+⟩$	$\begin{pmatrix} \frac{1}{\sqrt{2}} \\ i\frac{1}{\sqrt{2}} \end{pmatrix}$	$\frac{|0⟩+i\,|1⟩}{\sqrt{2}}$
	$i\,|-⟩$	$\begin{pmatrix} \frac{1}{\sqrt{2}} \\ -i\frac{1}{\sqrt{2}} \end{pmatrix}$	$\frac{|0⟩-i\,|1⟩}{\sqrt{2}}$

Table 14.5 Four Quantum Entanglement States in the Two-Qubit System, Commonly Known as Bell States

State Symbol	State Representation	State Symbol	State Representation
$\lvert\phi_+\rangle$	$\frac{1}{\sqrt{2}}(\lvert 00\rangle + \lvert 11\rangle)$	$\lvert\psi_+\rangle$	$\frac{1}{\sqrt{2}}(\lvert 01\rangle + \lvert 10\rangle)$
$\lvert\phi_-\rangle$	$\frac{1}{\sqrt{2}}(\lvert 00\rangle - \lvert 11\rangle)$	$\lvert\psi_-\rangle$	$\frac{1}{\sqrt{2}}(\lvert 01\rangle - \lvert 10\rangle)$

14.2.4 Entanglement theorem

Entanglement is a special type of superposition state where the given state cannot be represented as a product of other states, i.e., $\lvert\psi\rangle \neq \lvert\psi_1\rangle\lvert\psi_2\rangle$. In simpler terms, it is a special type of collective state where the measurement of one state inevitably disturbs the state of another unmeasured one. It was upon discovery called "spooky action a distance" to signify the absurdity in the concept. But since then, it has been a well-established phenomenon and represents a system that has no definite state of its own. Table 14.5 shows the maximally entangled states in a two-qubit system.

14.2.5 Impossibility theorems of quantum computing

Impossibility or no-go theorems emerge from the underlying principles of quantum mechanics and establish constraints on quantum computation.

14.2.5.1 No exponential information theorem

Quantum computers are exponential processing machines where n qubits will give us computational space equal to $2n$ bits, but we can only extract n bits of information after computation. This is a direct consequence of the measurement postulate.

14.2.5.2 No cloning theorem

A unitary transformation to copy an arbitrary (unknown) qubit state cannot exist, i.e., there exists no gate U such that $U\lvert 0\psi\rangle = \lvert\psi\psi\rangle$. This results in an important implication in security algorithms where an unknown state of the qubit cannot be copied by an eavesdropper without disturbing the superposition of the system.

14.2.5.3 No deletion theorem

It states that we cannot apply any transformation that results in loss of information i.e., $V|x,0,0\rangle = |x, f(x), g(x)\rangle$. The consequence of this condition is that every transformation in quantum computing should be reversible in nature.

14.3 GATES IN CIRCUIT MODEL OF COMPUTATION

The gate model or circuit model of computation is the most widely used programming paradigm of quantum computation. It is similar to gates in classical Boolean circuits where input is processed using gates of different logic. Gates are the method of implementing unitary transformation on qubit wave function in this paradigm of quantum computation. Gate can be mathematically defined as a matrix operation applied to the wave function to transform it into the desired qubit state. For instance, a quantum analog of the classical NOT gate can be symbolized as an X gate which transforms a zero qubit into one and vice versa. Table 14.6 provides the list of some common gates with their corresponding matrix representation and the action they invoke on the affected qubit [16].

14.4 NEED FOR QUANTUM COMPUTING IN SECURITY

The CIA triad (confidentiality, integrity, and availability) of data has always been under new threats, and mitigation techniques have also evolved over time, but today we are standing on the pavement of a new paradigm of computation which will have a twofold effect on the security of software systems in production and in-transit. The effect can be seen as discussed in subsections.

14.4.1 Quantum as a problem to classical cryptography

Classical cryptography mainly comprises symmetric and asymmetric encryption techniques. Techniques like Rivest-Shamir-Adleman, Data Encryption Standard, Advanced Encryption Standard, Diffie-Hellman, and Elliptic Curve Cryptography take advantage of the computational complexity of one-way operations grounded in mathematical theories. The presence of a robust fault-tolerant quantum system will challenge the dominance of cryptographic systems that work on the assumption of one-way computational complexities. Algorithms like Shor's and Grover's have certainly proven to be a threat to asymmetric and symmetric encryption, respectively. There are still some classes of algorithms called quantum-resistant algorithms like lattice-based algorithms or fully homomorphic algorithms that are away from the reach of quantum computers [17].

Table 14.6 Common Quantum Gates with Their Matrix Representation, Truth Table, and Symbol in the Circuit

Gate Name	Matrix	Truth Table	Symbol
X gate	$\begin{pmatrix} 0 & 1 \\ 1 & 0 \end{pmatrix}$	$X\|0\rangle = \|1\rangle$ $X\|1\rangle = \|0\rangle$	X
Y gate	$\begin{pmatrix} 0 & -i \\ i & 0 \end{pmatrix}$	$Y\|0\rangle = i\|1\rangle$ $Y\|1\rangle = -i\|0\rangle$	Y
Z gate	$\begin{pmatrix} 1 & 0 \\ 0 & -1 \end{pmatrix}$	$Z\|0\rangle = \|0\rangle$ $Z\|1\rangle = -\|1\rangle$	Z
H gate	$\frac{1}{\sqrt{2}}\begin{pmatrix} 1 & 1 \\ 1 & -1 \end{pmatrix}$	$X\|0\rangle = \|+\rangle$ $X\|1\rangle = \|-\rangle$	H
CX gate	$\begin{pmatrix} 1 & 0 & 0 & 0 \\ 0 & 1 & 0 & 0 \\ 0 & 0 & 0 & 1 \\ 0 & 0 & 1 & 0 \end{pmatrix}$	$CX\|00\rangle = \|00\rangle$ $CX\|01\rangle = \|01\rangle$ $CX\|10\rangle = \|11\rangle$ $CX\|11\rangle = \|10\rangle$	
CCX gate	$\begin{pmatrix} 1 & 0 & 0 & 0 & 0 & 0 & 0 & 0 \\ 0 & 1 & 0 & 0 & 0 & 0 & 0 & 0 \\ 0 & 0 & 1 & 0 & 0 & 0 & 0 & 0 \\ 0 & 0 & 0 & 1 & 0 & 0 & 0 & 0 \\ 0 & 0 & 0 & 0 & 1 & 0 & 0 & 0 \\ 0 & 0 & 0 & 0 & 0 & 1 & 0 & 0 \\ 0 & 0 & 0 & 0 & 0 & 0 & 0 & 1 \\ 0 & 0 & 0 & 0 & 0 & 0 & 1 & 0 \end{pmatrix}$	$CCX\|000\rangle = \|000\rangle$ $CCX\|001\rangle = \|001\rangle$ $CCX\|010\rangle = \|010\rangle$ $CCX\|011\rangle = \|011\rangle$ $CCX\|100\rangle = \|100\rangle$ $CCX\|101\rangle = \|101\rangle$ $CCX\|110\rangle = \|111\rangle$ $CCX\|111\rangle = \|110\rangle$	

14.4.2 Quantum as an opportunity for new cryptography systems

Techniques like One Time Pad are seen as unconditionally secure classical algorithms whose security is dependent on the production of randomness, which is shown to fail due to the application of pseudo-random numbers. In the NISQ era of computation, we are now able to generate actual randomness in the application of algorithms like Quantum One Time Pad, and the use of quantum channels gives us the advantage of detecting eavesdropping using the no-cloning theorem, which no classical counterpart is able to achieve. Quantum seals can also be used to restrict eavesdroppers at

Table 14.7 List of Quantum Cryptographic Algorithms

S. No.	Algorithm	Definition
1.	BB84 [21]	It is the first quantum cryptographic algorithm developed by Charles Bennett and Gillie Brassard using two orthogonal basis states (1984)
2.	BB92 [22]	Improved version of BB84 by Charles Bennett which can use non-orthogonal state (1992)
3.	E91 [23]	Cryptographic key distribution technique using entanglement of photon pairs (1991)
4.	SARG04 [24]	Developed in 2004 using four states instead of two as used in BB84
5.	S09 [25]	A multi-node key distribution protocol developed in 2009
6.	SSP [26]	Six state QKD protocol is similar to BB84, but it uses six states instead of two orthogonal states
7.	Kak's protocol [27]	Kak's three-stage protocol is an encryption algorithm that transmits the data itself in three stages instead of a key distribution
8.	QOTP [28]	Quantum One Time Pad is the extension of the classical one-time pad with help of quantum computers to generate randomness
9.	Quantum Teleportation [29]	It is a teleportation protocol using two classical bits to transmit one quantum data
10.	Superdense Coding [30]	It is the inverse of quantum teleportation using one quantum state to transmit two classical bits of data

the physical layer. The emergence of quantum algorithms has also evolved many quantum-inspired features enhanced in the existing algorithms like Quantum inspired TEA (tiny encryption algorithm) [18] and cascading discrete-time quantum walk [19,20].

In an overall sense, security in quantum computation comes from Heisenberg's principle of uncertainty giving rise to superposition and inviolability of the no-cloning theorem to replicate a non-orthogonal state.

Cryptographic applications of quantum computing have received the earliest attention in the field of quantum algorithms and have given some of the most sensational stories in the likes of Shor's and Grover's algorithms. Table 14.7 provides an overview of some cryptographic work in quantum computation.

14.5 IMPORTANCE OF QUANTUM COMPUTING IN SECURITY

The effect of quantum computers is not the same as the classical computer because quantum computers cannot be thought of as mere replacements for classical machines. Quantum machines have merit in a specific set of highly parallel optimization problems and hence are to be used so. The early adaptation of quantum computation has seen a shift toward cloud platforms with many companies and organizations providing QaaS (Quantum as a Service). Dedicated companies like Quantique, MagiQ, Quintessence Labs, and major players like IBM, Google, Microsoft, and D-wave are swaying the market toward cloud access. We have seen the evolution of application-specific ICs for quantum hardware in the NISQ era [31] and a plethora of open-source software frameworks. Moreover, it is much more practical to think of distributed quantum computing with different clusters of resources spread across. So, the objective has shifted from protecting data-in-transit to the complete protection of procedure, input, and output in such a distributed or decentralized environment. Blind quantum computation (BQC) is one of the first algorithms put forth by A.M. Childs [32] in 2005 that enabled the protection of data in remote locations. Universal blind quantum computation (BFK) [33] protocol is an extension of the idea put forth by Broadbent, Fitzsimons, and Kashefi in 2009 which works on the assumptions of a measurement-based model of quantum computation [34] and achieves universality using a brickwork state [14]. A review of the classification of the field of blind quantum computation and recent trends in the field can be found in a previous study [35].

14.6 CONCLUSION

Quantum computing has emerged as a field of interest for many disciplines in the ranks of physics, chemistry, finance, and computer science. Recent years have seen advancement in both the theoretical framework of functional quantum computers and the practical applicability of hardware systems in software advancements. With the increase in commercial hardware and simulators, exciting results are being published more frequently. Though we are far from fault-tolerant quantum computers, we certainly have entered a new generation of systems called Noisy Intermediate State Quantum Computers. These devices are helping us to solve short-term quantum algorithms and providing proof-of-concept for more powerful quantum devices. The techniques of quantum computing differ from classical computing. This chapter introduced the concept of quantum computing as an intuitive guide to computer scientists. It gives a familiar tone to the theory of superposition, and entanglement. It also gives a gentle introduction to circuit models of computation. It emphasizes the need for and importance of quantum technology in modern cryptography and justifies its place by giving a summary of new techniques of quantum security.

Acknowledgements This research is supported by a seed grant under IoE, BHU [grant no. R/Dev/D/IoE/SEED GRANT/2020-21/Scheme No. 6031]

REFERENCES

1. D. C. Marinescu, "The promise of quantum computing and quantum information theory—quantum parallelism (The Abstract of a Tutorial)," IEEE International Parallel and Distributed Processing Symposium, 2005.
2. P. Benioff, "Quantum mechanical Hamiltonian models of Turing machines," Journal of Statistical Physics, vol. 29, no. 3, pp. 515–546, 1982. Springer.
3. R. P. Feynman, "Simulating physics with computers," in Feynman and Computation, pp. 133–153, CRC Press, 1998.
4. J. Du, M. Shi, X. Zhou, Y. Fan, B. Ye, R. Han, and J. Wu, "Implementation of a quantum algorithm to solve the Bernstein-Vazirani parity problem without entanglement on an ensemble quantum computer," Physical Review A, vol. 64, p. 042306, September 2001.
5. D. R. Simon, "On the power of quantum computation," SIAM Journal on Computing, vol. 26, no. 5, pp. 1474–1483, 1997. SIAM.
6. P. Shor, "Algorithms for quantum computation: discrete logarithms and factoring," in Proceedings 35th Annual Symposium on Foundations of Computer Science (Santa Fe, NM, USA), pp. 124–134, IEEE Computer Society Press, 1994.
7. L. K. Grover, "A fast quantum mechanical algorithm for database search," in Proceedings of the Twenty-eighth Annual ACM Symposium on Theory of Computing—STOC '96 (Philadelphia, Pennsylvania, United States), pp. 212–219, ACM Press, 1996.
8. M. A. Nielsen and I. L. Chuang, Quantum Computation and Quantum Information, Cambridge; New York: Cambridge University Press, 10th anniversary ed., 2010.
9. A. Chatterjee, K. Phalak, and S. Ghosh, "Quantum error correction for dummies," April 2023. arXiv:2304.08678 [quant-ph].
10. T. Albash and D. A. Lidar, "Adiabatic quantum computation," Reviews of Modern Physics, vol. 90, p. 015002, January 2018.
11. R. Raussendorf, D. E. Browne, and H. J. Briegel, "Measurement-based quantum computation on cluster states," Physical Review A, vol. 68, no. 2, 022312, 2003.
12. P. Nimbe, B. A. Weyori, and A. F. Adekoya, "Models in quantum computing: a systematic review," Quantum Information Processing, vol. 20, p. 80, February 2021.
13. J. D. Hidary and J. D. Hidary, Quantum Computing: An Applied Approach, vol. 1., Cham: Springer, 2021. https://doi.org/10.1007/978-3-030-83274-2
14. M. M. Savchuk and A. V. Fesenko, "Quantum Computing: Survey and Analysis," Cybernetics and Systems Analysis, vol. 55, pp. 10–21, January 2019.
15. A. F. Kockum, "Quantum Optics with Artificial Atoms," PhD thesis, Applied Quantum Physics Laboratory, Department of Microtechnology and Nanoscience, Chalmers University of Technology, Göteborg, Sweden, 2014.
16. A. W. Cross, L. S. Bishop, J. A. Smolin, and J. M. Gambetta, "Open Quantum Assembly Language," 2017. https://doi.org/10.48550/arXiv.1707.03429

17. D. J. Bernstein and T. Lange, "Post-quantum cryptography," Nature, vol. 549, no. 7671, pp. 188–194, 2017. Nature Publishing Group.

18. W. Hu, "Cryptanalysis of TEA Using Quantum-Inspired Genetic Algorithms," JSEA, vol. 03, no. 1, pp. 50–57, 2010.

19. A. A. Abd El-Latif, B. Abd-El-Atty, M. Amin, and A. M. Iliyasu, "Quantum-inspired cascaded discrete-time quantum walks with induced chaotic dynamics and cryptographic applications," Scientific Reports, vol. 10, p. 1930, February 2020.

20. A. Kumar and S. Garhwal, "State-of-the-Art Survey of Quantum Cryptography," Archives of Computational Methods in Engineering, vol. 28, pp. 3831–3868, August 2021.

21. H. Bennett and G. Brassard, "Quantum cryptography: Public key distribution and coin tossing," Theoretical Computer Science, vol. 560, pp. 7–11, December 2014.

22. H. Bennett, "Quantum cryptography using any two nonorthogonal states," Physical Review Letters, vol. 68, pp. 3121–3124, May 1992.

23. A. K. Ekert, "Quantum cryptography based on Bell's theorem," Physical Review Letters, vol. 67, pp. 661–663, August 1991.

24. V. Scarani, A. Acın, G. Ribordy, and N. Gisin, "Quantum cryptography protocols robust against photon number splitting attacks for weak laser pulse implementations," Physical Review Letters, vol. 92, p. 057901, February 2004.

25. A. H. Serna, "Quantum key distribution protocol with private–public key," May 2012. arXiv:0908.2146 [quant-ph].

26. D. Bruß, "Optimal Eavesdropping in Quantum Cryptography with Six States," Physical Review Letters, vol. 81, pp. 3018–3021, October 1998.

27. S. Kak, "A three-stage quantum cryptography protocol," Foundations of Physics Letters, vol. 19, pp. 293–296, June 2006.

28. B. Schumacher and M. D. Westmoreland, "Quantum mutual information and the one-time pad," Physical Review A, vol. 74, p. 042305, October 2006.

29. S. Pirandola, J. Eisert, C. Weedbrook, A. Furusawa, and S. L. Braunstein, "Advances in quantum teleportation," Nature Photon, vol. 9, pp. 641–652, October 2015.

30. X. S. Liu, G. L. Long, D. M. Tong, and F. Li, "General scheme for superdense coding between multiparties," Physical Review A, vol. 65, p. 022304, January 2002.

31. J. Preskill, "Quantum computing in the NISQ era and beyond," Quantum, vol. 2, p. 79, August 2018. arXiv:1801.00862 [cond-mat, physics:quant-ph].

32. A.M. Childs, "Secure assisted quantum computation," QIC, vol. 5, no. 6, 2001. arXiv:quant-ph/0111046.

33. A. Broadbent, J. Fitzsimons, and E. Kashefi, "Universal blind quantum computation," in 2009 50th Annual IEEE Symposium on Foundations of Computer Science (Atlanta, GA, USA), pp. 517–526, IEEE, October 2009.

34. J. F. Fitzsimons, "Private quantum computation: an introduction to blind quantum computing and related protocols," NPJ Quantum Information, vol. 3, p. 23, June 2017.
35. M. Joshi, S. Karthikeyan, and M.K. Mishra, "Recent trends and open challenges in blind quantum computation," in I. Woungang, S.K. Dhurandher, K.K. Pattanaik, A. Verma, and P. Verma (eds.), Advanced Network Technologies and Intelligent Computing. ANTIC 2022. Communications in Computer and Information Science, vol. 1798. Springer, Cham, 2023. https://doi.org/10.1007/978-3-031-28183-9_34

Cyber-security technique for profound analytics of big data using AI

Mohd Haroon, Afsaruddin Khan, Manish Madhava Tripathi, Shish Ahmad, and Kalamuddin Ahmad

15.1 INTRODUCTION

Cyber-security problems have increased in businesses and organizations due to the increasing frequency and severity of cyber-attacks. The analysis of big data can play a very significant role in the detection and prevention of cyber threats. Artificial intelligence (AI) has proved to be a powerful tool for optimizing cyber security measures, offering the ability to automate processes, detect patterns, and learn from data [1].

As modern devices are generating this high volume of data by using advanced methods, the traditional security mechanism of using firewalls and antivirus is insufficient as well as inefficient to protect against advanced cyber-attacks. Big data analytics provides a way to analyse and interpret in real time the vast amount of data and is helpful to detect and respond to the threats as they occur [2,3].

With the computing power of AI and deep learning, machine learning, the big data can be easily analysed and the pattern of certain features of the data may easily detected, by which the potential threats can be easily located. This approach can be used to learn from historical data, to diagnose new threats, enabling organizations to stay ahead of evolving cyber threats. Big data analysis is very helpful for the organization to identify the vulnerability in the system to implement more security measures [3,4].

Big data analytics offers a promising solution to the challenges faced by organizations in detecting and preventing cyber threats. By analysing large volumes of data in real time, organizations can improve their ability to detect and respond to cyber-attacks, ultimately enhancing their overall cyber security posture [4,5].

15.2 BIG DATA ANALYTICS

Big data analytics is the topic of current research trends as data are generated in this digital era from several sources. The nature of these data is complex in nature. Big data analysis can involve the use of advanced data processing

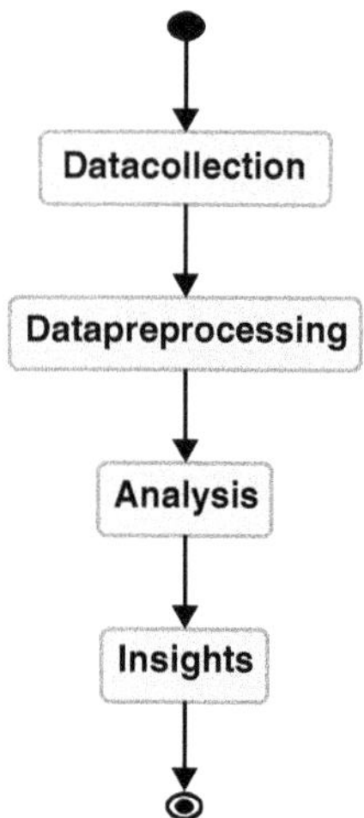

Figure 15.1 Process of Big Data Analytics.

algorithms and techniques, like data mining, data cleaning, machine learning, and Natural Language Processing (NLP), to extract meaningful insight from a huge amount of data [6–8].

Organizations can use big data analytics to gain valuable insights into customer behaviour, market trends, and business operations; this can also help us to make data-driven and immediate decision-making, which is depicted in Figure 15.1 [6].

Some of the key applications of big data analytics include:

Customer analytics: After the big data analysis customer behaviour and preferences can be analysed, enabling businesses to improve customer engagement and retention [8,9]. Operational analytics: Using big data analysis, business process can be monitored and optimized, thus helping organizations improve efficiency and reduce costs [8,9].

Fraud detection: Fraud can be easily detected and identified in financial transactions and other areas [10].

Predictive analytics: Big data analytics can predict future trends and outcomes so organizations can make proactive decisions and mitigate potential risks. However, big data analytics also come with some challenges, such as the need for advanced data processing techniques, privacy and security concerns, and the complexity of integrating data from multiple sources. Therefore, companies need to have a clear understanding of their data requirements and infrastructure to effectively use big data analytics [10].

15.2.1 Applications of deep learning in big data analytics

Deep learning can also be used in big data analysis as it can automatically learn multiple patterns and functions and extract relationships in large and complex data sets, which is shown in Figure 15.2. Following are some application areas for deep learning.

> **Image and video analysis:** Deep learning is widely used in image processing due to its ability to extract features and patterns. The above approaches are mainly used in the health industry, where a large amount of visual data needs to be analysed [11].
>
> **NLP:** NLP algorithms can be used to analyse large amounts of text data, generate meaningful insights, and identify various patterns. NLP can be used in sentiment analysis, text classification, and topic modelling [12,13].

In the field of recommender systems, deep learning can be used to predict products using association rules. It is used extensively by the Amazon and Flipkart commercial websites to recommend related products to customers. In agriculture, it is also useful to suggest crop types based on different climatic conditions and forecasts.

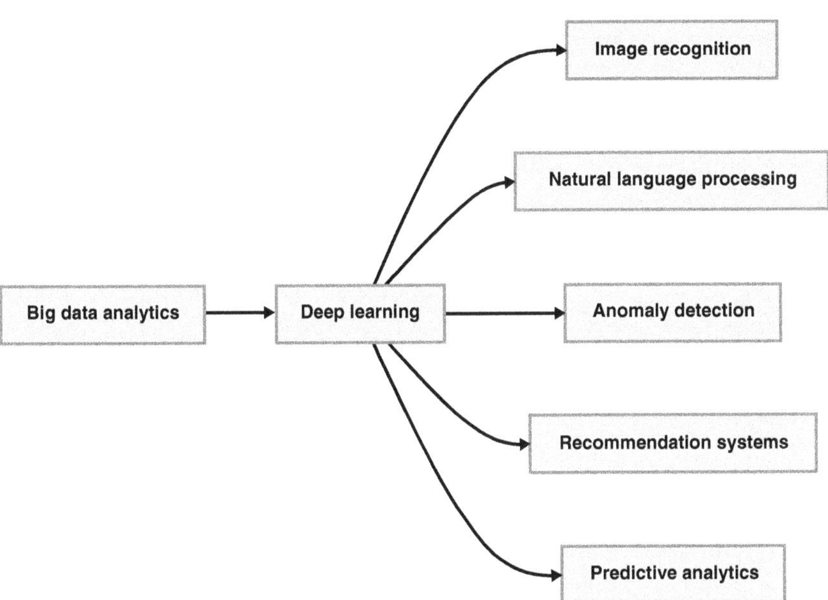

Figure 15.2 Deep Learning in Big Data Analytics.

Fraud detection: Deep learning algorithms can be used to analyse large amounts of variable data to identify patterns of suspicious behaviour. This allows businesses to more effectively detect and prevent fraud [12–14].

Predictive analytics: The feature of the predictive model given by deep learning can be used to forecast future trends and make outcome-based analyses of data. These approaches are very useful in the decision-making on data-driven and also gain competitive advantages.

Autonomous vehicles: Deep learning algorithms are used in autonomous vehicles to analyse sensor data and make decisions based on real-time input. This is critical for ensuring the safety and reliability of autonomous vehicles [14–16].

Deep learning has many applications in big data analytics and has the potential to revolutionize how organizations analyse and extract insights from large and complex data sets.

15.3 ARTIFICIAL INTELLIGENCE-BASED CYBER SECURITY APPROACH

In the highly technological age, AI techniques are very helpful for cyber security, which involves leveraging machine learning and other soft computing-based technologies to detect and prevent cyber threats. The machine learning approach includes advanced algorithms and techniques to analyse huge and large volumes of data, identify various patterns and different anomalies, and make a decision about the prediction for the potential of cyber threats shown in Figure 15.3 [17].

The AI-based system identifies the cyber threats automatically; this ability to identify the threats by the AI machine comes from pattern learning. After analysing the historical data the AI system can learn about the characteristics of different types of cyber-attacks, empowering them to categorize new and evolving threats in real time. This approach can also be used to automate certain aspects of cyber security, such as threat detection and response, reducing the workload on human security analysts [18].

Following are some AI-based approaches that can be used for cyber security purposes.

Machine learning: Machine learning is an AI-based algorithm, machine learning learns from data to forecast future events. Machine learning can detect patterns, features, and several kinds of anomalies in a large amount of data for the identification of potential threats. The mathematical model of the machine learning-based cyber technique and cyber security involved the use of statistical models and algorithms to identify and analyse the large amount of

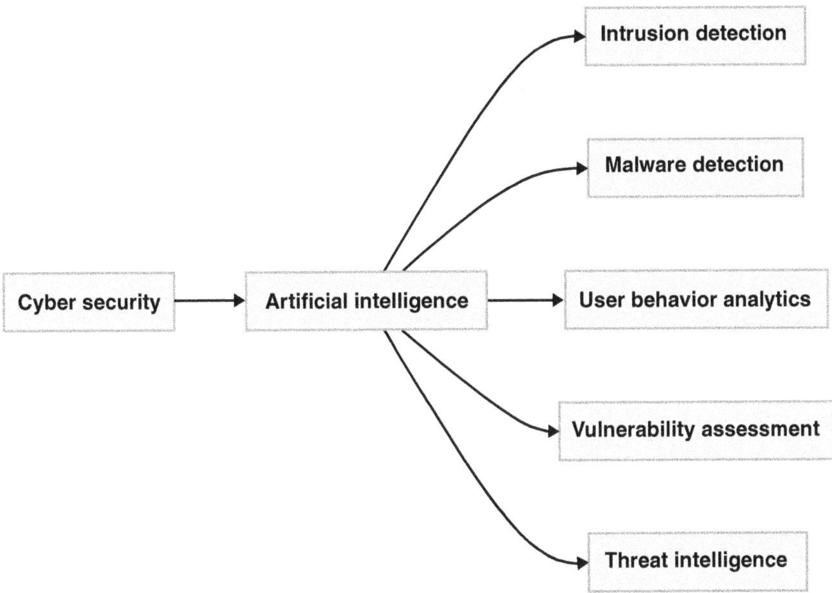

Figure 15.3 Artificial Intelligence-Based Cyber-Security Approach.

data to detect the cyber threats. A mathematical model consists of the following components [19–21].

Data pre-processing: Data pre-processing means making the data for analysis purposes. Data pre-processing means, cleaning the raw data transforming the raw data, and processing the raw data. In data processing, data redundancy can be minimized.

Feature selection: Feature selection allows you to select the most relevant features or variables from your dataset. Feature selection can minimize the dimensionality of your data and improve the efficiency and accuracy of your data analysis [22].

Model training: By machine learning, the various computing models can be trained for the recognition of patterns and various anomalies in the given data sets can be detected. This model is trained on labelled data (data previously classified as normal or malicious).

Model testing and validation: This involves testing the model against another test set dataset. After testing the data, you can evaluate the model's performance. The model performance can be assessed from the various performance matrices like recall, precision, and F1 score. In computing environments, this model can be deployed to detect and prevent cyber threats in real time.

Deployment: The deployment of the computing AI-based model only detects and protects the cyber threats in the real-time environment.

15.4 STATISTICAL DECISION THEORY [5]

In the case of quantitative output, we place ourselves as a random variable and probability space [5,23].

Let $x \in \mathbb{R}^p$ denote the real random input vector, and $y \in \mathbb{R}$ denote the real random output variable. With the joint distribution Dr(X, Y), now the function $f(x)$ is required to produce or predict Y. Some of the energy is going to loss, and the loss energy can be computed by the matrices of loss.

"$L(Y, f(X))$ for determining the error in the prediction, by most convenient square error loss $L(Y, f(X)) = (Y - f(X))^2$

This gives us the criteria of choosing f,

Common and convenient square error EPE $= E(Y - f(X))^2$

$$= \int [y - f(x)]^2 \, \mathrm{Dr}(dx, dy)$$

Estimated square error. By conditioning on input X, we can use the formula

$$\mathrm{EPE}(f) = ExE_{y|x}\left([y - f(x)]^2 \,|\, x\right)$$

After minimizing EPE $f(X) = \operatorname{argmin}_c Ey|x\left([y - c]^2 \,|\, X = x\right)$

Now the solution is $f(X) = E(Y|X = x)$

Thus, the best estimation of Y at point $X = x$ is the conditional means; the best solution can be measured by the average square error [5]".

Most of the frequent machine learning-based algorithms used in cyber security include:

Support vector machines (SVMs): The concept of support vector machine is based on a supervised learning algorithm. By the support vector machine, binary classification can happen.

Support vector classifier [5]: Now suppose the classes overlap in the feature space. One way to understand the overlaps is still to maximize the value M. Here allow the same point to the different side or wrong side of the margin. Now we are defining the slack variable $\xi = \left(\xi_1, \xi_2, \xi_{13}, \cdots - \xi_n\right)$. By the given equation, we can modify the constraint

$$y_i\left[x_i^\top \beta_i + \beta_0\right] \geqslant m - \xi_i \quad \text{or}$$
$$y_i\left(x_i^\top \beta + \beta_0\right) \geqslant M\left(1 - \xi_i\right)$$

$\forall i, \xi_i \geqslant 0, \sum_{i=1}^{n} \xi_i \leqslant$ constant. The above two different solutions provide us more natural solution, the first solution overlaps the actual distance from the margin, and the second solution overlaps the relative distance from the margin, which can change the width of the margin. The first solution is no convex optimization convex solution, and the second one is convex lead standard support vector classification [24,25].

The value ξ_i in the constraint $y_i \left[x_i^\top \beta_i + \beta_0 \right] \geqslant M(1 - \xi_i)$ is the supposed amount by which the estimation $F(x_i) = x_i^\top \beta_i + \beta_0$ is towards the wrong side of the defined margin. After the computing of all the sum $\sum \xi_i$, collect the total proportion amount by which the total prediction tends to the wrong side of the defined margin.

Here, β can be defined by $M = 1/\|\beta\|$

$$\min \beta \text{ subject to} \begin{cases} y_i \left(x^\top \beta + \beta_0 \right) > 1 - \xi_i \forall i \\ \xi_i \geqslant 0, \sum \varepsilon_i \leqslant \text{constant} \end{cases}$$

The suppose vector classifier is defined in a non-separable case, and the presence of a fixed length scale is 1 in the constraint $\left(x_i^\top \beta + \beta_0 \right) \geqslant \left(1 - \xi_i \right)$.

Inside the class boundary region, the identified point did not play a significant role in classifying the boundary. Identified points can be differentiated by non-discriminant analysis. The logistic regression is more similar to the support vector classifier. Using the concept of quadric programming solution of log range multiplier, it can be denoted mathematically:

$$\min \frac{1}{2} \|\beta\|^2 + c \sum_{i=1}^{N} \xi_i$$

Subject to $\xi_i \geqslant 0, y_i \left(x_i^\top \beta + \beta_0 \right) \geqslant 1 - \xi_i, \forall_i$

In the separable cases, the value of $C ==$ infinite.

The optimal Lagrange function is

$$L_p = \frac{1}{2} \|\beta\|^2 + C \sum_{i=1}^{N} \xi_i - \sum_{i=1}^{N} \alpha_i \left[y_i \left(x^\top \beta + \beta_0 \right) - \left(1 - \xi_i \right) \right] - \sum_{i=1}^{N} l_i \xi_i$$

which further minimized with respect to $\beta, \beta_0,$ and ξ_i, and the respective derivative is set to zero.

$$\beta = \sum_{i=1}^{N} \alpha_i y_i x_i$$

$$0 = \sum_{i=1}^{N} \alpha_i y_i$$

$$\alpha_i = C - \mu_i, \forall_i$$

Using the positive constraint $\alpha, \mu, \ \xi_i \geq 0$, we obtain the Lagrange dual objective function:

$$L_p = \sum_{i=1}^{N} \alpha_i - \frac{1}{2} \sum_{i=1}^{N} \sum_{i''=1}^{N} \alpha_i \alpha_i' y_i y_i' x_i^{\top} x_i'$$

We have to maximize the dual objective Lagrange function.

In addition to Karush–Kuhn–Tucker (KKT) conditions,

$$\alpha_i \left[y_i \left(x_i^{\top} \beta + \beta_0 \right) - \left(1 - \xi_i \right) \right] = 0$$

$$h_i \varepsilon_i = 0$$

$$y_i \left(x_i^{\top} \beta + \beta_0 \right) - \left[1 - \varepsilon_i \right] \geq 0$$

Now the solution β has the form $\beta = \sum_{i=1}^{N} \alpha_i y_i x_i$.

15.5 RANDOM FORESTS [5]

Random forest is an ensemble learning algorithm that can combine multiple decision trees to improve the accuracy of the system and the details are provided in Figure 15.4.

Trees are the ideal situation to bag the complex instruction, and they grow sufficiently in depth, given low bias the trees are notoriously noisy. Now each generated tree is captured and the expectation of the average of B, with such a tree resembling the expected tree, is identically sparred.

An average of B, random variable each with variance $\frac{1}{\sigma}$ and $\frac{\sigma^2}{B}$ if the variable is a simple identical distribution, but not required independence, the variance on an average is $\rho\sigma^2 + \frac{1-\rho}{B}\sigma^2$.

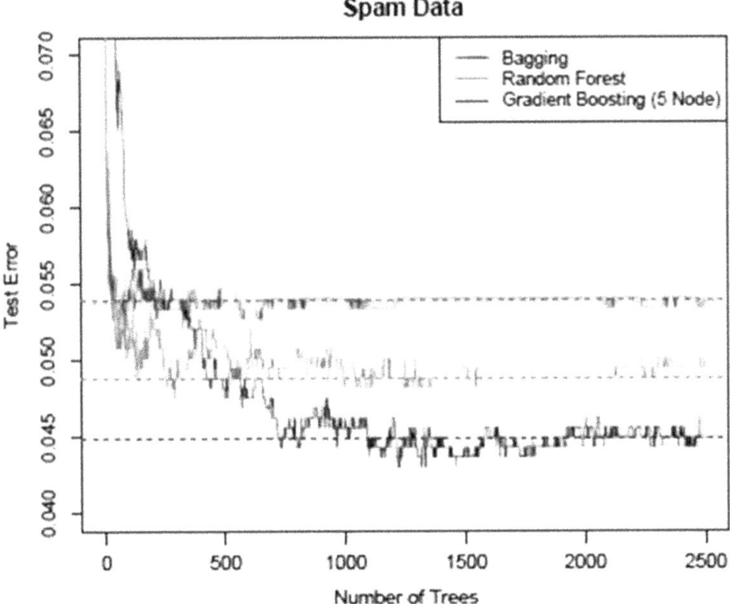

Figure 15.4 Random Forest.

Here, if B is increased, then the second term of the above expression disappears, but the first remains unchanged; the random forest that is meant to improve the variance reduction of the correlation between the tree $\{T(x;\Theta b)\}_1^B$

are grows, and the random forest predictors are $f_{rf}^B(x) = \dfrac{1}{B}\sum_{b=1}^{B}T(x;\theta b)$

Θb random forest tree in terms of split variables [26,27].

The different random forest and gradient boosting are applied for spam data. for the boosting of 5 nodes, and the number of trees chosen with 10-fold cross-validation of 2500 trees.

15.6 NEURAL NETWORKS

Using neural networks, nonlinear datasets and complex datasets can be processed. [27,28]. Analysing the huge and large amount of data and classifying the data and patterns and anomalies may indicate the possible attack with a mathematical model for cyber-security approach, intended to detect and prevent the cyber threats.

AI-based models can be trained on historical data and continuously updated to adapt to new and emerging threats [19].

15.7 NATURAL LANGUAGE PROCESSING

This approach is the use of artificial intelligence-based algorithms to analyse and interpret the human language. NLP can be used to analyse text-based data, like email and chat logs to identify potential threats [20].

In the mathematical model of the NLP, we can use AI-based algorithms and statistical models to analyse and interpret human language.

The NLP-based model consists of the following steps:

1. **Text pre-processing:** In text processing, we are normalizing the RAW data and cleaning the RAW data to ensure that the text data are in the format of data analysis. This step may involve removing stop words, stemming, and lemmatization.
2. **Feature extraction:** In the feature selection the AI-based algorithms convert the text data into the numerical format of data analysis. In this step, we can use techniques such as bag of words, term frequency, and inverse domain frequency to represent the text data.
3. **Model training:** In the AI subdomain machine learning-based algorithms are used to train the model to recognize the pattern and the relationship. This involves the use of machine learning algorithms to train a model to recognize patterns and relationships in the text data. The model is trained using labelled data (data that has been previously classified as either positive or negative).
4. **Model testing and evaluation:** Before the evaluation of the model, testing is required. Testing is done on a separate set of data (the data used for test purposes are known as testing). After that, the performance of the model can be evacuated. The performance of the model is typically evaluated using metrics such as accuracy, precision, recall, and F1 score.
5. **Deployment:** This involves deploying the model in a production environment to analyse and interpret text data in real time.

Some of the most commonly used machine learning algorithms in NLP include:

1. **Naive Bayes:** This is a probabilistic algorithm that is commonly used for text classification tasks.
2. **Support Vector Machines (SVMs):** These are supervised learning algorithms that are commonly used for text classification and sentiment analysis.
3. **Recurrent Neural Networks (RNNs):** These are deep learning algorithms that are capable of processing sequences of data, such as sentences and paragraphs.

The mathematical model for NLP is designed to analyse and interpret human language in order to extract meaning and insights from text data. The model is trained on labelled data and continuously updated to improve its performance and adapt to new and emerging trends in language use.

15.7.1 Deep learning

Deep learning can be used to synthesize and analyse the data coming from various sources; it is also used to analyse the complex data and datasets. Subsequently, deep learning can verify the network traffic, system logs, and the user's behaviours to identify the potential threats [21].

AI-based cyber-security approach offers a powerful way to detect and prevent cyber threats. By applying the cognitive view of machine learning and other AI technologies, the system models can enhance their ability to respond to cyber-attacks and protect their systems and data [22].

Mathematical model for deep learning can be used to analyse complex data, recognize the given pattern, and the relationship available between the data. Mathematical models of the system consist of the following steps:

Input layer: The input layer is the first layer of the neural network, and it is used to receive the input signal has the responsibility to receive the data.

Hidden layers: Hidden layers are the intermediate layers of the neural network; they are used to process the input data and extract the feature from the input data.

Output layer: This is the final and output layer of the neural network system; it is used to produce the output based on the input as well as the weight and bias applied to the neural network.

Activation function: The activation function is the mathematical function of the neuron; it is applied to the output function of each neuron. The activation function has to recognize the complex pattern.

Loss function: The loss function is the mathematical function. It is used to compute the difference between the predicted output and actual output. It is used to train the neural network by manipulating the weight and bias in the neural network model to minimize the loss.

Optimization algorithm: In the optimization of the neural network, the optimizing algorithms are used to minimize the loss by updating weight and bias on the neural network during the training time of the network. The optimization is based on gradient descent and involves adjusting weight and bias.

The most popular deep learning architectures are as follows:

Convolutional Neural Networks (CNNs): Using the convolutional neural network, the video and image recognition task is carried out.

Recurrent Neural Networks (RNNs): Recurrent neural networks are commonly used for NLP.

Generative Adversarial Networks (GANs): These are commonly used for augmentation and image generation purposes.

Mathematical model and statistical model of deep learning involve the use of the concept of neural network to recognize the pattern, image, and various kinds of data, and the relation between the data. This model can be trained by the labelled data and further optimized using various kinds of optimization algorithms to minimize the loss function. Most commonly, deep learning has been shown to be more effective for various kinds of tasks like image and speech recognition, natural language understanding and processing, and game playing [7, 23].

15.8 PROFOUND ANALYSIS FOR BIG DATA

Profound analysis of big data means analysing large and complex datasets to extract meaningful information using advanced technology and techniques. It consists of the following steps:

Data collection: In the data collection phase, large and complex, structured and unstructured data are collected from different sources, such as social media, sensor network, ad hoc network, mobile network, and transactional system [22–24].

Data pre-processing: In data pre-processing, data cleaning, data transformation happen. This is done to ensure to remove redundancy, missing values, outliers, and duplicates of characters, as well as the transformation of data into a suitable format. Data can be reproved for data analysis purposes.

Data integration: This involves integrating various data collected from various sources, and the purpose of this phase is to make the data unified.

Data analysis: This involves using advanced techniques and algorithms to analyse the data and extract meaningful insights. It may involve techniques such as data mining, machine learning, and NLP to identify patterns, relationships, and trends in the data.

Data visualization: In this phase, the result of the data can be represented in a visual format like a graph, chart, or dashboard.

This is done only to help the user to understand the results.

Profound analysis can be used in a variety of domains, like business, healthcare, finance, stock exchange, and senses. Profound analysis can immensely contribute to make a better decision, improve system efficiency, and a gain competitive edge. However, it also presents challenges such as data confidentiality, safekeeping, and ethical concerns, which need to be addressed to ensure that the benefits of big data analytics can be fully realized [29–30].

15.9 ARTIFICIAL INTELLIGENCE-BASED OPTIMIZATION METHOD FOR DEEP ANALYTICS OF BIG DATA [31]

AI-based optimization techniques for in-depth analysis of large amounts of data include using AI algorithms to reorganize the process of examining large and huge amount of complex datasets to provide meaningful insights and knowledge, which includes derivation. Escort typically involves the following steps:

Data pre-processing: This involves cleaning data and transforming it into a format that can be used for analysis.

1. Feature selection: Select the most appropriate features or variables from your dataset to reduce the dimensionality of your data and increase the efficiency of your analysis.
2. Model selection: This involves selecting the appropriate AI algorithm or model to perceive the data. The choice of model depends on the nature of your specific task and data.
3. Hyperparameter tuning: This is fine-tuning the parameters of an AI model to optimize its performance for a specific task or dataset.
4. Optimization: This refers to using AI optimization algorithms such as genetic algorithms and particle swarm optimization to find the optimal combination of hyper-parameters for the selected AI model.
5. Evaluation: This involves evaluating the performance of the optimized AI model against the validation set to ensure that it can be generalized to new data [25].

AI-based optimization technology for in-depth analysis of big data helps improve the accuracy, efficiency, and speed of analysis as depicted in Figure 15.5. One can also reduce the computational complexity of your analysis by automating the process of choosing the best hyper-parameters

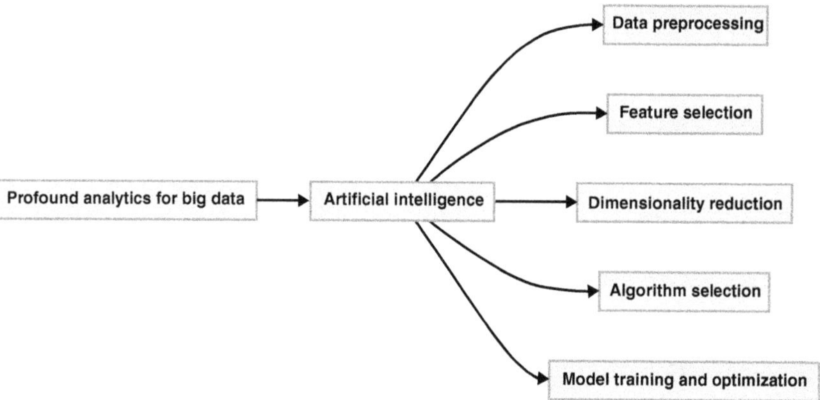

Figure 15.5 Artificial Intelligence-Based Optimization Method for Deep Analytics of Big Data.

for the AI model. Overall, AI-based optimization techniques have the potential to greatly enhance the value of big data analytics in various fields such as business, healthcare, and finance.

15.10 LYAPUNOV-BASED FEEDBACK CONTROL CYBER-SECURITY METHOD

Lyapunov-based feedback control is a commonly used technique in control systems to ensure stability and robustness. It can also be used to address cyber-security issues in control systems. The basic idea is to use the Lyapunov function to design controls that detect and defend against cyber-attacks and malicious activity.

Lyapunov-based feedback control principles can apply to cyber-security systems as follows:

System modelling: First, we need to model the control system and identify its vulnerabilities to cyber-attacks. This includes understanding the dynamics of your system and the attack vectors that can be exploited.

Threat analysis: This involves conducting an in-depth analysis of the cyber threats your system may be exposed to. This includes both insider threats (e.g. insider attacks) and external threats (e.g. network-based attacks).

Design Lyapunov functions: These functions capture the stability characteristics of control systems. The Lyapunov function should be able to quantify the performance and security aspects of the system. You also need to understand the impact of potential cyber-attacks on your system.

Attack detection: This involves developing algorithms that monitor system behaviour and detecting deviations and anomalies that may indicate cyber-attacks. This may entail comparing measured system output to expected behaviour based on Lyapunov functions.

Damage control strategy: Once an attack is detected, the control system should take appropriate action to mitigate the effects of the attack. This could be the adjustment of control laws or the implementation of defensive mechanisms to restore system stability and performance.

Robustness analysis: This involves performing a comprehensive analysis to assess control system robustness against various cyber-attack scenarios. This includes evaluating the system's resilience to different attack strategies and validating the effectiveness of Lyapunov-based feedback control approaches.

Below is a general overview of how Lyapunov-based feedback control can be used in the context of cyber security [16,18,29].

It is important to note that Lyapunov-based feedback control is just one approach to addressing cyber security in control systems. Other technologies such as anomaly detection, encryption, authentication, and intrusion detection systems can be combined with Lyapunov-based controls to provide a more comprehensive cyber-security solution. The specific implementation details and choice of algorithms depend on the characteristics of the control system and the types of cyber threats faced.

Let's consider a discrete time control system for simplicity.

15.10.1 System dynamics

System dynamics can be designed by the following mathematical equation:

$$x(k + 1) = f(x(k), u(k))$$

$x(k)$ represents the state of the system at time step k.
$u(k)$ is the control input at time step k.
$f(x(k), u(k))$ denotes the system subtleties function.

15.10.2 Lyapunov function

The Lyapunov function $V(x)$ is used to quantify the stability and security properties of the system. The following conditions must be met:

$$V(x) > 0 \text{ for all } x \neq 0$$

$$V(0) = 0$$

$$V(x(k + 1)) - V(x(k)) \leq -\alpha(x(k)) + \beta(u(k))$$

where $\alpha(x(k))$ is a positive-definite function that represents the potential impact of an attack or disturbance on the system's state.

$\beta(u(k))$ captures the effect of the control action in mitigating the impact of the attack or disturbance.

15.10.3 Control law

The control law aims to minimize the Lyapunov function $V(x)$ by appropriately choosing the control input $u(k)$. It can be formulated as:

$$u(k) = g(x(k), V(x(k)))$$

where $g(x(k), V(x(k)))$ is a control law function that determines the control input based on the system state and the Lyapunov function

15.10.4 Stability analysis

The stability of the system can be analysed by showing that the Lyapunov function $V(x)$ decreases with time. This can be expressed as: $V(x(k + 1)) - V(x(k)) \leq - \alpha(x(k)) + \beta(g(x(k), V(x(k))))$.

The specific form of the Lyapunov function, the choice of $\alpha(x)$ and $\beta(u)$, and the design of the control law $g(x, V)$ are intended to attack the nature of the desired dynamics and safety goals of the system. Detailed mathematical formulations and analyses are typically specific to individual control systems and their safety requirements.

15.11 CONCLUSION

Cyber-security technology based on AI for performing in-depth analysis of big data brings great benefits in improving the security and protection of digital systems. By combining the power of AI with advanced analytics, organizations can improve their ability to effectively detect and respond to cyber threats. This approach uses AI algorithms and techniques to analyse large amounts of data to identify patterns and identify anomalies that may indicate potential cyber-attacks or security breaches. The technology includes several key components such as intrusion detection, malware detection, user behaviour analysis, vulnerability assessment, and threat intelligence. The use of AI enables cyber-security systems to dynamically adapt and evolve, continuously learning from new data and improving their ability to detect and respond to emerging threats. AI algorithms can analyse network traffic, system logs, and user behaviour to detect suspicious activity, intrusions, and anomalous patterns that could indicate malicious intent. In addition, the integration of AI and big data analytics enables efficient and effective processing of large amounts of data, enabling real-time monitoring and detection of cyber threats. This approach makes it easier to automate tasks, reduces reliance on manual analysis, and allows security teams to focus on high-priority threats. By using AI-based cyber-security techniques, organizations can improve their overall security posture and reduce both incident response times as well as the risks associated with cyber-attacks. The combination of in-depth analytics and AI provides powerful defence mechanisms against evolving and sophisticated cyber threats. AI-based cyber-security techniques for in-depth analysis of big data are valuable approaches that enable organizations to proactively detect, prevent, and respond to cyber threats. By employing AI algorithms, organizations can achieve better threat detection, better incident response, and ultimately a stronger security posture in the face of ever-evolving cyber threats.

REFERENCES

1. Zhang, Y., Geng, P., Sivaparthipan, C. B., & Muthu, B. A. (2021). Big data and artificial intelligence based early risk warning system of fire hazard for smart cities. Sustainable Energy Technologies and Assessments, 45, 100986.
2. Barja-Martinez, S., Aragüés-Peñalba, M., Munné-Collado, Í., Lloret-Gallego, P., Bullich-Massague, E., & Villafafila-Robles, R. (2021). Artificial intelligence techniques for enabling Big Data services in distribution networks: A review. Renewable and Sustainable Energy Reviews, 150, 111459.
3. Kibria, M. G., Nguyen, K., Villardi, G. P., Zhao, O., Ishizu, K., & Kojima, F. (2018). Big data analytics, machine learning, and artificial intelligence in next-generation wireless networks. IEEE Access, 6, 32328–32338.
4. Srivastava, S., Haroon, M., &Bajaj, A. (2013, September). Web document information extraction using class attribute approach. In 2013 4th International Conference on Computer and Communication Technology (ICCCT) (pp. 17–22). IEEE.
5. Hastie, T., Tibshirani, R., & Friedman, J. (Eds.) The elements of statistical learning, trevor hastie, robert tishirani, jerome friedman, (2009). Springer.
6. Siddiqui, Z. A., & Haroon, M. (2023). Research on significant factors affecting adoption of blockchain technology for enterprise distributed applications based on integrated MCDM FCEM-MULTIMOORA-FG method. Engineering Applications of Artificial Intelligence, 118, 105699.
7. Najafabadi, M. M., Villanustre, F., Khoshgoftaar, T. M., Seliya, N., Wald, R., & Muharemagic, E. (2015). Deep learning applications and challenges in big data analytics. Journal of Big Data, 2(1), 1–21.
8. Khan, W., & Haroon, M. (2022). A pilot study and survey on methods for anomaly detection in online social networks. In Human-Centric Smart Computing: Proceedings of ICHCSC 2022 (pp. 119–128). Springer Nature Singapore.

9. Siddiqui, Z. A., & Haroon, M. (2022). Application of artificial intelligence and machine learning in blockchain technology. In Artificial Intelligence and Machine Learning for EDGE Computing (pp. 169–185). Academic Press.

10. Khan, R., Haroon, M., & Husain, M. S. (2015, April). Different technique of load balancing in distributed system: A review paper. In 2015 Global Conference on Communication Technologies (GCCT), Thuckalay, India (pp. 371–375). IEEE.

11. Xin, Y., Kong, L., Liu, Z., Chen, Y., Li, Y., Zhu, H., ... & Wang, C. (2018). Machine learning and deep learning methods for cyber-security. IEEE Access, 6, 35365–35381.

12. Khan, W., & Haroon, M. (2022). An unsupervised deep learning ensemble model for anomaly detection in static attributed social networks. International Journal of Cognitive Computing in Engineering, 3, 153–160.

13. Haroon, M., & Husain, M. (2015, March). Interest attentive dynamic load balancing in distributed systems. In 2015 2nd International Conference on Computing for Sustainable Global Development (INDIACom), New Delhi, India (pp. 1116–1120). IEEE.

14. Khan, W., & Haroon, M. (2022). An efficient framework for anomaly detection in attributed social networks. International Journal of Information Technology, 14(6), 3069–3076.

15. Ngiam, K. Y., & Khor, W. (2019). Big data and machine learning algorithms for health-care delivery. The Lancet Oncology, 20(5), e262–e273.

16. Sarker, I. H., Kayes, A. S. M., Badsha, S., Alqahtani, H., Watters, P., & Ng, A. (2020). Cyber-security data science: an overview from machine learning perspective. Journal of Big Data, 7, 1–29.

17. Haroon, M., Tripathi, M. M., Ahmad, T., & Afsaruddin. (2022). Improving the healthcare and public health critical infrastructure by soft computing: An overview. In Pervasive Healthcare: A Compendium of Critical Factors for Success (pp. 59–71). Springer.

18. Husain, M. S., & Haroon, D. (2020). An enriched information security framework from various attacks in the IoT. International Journal of Innovative Research in Computer Science & Technology (IJIRCST), 8(3), 2347–5552.

19. Rawat, D. B., Doku, R., & Garuba, M. (2019). Cyber-security in big data era: From securing big data to data-driven security. IEEE Transactions on Services Computing, 14(6), 2055–2072.

20. Khan, A. M., Ahmad, S., & Haroon, M. (2015, April). A comparative study of trends in security in cloud computing. In 2015 Fifth International Conference on Communication Systems and Network Technologies (pp. 586–590). IEEE.

21. Haroon, M., & Husain, M. (2013). Analysis of a dynamic load balancing in multiprocessor system. International Journal of Computer Science Engineering and Information Technology Research, 3(1).

22. Husain, M. S. (2020). A review of information security from consumer's perspective especially in online transactions. International Journal of Engineering and Management Research, 10(4). https://ssrn.com/abstract= 3669577

23. Sreedevi, A. G., Harshitha, T. N., Sugumaran, V., & Shankar, P. (2022). Application of cognitive computing in healthcare, cyber-security, big data and IoT: A literature review. Information Processing & Management, 59(2), 102888.

24. Khan, N., & Haroon, M. (2022). Comparativestudy of various crowd detection and classification methods for safety control system. Available at SSRN 4146666. http://dx.doi.org/10.2139/ssrn.4146666

25. Petrenko, S. A., Petrenko, A. S., & Makoveichuk, K. A. (2017). Problem of developing an early-warning cyber-security system for critically important governmental information assets. Network, 4, 7–8.

26. Siddiqui, Z. A., & Haroon, M. (2023). Research on significant factors affecting adoption of blockchain technology for enterprise distributed applications based on integrated MCDM FCEM-MULTIMOORA-FG method. Engineering Applications of Artificial Intelligence, 118, 105699.

27. Siddiqui, Z. A., & Haroon, M. (2022). Application of artificial intelligence and machine learning in blockchain technology. In Artificial Intelligence and Machine Learning for EDGE Computing (pp. 169–185). Academic Press.

28. Tripathi, M. M., Haroon, M., Khan, Z., & Husain, M. S. (2022). Security in digital healthcare system. In Husain, M.S., Adnan, M.H.B.M., Khan, M.Z., Shukla, S., Khan, F.U. (eds.), Pervasive Healthcare. EAI/Springer Innovations in Communication and Computing. Springer, Cham. https://doi.org/10.1007/978-3-030-77746-3_15

29. Truong, T. C., Zelinka, I., Plucar, J., Čandík, M., & Šulc, V. (2020). Artificial intelligence and cyber-security: Past, presence, and future. In Artificial Intelligence and Evolutionary Computations in Engineering Systems (pp. 351–363). Springer Singapore.

30. Khan, W., Haroon, M., Khan, A. N., Hasan, M. K., Khan, A., Mokhtar, U. A., & Islam, S. (2022). DVAEGMM: dual variational autoencoder with gaussian mixture model for anomaly detection on attributed networks. IEEE Access, 10, 91160–91176.

31. Martínez Torres, J., Iglesias Comesaña, C., & García-Nieto, P. J. (2019). Machine learning techniques applied to cyber-security. International Journal of Machine Learning and Cybernetics, 10, 2823–2836.

Chapter 16

A comprehensive examination of cloud forensics, including challenges and available solutions

Narayan P. Bhosale, Gaurav Chaudhari, Vijay Kumar Vishwakarma, Dhananjay Deshpande, and Siddharth Dabade

16.1 INTRODUCTION

A "cloud" is an elastic execution environment of resources involving multiple stakeholders and providing a metered service at multiple granularities for a specified level of quality (of service). According to the official NIST definition, "Cloud computing is a model for enabling ubiquitous, convenient, on-demand network access to a shared pool of configurable computing resources (e.g., networks, servers, storage, applications, and services) that can be rapidly provisioned and released with minimal management effort or service provider interaction."

Cloud computing has two models:

1. Deployment models

2. Services models

1. Deployment models are:
   a. Public cloud
   b. Internal or corporate cloud (private cloud)
   c. Multi-cloud (hybrid cloud)
   d. Community cloud

2. Service models are:
   a. Infrastructure as a service
   b. Platform as a service (PaaS)
   c. Infrastructure as a service (IaaS)

16.1.1 Cloud services

Table 16.1 shows the cloud-based services.

16.1.2 Digital forensics

Digital forensics is a specialized branch of forensics science, and it deals with a digital crime that occurs on a computer, mobile, or any other digital

DOI: 10.1201/9781003514312-16

Table 16.1 Tabular Representation of Cloud-Based Services (University of Teesside)

Features	Public	Private	Hybrid	Community
Provider	Service provider	Enterprise (Third)	Enterprise (Third)	Community
End-user	General public	Selected users	Selected users	Community users
Accessibility	Internet	Internet, VPN	Internet, VPN	Internet, VPN
Ownership	Service provider	Enterprise (Third)	Enterprise (Third)	Community

device. Cloud forensics (Chen et al., 2019) is a specialized field of digital forensics. Digital forensics is a technical field of forensics that deals with crime where any digital device such as a mobile, laptop, or computer is involved in a crime. Cloud forensics is a particular field of digital forensics (Shinde & Gurrala, 2023). Digital forensics is defined by Palmer 2001 as "The use of scientifically derived and proven methods toward the preservation, collection validation, identification, analysis, interpretation, documentation, and presentation of digital evidence derived from the digital sources to facilitate the reconstruction of the events found to be criminal or helping to anticipate unauthorized actions shown to be disruptive to planned operations."

16.1.3 Need for cloud forensics

Data put in the cloud is now known as secure and safe for storing data, but some issues are seen with this. There is a tempering of data, and if any digital crime occurs, cloud forensics needs to solve that crime and collect the evidence to solve that case. Continuous increase in crime in cloud platforms occurs, so need arises to solve that case of cloud forensics. Cloud forensics is the application of digital forensics in cloud computing as a subset of network forensics. Cloud forensics has three dimensions, which are named as legal, organizational, and technical dimensions.

16.1.4 Dilemmas in cloud forensics

We have a small amount of knowledge about cloud forensics after we learned about the cloud, cloud computing, deployment model, service model, and cloud forensics. Different types of dilemmas come in cloud forensics when we investigate the cloud. These are the general steps; we take when we do cloud forensics. There are four steps in cloud forensics, each stage with a different type of challenge.

Figure16.1 shows challenges in cloud forensics. It follows a cycle of various steps.

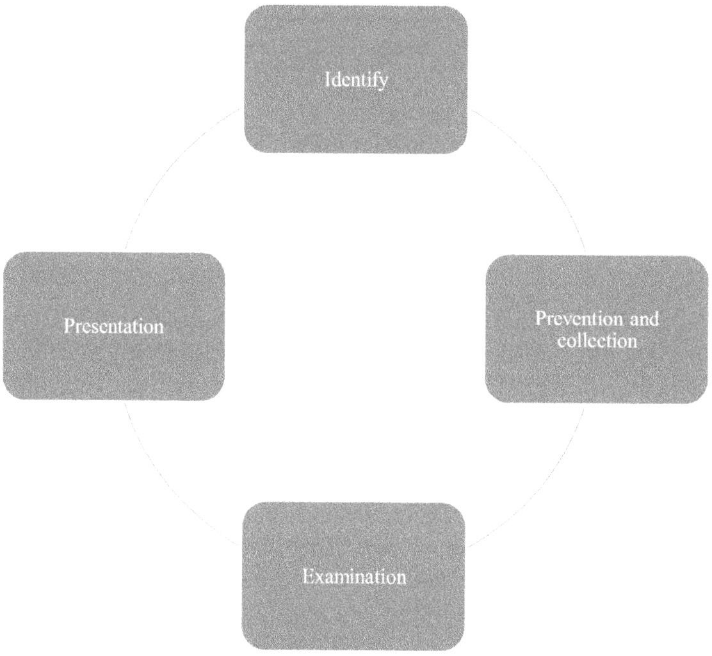

Figure 16.1 Challenges of Cloud Forensics.

16.1.5 Identify

In this stage, our priority is getting a log of the system where crime occurs. Log plays an essential role in solving any case in cloud forensics because without getting a log to investigate, they do not take any action against a criminal. In other words, we can say that a log is the backbone of cloud forensics. For solving any case, getting a log is an enormous issue for the investigator because many users use the same network resource for processing simultaneously. With the help of a log, it is confirmed whether a crime occurred or not. In cloud forensics, collecting the log is like getting the lock key. Fetching data from the cloud is very difficult due to its geographical distance. Cloud service provider (CSP) is also necessary for collecting evidence; with the help of CSP, any investigator gets the proof.

16.1.6 Prevention and collection

After accessing a log from Phase 1, we prioritize preventing data from tampering without harming its integrity. There are challenges in every face; this stage also comes with some challenges. The main difficult task for this stage is securing the data. It means that violating the data is prevented if the data is violating that it is not accepted in court—the maintenance of custody in the whole investigation. Every piece of evidence should be prevented from the first day of the investigation until the day of the decision. If the chain is mismanaged, broken, or interrupted, it is also affected by justice.

16.1.7 Examination

Forensics tools are required to analyze the data to solve any case in cloud forensics. Many tools are present for forensics, but all of these tools are for digital forensics. Some specific tools are only used for cloud investigation, which are used to examine cloud forensics. With the help of tools and their knowledge, investigators want to solve the case, but this is also true that no specific tools were designed for the cloud forest. Some tools that we discuss are not for cloud forensics that are used in digital forensics after seeing their roles in cloud forensics, and these tools are worked in the use of cloud forensics.

16.1.8 The volume of data

After getting a tool, the next major problem is that a large amount of data is present in the cloud. Due to this, it takes much time to monitor that data. The data center is also massive and takes much time to invest in. Data were increasing per second. Many techniques are also used to get appropriate evidence.

16.1.9 Encryption

Data stored in a cloud made is in encrypted form for securing data in the cloud platform. After getting that data decrypted because it is the encrypted format in this format, anyone cannot read this. Getting a valid key is a big task at this stage. Then analyze the data. It takes much time for the investigator to get good evidence.

16.1.10 Presentation

The data we get from all the stages does not mean our case is solved. The main essential and last challenge has come in this phase. The information we get is appropriately set because the jury has a piece of less technical knowledge.

They have some technical expertise in cybercrime, but if we talk about cloud forensics. They have very, significantly less knowledge in this field. This is a very new topic for the jury and our society. The inability to "go back" and obtain the original again is a unique issue that challenges cloud forensics investigations from an authenticity standpoint.

At this stage, the documentary is made from all the evidence from all the phases. The data we get from all the phases are set in a proper manner. It is presented in a straightforward word and understandable format. The person who has less technical knowledge will also understand the whole scenario. The evidence is not a temper or changes throughout the investigator.

Table 16.2 (Isaac Abiodun et al., 2022) presents discussion on the whole procedure of the investigation process in cyber forensics. The author showed the whole procedure to investigate step-wise and comply with the legal process in cyber forensics. The authors put all the points ethically.

16.2 LITERATURE SURVEY

Cloud computing services are improving, and their use is increasing daily. Today it is a very famous topic in the field its world, and everyone knows this topic which is connected to this area. There were fewer resources before introducing cloud computing, or we can say that limited services were present for users. After coming of cloud computing, various types of services come for a customer in the same network. Many companies are moving their services to the cloud (Simou et al., 2016), so with the continuous increase in shifting data in cloud computing, digital crime also increases in the cloud environment. According to a report from McAfee publisher, the cost of a cyber-crime-related activity is $500 billion every year, more than the cost of drug traffic globally (Alex & Kishore, 2017).

The main reasons for shifting to cloud computing are backup and data restoration, better cooperation, magnificent accessibility, reasonable maintenance cost, mobility, service in the pay-per-user model (Zeng, 2019), unlimited storage capacity, and data security. Cloud forensics is a specialized area that specializes in investigating and reading digital evidence within cloud computing environments (Akter & Rahman, 2023). As more companies undertake cloud offerings for their infrastructure, programs, and facts storage, the need for powerful cloud forensics has become increasingly critical (Isaac Abiodun et al., 2022). This comprehensive exam will explore the challenges related to cloud forensics and speak about some available solutions.

The focus of the traditional digital forensic inquiry procedure has always been on the aftermath of the occurrence. The cloud may not be able to handle this lengthy process. The current study looks into how the standard digital forensic investigation procedure can be streamlined and updated to fit cloud computing environments better. When an inquiry is necessary, this strategy will expedite the collection of relevant evidence, enabling the investigator to begin the analysis and inspection immediately.

Table 16.2 Investigation Process in Cyber Forensics (Isaac Abiodun et al., 2022)

Identification	Preservation	Collection	Examination	Analysis	Presentation	Admissibility
Identify the crimes and incidents	Managing the case	Conservation or protection	Conservation or protection	Preservation or protection	Documentation and record-keeping	Presentation
Digital signature	Image development	Accepted strategy	Trailing and tracking	Trailing and tracking	Expert opinion	Comply with the constitution
Profile Identification	Custodial chain	Software compatibility	Methods for validation or verification	Analyzing data, empirically	Evidence clarification	No uncertainty in the evidence
Abnormal discovery	Time synchronization	Hardware compatibility	Sieving the procedures	Protocols and algorithms used	Effect of the mission	Testimony from expert
Constructive comments and criticisms	Notifying the authority	Legal authorization	Pattern matching	Data gathering and analysis	Worthy recommendation	Worthy recommendations
Surveillance and vigilance	-	Lossless data compression	Listing data sources	Events, timeline, history	Data interpretation	Data interpretation
Auditing	-	The survey, sampling reduction of data, and recovery	Hidden and lost data extraction	Link special AI	-	Case proving

In distributed computing, which many researchers study, cloud computing represents an alternative paradigm. In this setting, it is clear that it is vital to understand the precise location, timing, and method of data processing and storage. Cloud forensics presents more challenges than traditional digital forensics because data is not stored (Patrascu & Patriciu, 2015) in a single location, and virtualization technologies are used. Describe a new technique for monitoring activity in data centers and cloud settings using a safe cloud forensic framework. The author discusses the architecture, and this framework must support and describe how it may be implemented on top of new or existing cloud computing implementation.

For the incident tracking to be finished, the log files' integrity must be guaranteed. To do this, the authors (Wang et al., 2019) suggest a public approach based on a third-party auditor for confirming the accuracy of cloud logs. The combined log block tags (Achar, 2021) use the traditional Merkle hash tree structure to create the root node that will be stored in the blockchain and prevent tampering with the log data. The suggested system also prevents any log content from being exposed during a public audit. The analysis and experimental findings support the claim that the proposed method may successfully conduct the security audit of cloud logs while requiring less computing overhead than previous methods.

The forensic investigator must trust the CSP (Achar, 2021), who gathers the logs from various sources within the cloud environment. However, untrusted cloud stakeholders and hostile outsiders can work together to change (Ganesh et al., 2022) the logs after the fact and stay undetectable. As a result, the reliability of the logs provided for forensics may be in doubt. The author(s) (Rane & Dixit, 2019) suggested secure logging-as-a-service assisted by forensic-aware blockchain for a cloud environment to securely store and process logs by addressing multi-stakeholder collusion problems and maintaining integrity and confidentiality. The unchangeable aspect of blockchain technology is used to guarantee the integrity of logs. Only BlockSLaaS, which protects log confidentiality, allows Cloud Forensic Investigator (CFI) (Trivedi et al., 2023) to access the logs for forensic inquiry.

Cloud computing is a technological solution that takes all of these into account. Furthermore, many small and medium-sized businesses were drawn in by recent improvements in cloud computing. However, the level of protection and privacy (Achar, 2021) offered to the tenant's data is not apparent or appropriate. Attacks on the cloud in the present day support this claim. By undertaking forensics, one might take a reactive strategy to handle a cloud event that has already occurred. Cloud forensics, however, is still a relatively new field. The National Institute of Standards and Technology (NIST) (Jahankhani & Hosseinian-Far, 2017) published a draft in 2014 that outlines several technological, architectural, organizational, and legal difficulties that arise when conducting forensics in a cloud setting. The authors (Kumar Raju & Geethakumari, 2018) focus on event reconstruction as one

Table 16.3 Overview of the Existing Methods (Vadetay Saraswathi Bai, 2023)

Author (s)	Evidence Collection Techniques	Forensics Analysis	Framework Proposed	Implementation Details	AI/ML	Trust Provenance
	Yes	Yes		Yes	Yes	Yes
Zawoad et al. (2013)	Yes		Yes			
Povarand Geethakumari (2014)		Yes	Yes		Yes	Yes
Ahsan et al. (2021)	Yes	Yes		Yes		
Almulla et al. (2013)		Yes	Yes	Yes	Yes	Yes
Sampana (2019)	Yes	Yes	Yes			
Hemdanand Manjaiah (2018)				Yes	Yes	Yes
Battistoni et al. (2016)	Yes	Yes		Yes	Yes	
Kao (2016)		Yes	Yes	Yes		Yes
Roussev et al. (2016)		Yes		Yes	Yes	

of the technical difficulties. Cloud virtual machine artifacts are regarded as achieving the same.

Table 16.3 shows the summarized version of the existing methods for cloud forensics.

Challenges in cloud forensics

Multi-tenancy: Cloud service vendors (CSPs) commonly share physical resources among multiple users (Awuson-David et al., 2021), resulting in challenges in setting apart and maintaining digital evidence precise to a particular user or instance.

1. Dynamic and elastic nature: Cloud environments are enormously dynamic, with assets being hastily provisioned, scaled up or down, and decommissioned (Shaik & Baskiyar, 2021). This dynamic nature complicates proof series and upkeep efforts, as these methods can misplace or alter crucial facts.
2. Lack of bodily management: Unlike conventional on-premises environments, investigators have restricted bodily management over the infrastructure and storage gadgets in cloud environments. This issue can affect the ability to perform specific forensic tasks, including direct memory acquisition.

3. Data segregation and encryption: Cloud carriers regularly use facts and segregation techniques to make particular data privacy and safety, which could make it hard to get entry to and examine precise statistics without proper authorization (Tiwari et al., 2019). Encryption practices carried out by using CSPs may also complicate the recovery and evaluation of encrypted records.

4. Jurisdictional issues: Cloud offerings can be hosted in records facilities in distinctive jurisdictions (Siry, 2019), making it complicated to navigate felony and jurisdictional demanding situations when engaging in investigations across worldwide limitations.

5. Logging and auditing: Cloud service carriers might not provide comprehensive logging and auditing mechanisms, or they will hold logs for restrained periods (Ljubuncic & Litterer, 2019). This loss of precise and long-term logging can avoid forensic investigations.

16.2.1 The solution in cloud forensics

One of the biggest challenges for cloud forensics is getting a log from the cloud environment. Both users and investigators have no control over the cloud environment. All the power is in the hands of the CSP—the biggest failure is not solving the case related to a cloud because of not getting a log. There is no control of the user and investigator in the cloud environment. All the power is on hand, CSP. This CSP's secure-logging-as-a-service (SecLaas) mechanism allows the storage of VM logs. It provides access to forensics investigators and prevents the confidentiality of their users.

A log-based approach makes digital forensics easier on cloud computing. It reduces the complexity of forensics for non-repudiation of behaviors on the cloud. The team of investigators must be masters of their field. Cloud service providers also hire a forensics investigator. When any crime occurs in a cloud environment, the person with a piece of relevant knowledge in this field can collaborate with the team and help solve a case, give the training and make an expert in cloud forensics. He/she better understands the cloud environment, its architecture, storage technology, and the use of tools related to cloud forensics.

In most cases, the researcher saw that enough forensics in cloud forensics, there are no specific tools, so designing particular tools for cloud forensics help to solve any case. In the forensics team, one person is a master in cryptography because all the data we get from a cloud is an encryption form, so decrypting that data with crypto graphical knowledge is essential. Due to this, it takes less time to investigate crime. The jury hearing the case has more knowledge of the cybercrime file. So, the team explains the issue in a more technical aspect.

16.2.2 Available solutions in cloud forensics

Legal and coverage concerns: Organizations should establish clear policies and agreements with cloud carrier companies to outline the obligations and expectations regarding proof handling, facts retention, and incident reaction. Legal experts with cloud computing and virtual forensics information must be worried about growing and reviewing these agreements.

Incident reaction-making plans: Organizations must have well-defined incident reaction plans that encompass cloud-precise strategies. These plans should outline the stairs to be taken during a safety incident, maintaining virtual evidence, contacting CSPs, and concerning appropriate forensic professionals.

Cloud-unique forensic tools: Specialized forensic tools designed for cloud environments are emerging to cope with the demanding situations of cloud forensics (Raghavendra et al., 2022). This equipment frequently provides features like faraway acquisition, integration with cloud carrier company application programming interfaces, and analysis of cloud-unique artifacts.

Memory forensics: Traditional disk-based forensic techniques might also have limitations in cloud environments, but memory forensics can be a precious approach (Prachi & Kumar, 2022). Analyzing volatile reminiscence can help recover manner records, encryption keys, and other vital artifacts to research.

Collaboration with CSPs: Establishing relationships and channels of conversation with cloud carrier providers can significantly help in cloud forensics investigations (Benjula Anbu Malar & Prabhu, 2020). CSPs can provide vital records, assist with proof renovation, and provide insights into their infrastructure and logging practices.

Data class and encryption key management: Properly classifying information primarily based on its sensitivity and enforcing appropriate encryption key control practices can resource in information recuperation and evaluation at some point in forensic investigations (Kumar & Goyal, 2019). Strong critical control practices ensure investigators can get admission to encrypted facts while vital.

Training and know-how: Cloud forensics is a complex and unexpectedly fast evolving discipline (Dacorogna & Kratz, 2023). Organizations must spend money on schooling virtual forensics groups to increase specialized capabilities and stay updated with today's improvements, exceptional practices, and prison necessities in cloud forensics. It is important to note that each cloud environment and research is precise, and answers must be tailor-made to precise instances. Collaboration among digital forensics specialists, criminal professionals, and cloud provider vendors is essential to overcoming challenges and carrying out effective cloud forensics investigations.

16.3 STATISTICAL STUDY

16.3.1 Dataset

The dataset below shows the year-on-year increase in cybercrime by a ratio of 1200. It seems that the cost to resolve those cyber incidents is also increasing by $1.2 billion, year on year basis. This cost and an increasing number of cyber incidents can give headaches to any cybercrime unit/agency in the world.

Table 16.4 presents the statistics of the world cybercrime and the increasing cost for measures taken to resolve the related crime. This table displays the rising trend in both the overall cost and the number of incidents from 2013 to 2022. According to the data both the frequency and the financial costs of cybercrime have increased significantly throughout this time. It is important to remember that these numbers only reflect reported occurrences and costs; as a result, undiscovered or underreported instances may result in an even greater impact of cybercrime. To reduce the risks posed by cybercrime, businesses and people should continue to emphasize cybersecurity measures.

The graph in Figure16.2 represents the increasing number of cybercrimes throughout the year. This bar graph displays the rising trend in the number of incidents from 2013 to 2022.

16.4 CONCLUSION

We still have inadequate knowledge about cloud forensics after we learned about the cloud, cloud computing, deployment model, service model, and cloud forensics. Some tools that we discuss are not for cloud forensics tools that are used in digital forensics after seeing their roles in cloud forensics, and

Table 16.4 Dataset for Worldwide Increasing Cybercrime Year on Year [NCRB]

Year	Number of Incidents	Total Cost
2013	1,200	$1.2 billion
2014	2,400	$2.4 billion
2015	3,600	$3.6 billion
2016	4,800	$4.8 billion
2017	6,000	$6.0 billion
2018	7,200	$7.2 billion
2019	8,400	$8.4 billion
2020	9,600	$9.6 billion
2021	10,800	$10.8 billion
2022	12,000	$12.0 billion

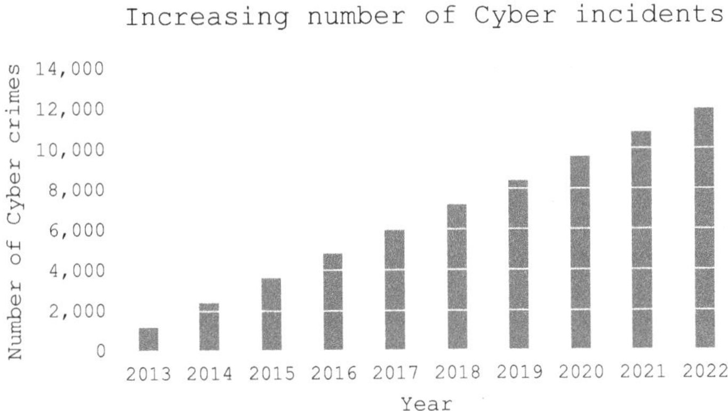

Figure 16.2 Statistics for Cybercrime in the World.

these tools are worked in the use of cloud forensics. A better understanding of the cloud environment, its architecture, storage technology, and the use of tools related to cloud forensics is essential. In most cases, the researcher observed enough forensics in cloud forensics; there are no specific tools, so designing particular tools for cloud forensics helps to solve any case. There is a tempering of data, and if any digital crime occurs, cloud forensics needs to solve that crime and collect the evidence to solve that case. As a result of the ongoing rise in crime on cloud platforms, it is necessary to define cloud forensics as the use of digital forensics in cloud computing as a subset of network forensics to resolve such cases. These are the general steps that we take when we do cloud forensics. There are four steps in cloud forensics, each with a different challenge. Our priority is getting a log of the system where crime occurs. In cloud forensics, collecting the log is like getting the lock key. CSP is also necessary for collecting evidence; with the help of CSP, any investigator gets the proof. The CSP has no legal responsibility to give such information if the service level agreement does not specify what kind of procedural forensics data should be provided to the customer. Some tools that we discuss are not for cloud forensics tools that are used in digital forensics after seeing their roles in cloud forensics, and they are worked in the use of cloud forensics. Seeing huge volume of data after getting a tool, the next major problem is that a large amount of data present in the cloud is to be studied. Encryption data stored in a cloud is encrypted for secure data in the cloud platform. As a result, the reliability of the logs provided for forensics may be in doubt. Due to sharp increase in crimes involving cloud computing services, the field of cloud forensics has become essential. As there are currently no dedicated tools for cloud forensics, there is a pressing need for experts in the field to create solutions that are specifically designed for cloud environments. Additionally, the success of forensic investigations depends on knowing the fundamentals of data collecting and collaborating closely with CSPs. The sheer amount of data and encryption, however, might pose additional challenges and possibly jeopardize the accuracy of forensic records. Therefore, improving cloud-forensic techniques and encouraging collaboration between customers and CSPs are essential steps to more effectively combating cybercrime.

16.5 FUTURE SCOPE

Most infrastructure, such as networking hardware and servers, has been replaced by virtual server applications designed to specifically address the needs of the modern business landscape. The fact is that any party can deploy and control third-party resources and data and that the customer can host them from any location are both drawbacks. Applications and data are stored on several hard drives located throughout the world. Thus, optical forensic artifacts have a non-uniform characteristic. The fact that automated inquiry technology and methods range in terms of cost, level of sophistication, and accessibility is another distinction. Most digital forensic tools are incompatible as they use a variety of hardware and software, which restricts what they can achieve.

Depending upon the categories of cloud service utilized, forensic data may or may not be accessible. Clients of IaaS get free access to the information required for forensic analyses. SaaS consumers can have restricted or no access to such data simultaneously. The creation of the chronology in forensics activities requires time synchronization. In a cloud environment, time synchronization is challenging as relevant data is dispersed across numerous physical computers and locations. The expansion of endpoints is becoming increasingly challenging as several resources are associated with the cloud. Multiple instances coordinating on the physical system can be divided from one another in the cloud because of virtualization. Because of this limitation, isolating resources during a forensics investigation might be challenging without endangering the privacy of other users who share the infrastructure.

REFERENCES

Achar, S. (2021). A Comprehensive study of current and future trends in cloud forensics. *Asian Journal of Humanity Art and Literature*, 8, 95–104. https://doi.org/10.18034/ajhal.v8i2.651

Ahsan, M. A. M., Wahab, A. W. B. A., Idris, M. Y. I. Bin, Khan, S., Bachura, E., & Choo, K.K. R. (2021). CLASS: Cloud log assuring soundness and secrecy scheme for cloud forensics. *IEEE Transactions on Sustainable Computing*, 6(2), 184–196. https://doi.org/10.1109/TSUSC.2018.2833502

Akter, S. S., & Rahman, M. S. (2023). Cloud Forensic: Issues, Challenges and Solution Models. https://doi.org/10.1142/13337

Alex, M. E., & Kishore, R. (2017). Forensics framework for cloud computing. *Computers & Electrical Engineering*, 60, 193–205.

Almulla, S., Iraqi, Y., & Jones, A. (2013). A distributed snapshot framework for digital forensics evidence extraction and event reconstruction from cloud environment. 2013 IEEE 5th International Conference on Cloud Computing Technology and Science, 699–704. https://doi.org/10.1109/Cloud Com.2013.114

Awuson-David, K., Al-Hadhrami, T., Alazab, M., Shah, N., & Shalaginov, A. (2021). BCFL logging: An approach to acquire and preserve admissible digital forensics evidence in the cloud ecosystem. *Future Generation Computer Systems*, 122, 1–13. https://doi.org/10.1016/j.future.2021.03.001

Battistoni, R., Di Pietro, R., & Lombardi, F. (2016). CURE—Towards enforcing a reliable timeline for cloud forensics: Model, architecture, and experiments. *Computer Communications*, 91–92, 29–43. https://doi.org/10.1016/j.com com.2016.03.024

Benjula Anbu Malar, M. B., & Prabhu, J. (2020). A distributed collaborative trust service recommender system for secure cloud computing. *Transactions on Emerging Telecommunications Technologies*, 31(12). https://doi.org/10.1002/ ett.3991

Chen, L., Takabi, H., & Le-Khac, N. (Eds.). (2019). *Security, Privacy, and Digital Forensics in the Cloud*. Wiley. https://doi.org/10.1002/9781119053385

Dacorogna, M., & Kratz, M. (2023). Managing cyber risk, is a science in the making. *Scandinavian Actuarial Journal*, 1–22. https://doi.org/10.1080/03461 238.2023.2191869

Federici, C. (2013). AlmaNebula: A computer forensics framework for the cloud. *Procedia Computer Science*, 19, 139–146. https://doi.org/10.1016/ j.procs.2013.06.023

Ganesh, N. S. G., Venkatesh, N. G. M., & Prasad, D. V. V. (2022). A Systematic Literature Review on Forensics in Cloud, IoT, AI & Blockchain (pp. 197–229). https://doi.org/10.1007/978-3-030-93453-8_9

Hemdan, E. E.D., & Manjaiah, D. H. (2018). CFIM: Toward Building New Cloud Forensics Investigation Model (pp. 545–554). https://doi.org/10.1007/978-981-10-3812-9_56

Isaac Abiodun, O., Alawida, M., Esther Omolara, A., & Alabdulatif, A. (2022). Data provenance for cloud forensic investigations, security, challenges, solutions and future perspectives: A survey. *Journal of King Saud University— Computer and Information Sciences*, 34(10), 10217–10245. https://doi.org/ 10.1016/j.jksuci.2022.10.018

Jahankhani, H., & Hosseinian-Far, A. (2017). Challenges of Cloud Forensics (pp. 1–18). https://doi.org/10.1007/978-3-319-54380-2_1

Kao, D.Y. (2016). Cybercrime investigation countermeasure using created-accessed-modified model in cloud computing environments. *The Journal of Supercomputing*, 72(1), 141–160. https://doi.org/10.1007/s11227-015-1516-7

Kumar, R., & Goyal, R. (2019). On cloud security requirements, threats, vulner-abilities and countermeasures: A survey. *Computer Science Review*, 33, 1–48. https://doi.org/10.1016/j.cosrev.2019.05.002

Kumar Raju, B. K. S. P., & Geethakumari, G. (2018). Timeline-Based Cloud Event Reconstruction Framework for Virtual Machine Artifacts (pp. 31–42). https://doi.org/10.1007/978-981-10-3376-6_4

Ljubuncic, I., & Litterer, T. (2019). System Administration Ethics. Apress. https://doi.org/10.1007/978-1-4842-4988-8

Patrascu, A., & Patriciu, V.V. (2015). Logging for cloud computing forensic systems. *International Journal of Computers Communications & Control*, 10(2), 222. https://doi.org/10.15837/ijccc.2015.2.802

Povar, D., & Geethakumari, G. (2014). A Heuristic Model for Performing Digital Forensics in Cloud Computing Environment (pp. 341–352). https://doi.org/10.1007/978-3-662-44966-0_33

Prachi, & Kumar, S. (2022). An effective ransomware detection approach in a cloud environment using volatile memory features. *Journal of Computer Virology and Hacking Techniques*, 18(4), 407–424. https://doi.org/10.1007/s11416-022-00425-2

Raghavendra, S., Srividya, P., Mohseni, M., Bhaskar, S. C. V., Chaudhury, S., Sankaran, K. S., & Singh, B. K. (2022). Critical retrospection of security implication in cloud computing and its forensic applications. *Security and Communication Networks*, 2022, 1–19. https://doi.org/10.1155/2022/1791491

Rane, S., & Dixit, A. (2019). BlockSLaaS: Blockchain Assisted Secure Logging-as-a-Service for Cloud Forensics (pp. 77–88). https://doi.org/10.1007/978-981-13-7561-3_6

Roussev, V., Ahmed, I., Barreto, A., McCulley, S., & Shanmughan, V. (2016). Cloud forensics—Tool development studies & future outlook. *Digital Investigation*, 18, 79–95. https://doi.org/10.1016/j.diin.2016.05.001

Sampana, S. S. (2019). FoRCE (Forensic Recovery of Cloud Evidence): A digital cloud forensics framework. 2019 IEEE 12th International Conference on Global Security, Safety and Sustainability (ICGS3), 212–212. https://doi.org/10.1109/ICGS3.2019.8688215

Shaik, S., & Baskiyar, S. (2021). A scalable approach to service placement in fog/cloud environments. 2021 IEEE International Performance, Computing, and Communications Conference (IPCCC), 1–8. https://doi.org/10.1109/IPCCC51483.2021.9679396

Shinde, S. M., & Gurrala, V. R. (2023). Securing trustworthy evidence for robust forensic cloud-blockchain environment for immigration management with improved ECC encryption. *Expert Systems with Applications*, 229, 120478. https://doi.org/10.1016/j.eswa.2023.120478

Simou, S., Kalloniatis, C., Gritzalis, S., & Mouratidis, H. (2016). A survey on cloud forensics challenges and solutions. *Security and Communication Networks*, 9(18), 6285–6314. https://doi.org/10.1002/sec.1688

Siry, L. (2019). Cloudy days ahead: Cross-border evidence collection and its impact on the rights of EU citizens. *New Journal of European Criminal Law*, 10(3), 227–250. https://doi.org/10.1177/2032284419865608

Tiwari, K., Sharma, P., & Tiwari, R. (2019). Cloud security intensification by client-side data segregated and encryption. *International Journal of Computer*

Sciences and Engineering, 7(5), 1672–1676. https://doi.org/10.26438/ijcse/v7i5.16721676

Trivedi, D., Boudguiga, A., & Triandopoulos, N. (2023). *SigML: Supervised Log Anomaly with Fully Homomorphic Encryption* (pp. 372–388). https://doi.org/10.1007/978-3-031-34671-2_26

Vadetay Saraswathi Bai, T. S. (2023). A systematic literature review on cloud forensics in cloud environment. *International Journal of Intelligent Systems and Applications in Engineering*, 11(4s SE-Research Article), 565–578. https://ijisae.org/index.php/IJISAE/article/view/2727

Wang, J., Peng, F., Tian, H., Chen, W., & Lu, J. (2019). Public Auditing of Log Integrity for Cloud Storage Systems via Blockchain (pp. 378–387). https://doi.org/10.1007/978-3-030-21373-2_29

Zawoad, S., & Hasan, R. (2013). Cloud forensics: A meta-study of challenges, approaches, and open problems. https://doi.org/10.48550/arXiv.1302.6312

Zeng, X. (2019). *Management of Service Level Agreements for Big Data Analytics Applications in Cloud: A Layer-Based Study*. Australia: The Australian National University.

Chapter 17

Pivotal role of cyber security in Internet of Things-based smart cities

An analytical framework for safe smart cities

Saurabh Srivastava, Tasneem Ahmed, and Nayyar Ali Usmani

17.1 INTRODUCTION

The term "smart cities" refers to technologically based solutions that improve citizen satisfaction, foster better communication with the government, and advance sustainable development (Ismagilova et al., 2020). When the development process for managing the important assets, resources, and the flow of real-time processes balances and compares economic, social, and environmental variables, the city can be considered smart (Ismagilova et al., 2020). Smart cities are evolving around the world, and information technology is crucial to their development as well as to the creation of new social and economic opportunities (Elmaghraby & Losavio, 2014). Smart cities take on new dimensions because of the widespread use of information and communication technology (ICT) in urban infrastructure. ICT use in urban areas boosts both the local economy and the standard of living for residents. Smart cities aim to manage a sizable populace and their requirements effectively. The newest technologies, including networks, mobile, cloud computing, electronic equipment, IoT devices, etc., are all used in smart cities (Hamid et al., 2019).

Smart cities rely on the Internet of Things (IoT) to supply services across multiple industries, thereby facilitating human lives. The role of ICT in smart cities includes information gathering, data storage and access, and distribution of information. The role of IoT started in smart cities as a set of technologies that are capable to access and gather data from different sources such as wire, wireless, and internet networks. The graphical and mathematical models are added to IoT systems by locating the privacy and security, people, and sensors but the methodology of the system is not discussed with others (Hamid et al., 2019). Because the distributed architecture can monitor issues linked to technologies and the extent of smart cities, the IoT is a significant factor in urban areas and plays a significant role in the development of smart cities. The IoT and cloud computing are currently working together and creating a new dimension in the smart city

DOI: 10.1201/9781003514312-17

known as the cloud of things (CoT) (Bibri, 2018; Hamid et al., 2019). The IoT is gaining ubiquity in terms of devices, and it is becoming ubiquitous in terms of services. Attacks against IoT devices and services are becoming more frequent. Although IoT-based devices and services are not immune to cyberattacks, it is imperative to have a security mechanism that can comprehend cyberattacks on IoT infrastructure and the thread, as IoT will become increasingly integrated into our daily lives (Abomhara & Køien, 2015; González-Zamar et al., 2020). Security is characterized as a procedure that can guard an object against physical harm, theft, unauthorized access, and loss to preserve the object's high level of confidentiality and integrity (Abomhara & Køien, 2015). Budgeting for security is necessary because it is an expensive activity. To guarantee that smart cities continue to succeed, security concerns are legitimate and require attention. In smart cities, privacy and security concerns are essential since they improve and optimize the quality of life for locals (Hamid et al., 2019).

In the 1990s, only 13% of world population lived in urban areas as per a report, and now approximately 70% population lives in urban areas. Water usage is also increasing day by day, and it is two times greater than the population. The smart city focuses on efficiency, sustainability, a wide role in decision making, ICT solutions, and providing various types of services. The smart city is a significant phenomenon for community growth and economic innovation, as well as a fusion of contemporary technology innovation (Szabo, 2019; Huang & Nazir, 2021). When a city uses contemporary technology for resource management, traffic control, transportation, disaster security and response, and other aspects of city administration, it can be considered smart (Alibasic et al., 2016). Smart city designers make the most of contemporary technologies like sensors, cloud computing, networks, IoT, and machine learning to allow various smart city components to communicate with the network architecture (Ismagilova et al., 2020). IoT-based applications are not very dependable due to numerous problems and concerns related to cyber security. An analytical model that characterizes various cyber concerns in terms of smart cities has been proposed in this study.

17.2 CYBER SECURITY'S ROLE IN SMART CITIES

The urban population is increasing all over the world, and interconnectedness is also increasing which highlights the importance of cyber security in smart cities (Kalinin et al., 2021). A single cyber-attack can affect a system that contains a wide variety of features such as home applications, medical devices, hospitals, and air defense systems. Cyber-attacks always target the smart grid, and cyber threats have increased in the last few years. In the current era, malicious software (malware) acts as a bomb that is designed to damage, destroy, or take control of the complex systems that control the operations of the smart grid (Alibasic et al., 2016). Smart cities provide many potential advantages due to technological advancements, including enhanced management, personal security, and energy efficiency. Due to their reliance on contemporary technology, smart cities increase their vulnerability to cyber-attacks (Alibasic et al., 2016). A smart city is outfitted with a wide range of electronic gadgets, machinery, and technologies that make life in the city smarter for its residents and increase the accessibility and usefulness of its many features. As the population density inside the urban area increases, new infrastructure and services are required to meet the growing needs of the citizens. In the past few years, it has become extremely difficult or impossible to combine digital devices that can essentially plan the development of a city and gather information for daily management activities (Talari et al., 2017; Komninos & Mora, 2018). ICT amenities are included in the smart city to raise the standard of living. A city can become smarter in six different area: the environment, citizens, economy, living mobility, and governance. The infrastructure and government utilities of public, private, and civil organizations are included in smart governance (Hamid et al., 2019). Smarter citizens are those who are aware of the latest technology, and who also use e-skills for general problems. The smart economy includes e-business, e-commerce, production, etc. Smart mobility defines transparency, logistic systems, and traffic. Smart living defines a superior lifestyle with reasonable consumption and stable behavior. The smart environment includes water, energy, and waste management for a clean environment, and a controlled population (Hamid et al., 2019).

Global urban population growth is driving smart city development initiatives to leverage IoT to create more intelligent, appropriate, and efficient solutions. IoT-based gadgets are self-diagnosing and have the ability to anticipate future malfunctions or maintenance requirements. A very intricate, interconnected network of platforms, systems, devices, and users exists in smart cities (Szabo, 2019). Globally, the concept of security has always aimed to ensure people's safety and give them confidence that their rights, property, or lives will not be violated. The connected world is detrimental from various angles as well, including administrative, technical, and financial ones. The legal, political, and social environments of the cities must be taken into consideration when weighing the trust factors (Elmaghraby & Losavio, 2014). New networked systems for automation, monitoring, and control are changing city infrastructure and security. These systems are also utilized for disaster recovery, emergency response, and water and sanitation. Cybersecurity is a subset of computer systems; it deals with information processing, data exchange channels, and sanction violations, which are illegal. Cybersecurity primarily faces two challenges: security and privacy. Systems that gather information and respond quickly when needed are protected by privacy from ongoing security threats, while security keeps an eye out for attacks brought on by the physical distribution of services (Elmaghraby & Losavio, 2014; Ma, 2021).

IoT devices can connect to other IoT devices and ICT systems via a variety of technologies, including wireless, Wireless Local Area Network (WLANs), cellular (3G, LTE), and others. IoT entities and services are readily integrated into service-oriented architecture (SOA) and service science (Abomhara & Køien, 2015). The primary issue with IoT services and systems is data security. Not only can a user access data in IoT systems and services, but authorized objects or people can also access data. Because the IoT environment is so pervasive, privacy is a major concern for IoT-based systems and services. In many research projects, including those involving data collection, sharing, management, and security, privacy is a sensitive topic (Abomhara & Køien, 2015). Cyber-attacks on IoT devices are on the rise and have the potential to wipe out individuals and companies. Initial settings make it easy to operate IoT devices that are connected to the internet but have not been protected (Neshenko et al., 2020). IoT devices urgently require security measurement techniques. The fastest-growing area of modern technology is made up of smart systems and services, which are connected via the internet and have sophisticated features and functions. The value of IoT is frequently increasing and in 2022 it was projected at approximately 83 billion dollars (Szabo, 2019).

17.3 INTERNET OF THINGS ROLE IN SMART CITIES

IoT is defined as a physical system designed to enhance smart city infrastructure, performance, and security, particularly by having minimal processing and storage power (Talari et al., 2017). The perception, network, and application layers make up the three layers of the IoT. A collection of internet-enabled devices that can gather and share data with other devices via the internet communication network is referred to as the perception layer. The perception layer consists of Global Positioning System (GPS), cameras, radio frequency identification (RFID), and other sensors. The network layers carry the data that is transmitted from the perception layer to the application layer. Wi-Fi and Bluetooth are two examples of short-range network communication technologies that are used by the IoT to transfer data from various perceptual devices to adjacent gateway-based communication parties. Long-distance applications use internet-based technologies like wireless, cellular services, and programmable logic controller (PLC) to transport information. The platform where data is received and processed is called the application layer. Power system monitoring, smart cities, and smart homes are a few examples of applications (Talari et al., 2017). The internet is growing and spreading fast nowadays; formally independent communication devices and services are connected to the internet (Szabo, 2019; RP et al., 2021).

The ability to provide crucial information to citizens connected by multiple devices, including wired and wireless internet services, is defined by the IoT (Zheng & Lee, 2004; Hamid et al., 2019). According to recent studies, 1.6 billion IoT devices and modules were used in 2017, which is approximately 39% more than the number of IoT devices used in 2015 (Hamid et al., 2019). Digital security investment in smart cities is lacking which will create future vulnerabilities in the IoT systems (Szabo, 2019). The most common use case for IoT systems in the current era of advanced technology is machine-to-machine (M2M) communication. These days, M2M is widely used to control and monitor users in a variety of industries, including water, oil, management, retail, public service, transportation, and health (Abomhara & Køien, 2015). According to reports, by 2020, M2M applications will have 12 billion connections and have brought in over 714 billion euros in income (Abomhara & Køien, 2015). The possibility of interconnecting devices with the internet is increasing when it starts the adoption of the IoT paradigm, and similar in the smart city environment. Numerous applications are developing that are based on the real concern of smart cities, and these are offering advanced services to citizens and city administrators (Righetti et al., 2018). The IoT is becoming increasingly popular and overcoming several challenges for real-time applications. Artificial intelligence is widely used in big data analytics that offers accurate real-time data analysis (Sharma et. al., 2021). New network architectures like edge computing and fog computing can be used by IoT devices to collect data from metropolitan areas and deliver real-time management information (Ferreira et al., 2021). Nevertheless, there are countless IoT mobile devices in use today that might replace the functionality offered by specialist equipment. About 75.44 billion people will be using mobile IoT devices by 2025, including laptops, Android-powered smartphones, iPhones, iPads, and other gadgets (Wang et al., 2022). Data transport and acquisition utilizing IoT-enabled smart device technologies were assisted by the usage of communication technologies. Industries may now save time and money on maintenance costs if a machine breaks down due to unforeseeable circumstances because of the real-time data storage, processing, and analysis made possible by wireless sensor networks (WSNs) and cloud computing (Khalil et al., 2022).

The IoT is a rapidly expanding technology that has a significant impact on the social and business environments due to its frequent development and capacity to provide a wide range of services. With increased complexity comes an increase in the amount of threats and other cyber-attacks. Not only number of attacks increasing but the tools that can help in cyber-attack are also increasing which makes the attackers more sophisticated, efficient, and effective (Abomhara & Køien, 2015). In an insecure IoT environment, trust plays a crucial role in secure communication. In the context of the IoT, trust can be characterized as both user-perceived trust in the system and trust in the messages exchanged between entities (Abomhara & Køien, 2015). According to Koin, various system components, including hardware, sensors, processors, memory, actuators, and software resources like operating systems and hardware-based software, determine how reliable IoT systems and services are. Human life is made easier by IoT applications, but security concerns are consistently at play (Goyal et al., 2018). A few locations make the linked systems extremely susceptible to cyber-attacks.

- As most IoT systems are managed by humans, it is simple for attackers to get physical access to the system (Abomhara & Køien, 2015).
- As most IoT systems interact via wireless networks, it is easier for an attacker to eavesdrop and collect private information (Abomhara & Køien, 2015).
- The minimal power and processing resource capabilities of most IoT devices prevent them from supporting complicated societal procedures and increase the possibility of an attacker gaining physical access to the system (Abomhara & Køien, 2015).mart

17.4 ANALYTICAL FRAMEWORK FOR SAFE AND SECURE SMART CITIES

The USA coined the term "smart city," meaning that a city is considered smart if it achieves appropriate economic development and makes balanced investments in both traditional and digital infrastructures. The various social capital and networked elements that make up a smart city can exchange data continuously. Intelligent networks, cloud platforms, IoT sensors, building automation systems, and intelligent cars are some of the components that make up a smart city (Szabo, 2019; Qasem & Al Mobaideen, 2019). Large volumes of data are processed by sophisticated and intelligent systems in smart cities. These systems gather and produce various kinds of data, including common roots, sensor data, and location data, which are useful for a variety of applications (Szabo, 2019). The urban strategist "Boyd Cohen" divides the smart city into six different components such as smart economy, smart environment, smart government, smart living, smart mobility, and smart citizens (Alibasic et al., 2016). The smart city components are shown in Figure 17.1.

A smart economy relies on entrepreneurship, innovation, and the productivity of interconnectedness on a local and global scale. The supply and demand policies for ICT-based systems, e-government apps, and open data transparency are created by the smart government. The well-being, prosperity, contentment, and security of residents in smart cities are all aspects of smart living. Mixed modalities, clean and non-motorized options, and an ICT-connected environment are all components of smart mobility. It is creative and always focused on achieving academic excellence, and intelligent

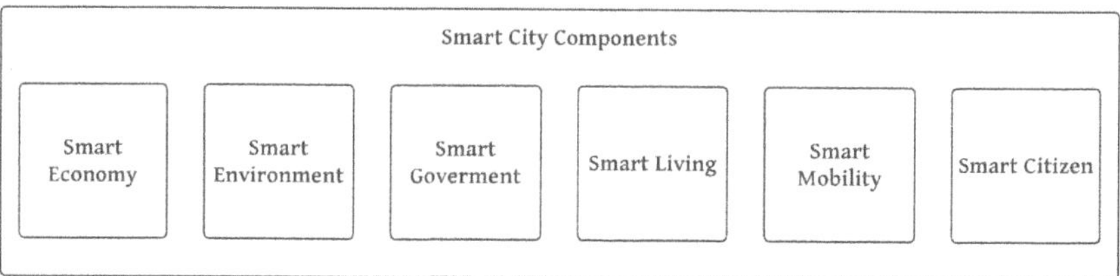

Figure 17.1 Smart City Components.

citizens (Alibasic et al., 2016). The term smart city can define a city that has properly balanced the economy, mobility, living standard, environment, etc. and a smart city has a lot of features that make citizen's life easy. The concept of a smart city cannot be implemented without using technology (Joshi et al., 2016). Modern technology such as ICT, IoT, and digital and information technologies are playing an important role in developing smart cities. Numerous devices and services are launched per day with new and advanced features that make human life easy (Samih, 2019). In the current era, no device or object can be fully secure. The backbone of smart applications is modern technologies, devices, and sensors, and the smart devices are not so reliable because of several cyber concerns behind them. The hackers perform different kinds of activities such as attacks, threats, etc., to take control of smart devices. Different shades of smart applications (positive and negative) are shown in Figure 17.2.

17.5 MODERN TECHNOLOGIES IN SMART CITIES

The smart city cannot be imagined without modern technologies. A smart city uses different techniques such as IoT, ICT, and Cloud to enhance the quality and performance of urban area services at low cost with resource consumption. The IoT has made it much easier to create real-time apps for smart cities. In this study, some smart IoT-based technologies that are very useful in developing applications for smart cities are discussed. The useful technologies of IoT that are frequently used to develop real-time applications of smart cities are shown in Figure 17.3.

- RFID comprises tags and readers, which are essential components of the IoT. It offers a variety of smart applications, such as asset management, parking lots, healthcare applications, and item tracking and localization (Talari et al., 2017).
- Near field communication (NFC) is a technology that is utilized in smartphones in particular for bidirectional short-range communication. NFC-enabled smartphones are utilized as wallets that let users

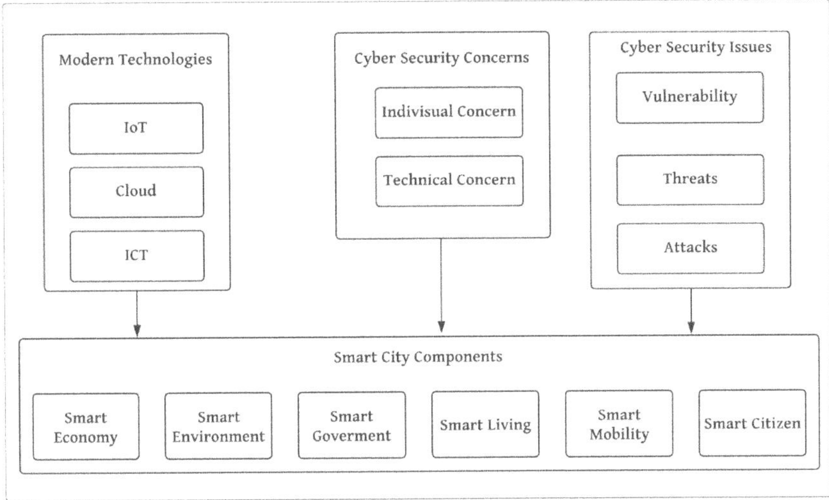

Figure 17.2 Different Shades of Smart Applications.

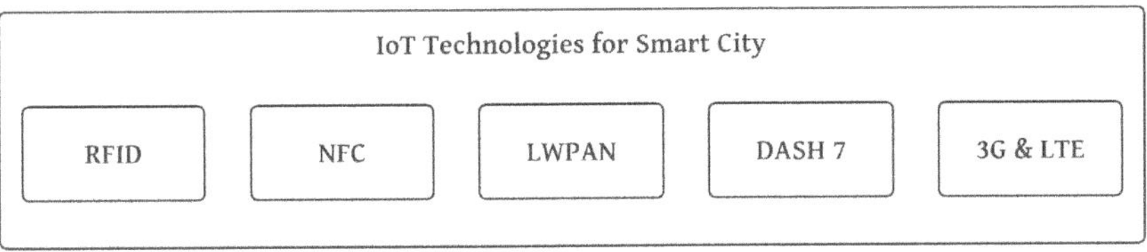

Figure 17.3 IoT-Based Technologies for Smart Cities.

use their phones as credit/debit cards, bank cards, access control cards, transit cards, and more. The NFC is bidirectional and useful for a variety of tasks, including document sharing, multimedia, and data transfer across devices (Talari et al., 2017).

- A short-range radio technology known as Low Rate Wireless Personal Area Network (LWPAN) has a long range of roughly 10 to 15 kilometers. LWPAN devices use incredibly little energy, and a battery's lifespan is about 10 years (Talari et al., 2017).

17.6 CYBER SECURITY CONCERNS IN SMART CITIES

The users are sharing their personal and professional information from different platforms to take advantage of the latest services. Hackers perform different kinds of activities to collect users' information, and the secret

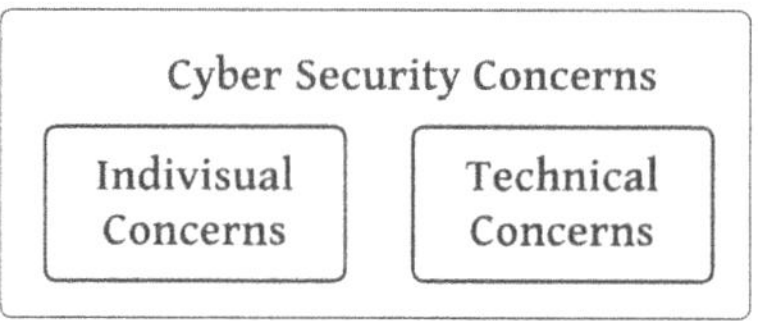

Figure 17.4 Cyber Security Concerns.

information can act as a key to unlock the user's confidential objects such as banking systems, e-mail, social media, etc. These are the major concerns of cyber security in IoT-based applications for smart cities. Hackers perform various kinds of unethical activities to take control of the system and steal and destroy individual, and organizational information. In this study, the cyber concern is categorized into two phases that are individual concerns and technical concerns, as shown in Figure 17.4. Both types of cyber security concerns are discussed in detail.

17.6.1 Individual concern

Everyone is sharing their personal and professional information on social media and other platforms nowadays. The hackers are using information that is shared by a citizen to harm the citizen. The individual perspective concerns are related to citizens' safety, communication, transportation, banking, and finance. The smart city provides a platform of services for the citizens that are based on social issues that are raised by them. The smart city improves the quality of many areas including health, communication, banking, finance, etc. (Hamid et al., 2019). Some important issues on behalf of individual perspectives are defined below.

- The primary concern about the health sector is that hackers can change the critical information of patients.
- The user can frequently share health-related and personal data through different platforms; therefore, the data is very crucial and needs a high-security mechanism to secure individual data.
- Many service providers such as hospitals and other social networks are selling personal and professional information about the individual such as health and social data that can create problems in individual personal and professional life so need an act that can monitor these kinds of unethical practices.
- The social cloud server can be vulnerable to an unauthorized entity. The organization must share information authentically via the cloud.

- The telecommunication infrastructure is important as well as vulnerable in smart cities. Several financial and government actions are carried out via telecommunication and wireless networks.
- Bluetooth, wireless networking, and cloud computing consider privacy and security issues; therefore, there is a need for a powerful security mechanism that can monitor unethical activities.
- The privacy of citizens must be guaranteed by the smart city infrastructure. The privacy concerns related to the social network depend on the information that is available on the social media platforms of the individual citizen.
- Smartphones affect the security of smart cities because they contain smart applications that use GPS location, Bluetooth, Wi-Fi, etc., and threats can come from different sources such as social media, unauthorized applications, Bluetooth, Wi-Fi, etc.

17.6.2 Technical concern

Numerous applications exist that provide many services that make human life easy. Everyone tries to take more benefits from it but a little mistake by users harms them. Hacker performs different activities to take control of the user or unauthorized access. Some technical devices can be used for monitoring, security, or avoiding unauthorized access.

- Biometrics is used to monitor unauthorized access; Biometrics is capable of automatically recognizing a citizen by the behavior of their biometric characteristics. There are two types of biometric characteristics: physical and behavioral. Biometrics plays an important role in resolving several security issues in the smart city, especially things of malicious attacks, data theft, and online fraud.
- Smart grids have a wide role in smart cities because the smart grid can take responsibility for power development. The smart grid includes enabled sensors with networks that can be managed remotely. The applications of the smart grid include smart homes, smart appliances, and infrastructure.

17.7 CYBER SECURITY ISSUES

Smart city deals with modern technologies and different kinds of other resources such as sensors, wireless networks, etc. In the current era, there is not any object that is fully secured. Hackers and crackers have become more advanced because the latest hacking and cracking tools make their tasks easy. IoT-based applications are not so reliable because of lots of cyber security issues. The cyber security issues have been categorized into three phases, i.e. vulnerability, threats, and attacks, as mentioned in Figure 17.5.

Figure 17.5 Cyber Security Issues.

17.7.1 Vulnerability

Vulnerability defines a weakness in a system or design of a system that allows an attacker to access authorized data, execute commands, and conduct different types of attacks such as denial of service attacks, etc. Exposure is a concern or mistake in the configuration of the system that can allow an attacker to organize unethical activities such as information gathering, etc. (Abomhara & Køien, 2015).

17.7.2 Threats

Threats are software programs that take advantage of a security weakness in the system and harm it. The threats can be divided into natural threats and human threats. Natural threats include floods, earthquakes, and computer systems that are damaged by fire cause. Human threats are those conducted by humans such as malicious threats. Human threats can be categorized into two categories structured threats and unstructured threats (Abomhara & Køien, 2015).

- Structure threats include threats such as people knowing about the vulnerabilities of the system and understanding the developed code and script (Abomhara & Køien, 2015).
- Unstructured threats consist of the most inexperienced individuals who use hacking tools (Abomhara & Køien, 2015).

17.7.3 Attacks

Attacks are the procedures or programs that are used to harm systems or disturb normal operations by exploiting vulnerabilities through different tools and techniques. The attackers can perform their activities from many sources such as active networks and passive networks (Abomhara & Køien, 2015). Some cyber-attacks that are frequently used by hackers are mentioned in Table 17.1.

Today, everyone is sharing their personal and professional information on different portals. A citizen's shared information is being used by

Table 17.1 List of Frequently Used Cyber-Attacks by Hacker

Cyber Attacks	Objective
Physical attacks	Physical attacks are sort of attacks tempered with a hardware component (Abomhara & Køien, 2015)
Reconnaissance attacks	The unapproved identification and mapping of systems, services, and vulnerabilities is known as a reconnaissance attack (Abomhara & Køien, 2015).
Denial of service (DoS)	The goal of denial-of-service (DoS) attacks is to prevent users from accessing computers and networks (Abomhara & Køien, 2015). DoS attacks can cause a network or machine to crash, rendering the user unable to access it (Hamid et al., 2019).
Access attacks	Access attacks can define that an unauthorized person gets access to networks and devices that have no right to access them. The access attacks can be categorized into two types that are physical access and remote access (Abomhara & Køien, 2015).
Data mining attacks	Data mining attacks enable attackers to discover information that is not anticipated in a certain database (Abomhara & Køien, 2015).
Cyber espionage	Cyber espionage can be done by using creaking techniques and malicious software to steal and gather secret information about individuals, organizations, and governments (Abomhara & Køien, 2015).
Eavesdropping	Eavesdropping is defined as listening to a conversation between two parties (Abomhara & Køien, 2015). This exploit takes data by listening in on network communications without actively participating. The most frequent threat to RFID systems is eavesdropping (Hamid et al., 2019). This attack can be categorized into two types: directly listening and sniffing of the data.
password-based attacks	The password-based attacks can be categorized in two ways that are dictionary attacks and brute force attacks. The dictionary attacks trying different possible combinations of numbers and letters to predict the user's password. The brute force attack uses different cracking tools to find all possible combinations of passwords to unlock the system (Abomhara & Køien, 2015).
Jamming attack	The jamming attack disturbs communication in the wireless environment, and wireless devices may be vulnerable to this attack (Hamid et al., 2019).
Spoofing attack	A security risk known as "spoofing" is when an attacker uses a link to gather the IP address and MAC address of a network that is vulnerable (Hamid et al., 2019).
Cryptanalysis attack	Cryptanalysis attack is used to break cryptographic security system and can get access to encrypted data and messages. There are many techniques such as cipher-text attacks, differential middle attacks, integral cryptanalysis attacks, and dictionary attacks (Hamid et al., 2019).
Spyware Attack	Malware with the ability to gather data about an individual, community, or institution is known as spyware (Hamid et al., 2019). A user's moment can be tracked by the device's unique identification number (UID) (Abomhara & Køien, 2015).

hackers to harm the citizen. Concerns from the person's perspective are related to citizen safety in areas like banking, transportation, and communication. The smartphone of the user contains a lot of things of the user that can be hacked and used to harm the user. In this study, an analytical framework has been proposed that can be used to build safe and secure applications for smart cities. The block diagram of a smart city is presented in Figure 17.6.

The proposed model divides the application of IoT into three major units: input unit, processing unit, and output unit. The input unit collects the data from a smartphone user using the graphical user interface (GUI) and embedded sensor. Security-1 receives the input from the input unit, converts it into a single string with a request key, and encodes the string using a cryptography algorithm. The encoded string interacts with the processing unit to perform a specific task. The processing unit decodes the string and fetches the required dataset from the centralized database according to the operation for performing a task, while the processing unit forwards the obtained result with the request key to Security-2 after the successful completion of performing a task. Security-2 encodes the result and forwards it to the output unit. The output unit receives the encoded result, and again decodes it and serves it as a result to the user's smartphone. Centralized may be a collection of different kinds of datasets that help in the processing phase to obtain a better result. The accuracy of IoT applications depends on the algorithm that is used to extract information from the dataset, and the dataset itself. For building a centralized dataset, the data can be gathered from smartphone sensors, websites, social media, repositories and warehouses, and transaction and moment record details of the user.

Many repositories are available that provide sample datasets for developing and validating the model. Some popular repositories are Kaggle,

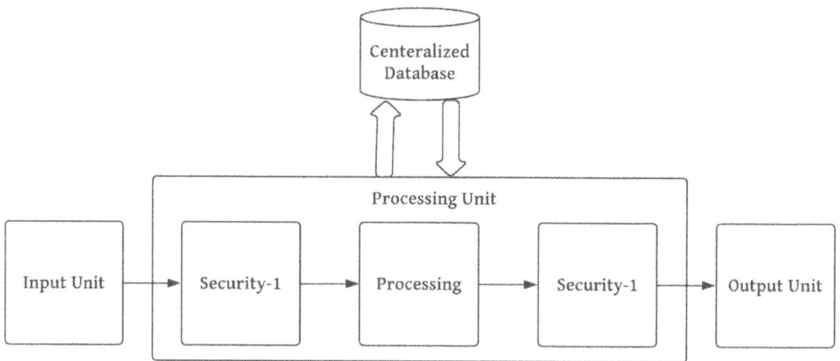

Figure 17.6 An Analytical Framework to Build Safe and Secure Applications for Smart Cities.

UCI Machine Learning Repository, Visual Data, The Big Bad NLP Database, etc. Security-1 and Security-2 used the cryptography algorithm to encode and decode the data and information (input and output). Cryptography is a method that protects data and information through programming codes. The term cryptography is a combination of two words that are "crypt" and "graph." The meaning of crypt is hidden and graph means "writing" hence the meaning of cryptography is hidden writing. Various cryptography algorithms are available that can be used in Security-1 and Security-2, but in this study, we will focus on DES, 3DES, and AES algorithms because these have been used by the US Government frequently. This algorithm is capable of protecting highly sensitive data. The proposed model is capable of verifying and validating the user using authentication that helps to provide information to the genuine user, the request key helps the model to verify and validate the user. Everyone cannot read the encoded data and information; only the authentic person can read the data and information. In the future, the proposed model will help develop safe and secure IoT-based smart cities and can effectively and efficiently be used for smart city applications such as traffic monitoring and intelligent transportation systems.

The current era is technology-oriented; several applications are being launched every day which claim that they are making human life easy. The citizens of the modern era have become lazy and are trying to take more and more benefits from different sources without their effort. Nothing is free in this world; everyone is trying to take advantage of others. Applications and services take the personal and professional information of the user, and they are utilizing it for their purpose or they are selling the user's information to other organizations for their benefit. This study covers various viewpoints on IoT-based applications for smart cities, including cyber security difficulties and concerns, as well as the technologies utilized to create real-time applications for smart cities. Modern technologies such as IoT, ICT, Cloud, etc. are playing an important role in developing real-time applications for smart cities, and the IoT is the backbone of smart city applications. IoT-based technologies such as RFID, NFC, LWPAN, Dash 7, etc. are frequently used in smart city applications. IoT-based applications are not so reliable because of lots of cyber security concerns and issues behind them. Cyber security concerns have been categorized into two types such as individual and technical concerns and cyber security issues include vulnerabilities, threats, and attacks.

17.8 CONCLUSION

Cyber security is the major concern of IoT-based applications for smart cities. The IoT-based applications run on the internet and various techniques and tools exist that provide the easiest way to access data from different servers. The hackers perform different kinds of activity to get control of the user or organization system, and they can destroy, and manipulate the data which are stored on the server. The user shares their personal and professional information on social media and other platforms to get attention from others or take benefits from applications. The shared data by the user can become the key for the hacker to unlock the user's system. Some organizations are collecting users' data and providing it to others for money and other benefits. Hence, there is an urgent need for a law that ensures that the users' information cannot be bought for money, the user information must be used to provide a good experience to the user. Technical equipment such as biometrics, smart grids, and other smart devices must be used to monitor unauthorized access. Many organizations are collecting information about users through applications and surveys, and they are selling users' information for their benefit. This information helps the hackers to collect confidential information of the user, hacker performs different activities such as phishing and other theft and attacks to gather confidential information. Users share their personal and professional information on social media and other platforms, Sometimes the data is very crucial and needs a high-security mechanism to secure user data. The collected data of the user are stored on the servers such as the web and cloud. The hackers try different activities to get access to the server in an unauthorized way to steal, manipulate, or destroy user, and organization information. We need a law that can ensure that the user information cannot be shared with others, and the user information will be used only to understand the user behaviors which can help to enhance the quality of applications. Also, we need a high-level security mechanism that includes the latest technologies and devices and can ensure that data that are traveling from one node to another are secured, and the server where all the confidential data is stored must be highly secure.

BIBLIOGRAPHY

Abomhara M., Køien G.M. (2015). Cyber security and the Internet of Things: Vulnerabilities, threats, intruders and attacks. Journal of Cyber Security, 4(1): 65–68.

Alibasic A., Al Junaibi R., Aung Z., Woon W.L., Omar M.A. (2016). Cybersecurity for smart cities: A brief review. International Workshop on Data Analytics for Renewable Energy Integration, 22–30.

Andrade R.O., Yoo S.G., Tello-Oquendo L., Ortiz-Garcés I. (2020).A Comprehensive study of the IoT cybersecurity in smart cities. IEEE Access, 8: 228922–228941.

Bibri S.E. (2018). The IoT for smart sustainable cities of the future: An analytical framework for sensor-based big data applications for environmental sustainability. Sustainable Cities and Society, 38, 230–253.

Elmaghraby A.S., Losavio M. (2014). Cyber security challenges in Smart Cities: Safety, security and privacy. Cairo University Journal of Advanced Research, 5(1): 491–497.

Ferreira C.M., Garrocho C.T., Oliveira R.A., Silva J.S., Cavalcanti C.F. (2021). IoT registration and authentication in smart city applications with blockchain. Sensors, 21(4): 1323.

González -Zamar M.D., Abad –Segura E., Vázquez-Cano E., López-Meneses E. (2020). IoT technology applications-based smart cities: Research analysis. MDPI Electronics, 9(8): 1–36.

Goyal K.K., Garg A., Rastogi A., Singhal S., (2018). A literature survey on Internet of Things (IoT). International Journal of Advanced Networking and Applications, 9(6): 3663–3668.

Hamid B., Jhanjhi N.Z., Humayun M., Khan A., Alsayat A. (2019). Cyber security issues and challenges for smart cities: A survey. 13th International Conference on Mathematics, Actuarial Science, Computer Science and Statistics (MACS), 1–7.

Huang C., Nazir S. (2021). Analyzing and evaluating smart cities for IoT based on use cases using the analytic network process. Hindawi Mobile Information Systems, 1–13.

Ismagilova E., Hughes L., Rana N.P., Dwivedi Y.K. (2020). Security, privacy and risks within smart cities: Literature review and development of a smart city interaction framework. Information Systems Frontiers, 1–22.

Kalinin M., Krundyshev V., Zegzhda P. (2021). Cybersecurity risk assessment in smart city infrastructures. MDPI Machines, 9(78): 1–19.

Khalil U., Malik O.A., Hussain S. (2022). A blockchain footprint for authentication of IoT-enabled smart devices in smart cities: State-of-the-art advancements, challenges and future research directions. IEEE Access, 10: 76805–768023.

Komninos N., Mora L. (2018). Exploring the big picture of smart city research. Scienze Regionali, 17(1): 33–56.

Ma C. (2021). Smart city and cyber-security; technologies used, leading challenges and future recommendations. Energy Reports, 1–14.

Neshenko N., Nader C., Bou-Harb E., Furht B. (2020).A survey of methods supporting cyber situational awareness in the context of smart cities. Journal of Big Data, 7(92): 1–41.

Qasem M.H., Al Mobaideen W. (2019). Heterogeneity in IoT-based smart city designs. International Journal of Interactive Mobile Technologies, 13(12): 210–225.

R.P. Janani, K. Renuka, A. Aruna, K.N. Lakshmi (2021). IoT in Smart Cities: A Contemporary Survey. Global Transitions Proceedings.

Righetti F., Vallati C., Anastasi G. (2018). IoT applications in smart cities: A perspective into social and ethical issues. IEEE International Conference on Smart Computing, 387–392.

Samih H. (2019).Smart cities and Internet of Things. Journal of Information Technology Case and Application Research, 21(1): 3–12.

Sharma A., Podoplelova E., Shapovalov G., Tselykh A., Tselykh A. (2021). Sustainable smart cities: Convergence of artificial intelligence and blockchain. Sustainability, 13(23): 13076.

Sujataa J., Sakshamb S., Tanvic G., Shreyad. (2016). Developing smart cities: An integrated framework. 6th International Conference on Advances on Computing & Communications, ICACC 2016, 6–8 September 2016, Cochin, India, 902–909.

Szabo Z. (2019). The effect of Globalization and cyber security on smart cities. Interdisciplinary Description of Complex Systems, 17(3): 503–510.

Talari S., Shafie-khah M., Siano P., Loia V., Tommasetti A., Catalao J. P. S. (2017). A review of smart cities based on the Internet of Things concept. MDPI Energies, 10(421): 1–23.

Wang J., Zhu J., Zhang M., Alam I., Biswas S. (2022).Function virtualization can play a Great Role in Blockchain Consensus. IEEE Access, 18(10): 59862–59877.

Zheng J., Lee M.J. (2004). Low rate wireless personal area networks for public security. IEEE 60th Vehicular Technology Conference, 2004. VTC2004-Fall. 2004, 4568–4572.

Chapter 18

Securing information systems through quantum computing

Grover's algorithm approach

Mohd Nadeem, Amal Krishna Sarkar, and Mohammed Ishrat

18.1 INTRODUCTION

The life cycle of the software comprises several phases, such as assistance for needed design, planning, coding, testing, and troubleshooting. Upkeep is seen as the final step of progress [1]. Software functionality can be identified as the circumstances under which programming is still useful or feasible. To achieve viability, the utility of the programme should be strong [2]. Sturdiness in software is determined by the timeframe that the administration provides for programming. In the mid-21st century, the increasing adaptive climate created new problems for everyone, even for products [3, 4]. In the age of quantum computing, quantum safety is the focus. The pace of quantum technology development is exponential right now. The complete quantum processor, the Sycamore Processor, which designs a quantum circuit in 200 seconds, has been successfully built by a group of scientists who developed it in 10,000 years using a classical supercomputer [5]. The present encryption and security structure of many networks, online applications, software and software, financial encryption structures, and security in defence and everything related to computer networks are involved in the creation of the quantum processor. In traditional computers, the current security techniques are symmetrical and asymmetrical. The same key is utilised for the encryption and decryption of data in a symmetrical method. Different keys are utilised throughout the asymmetric method. The security is completely reliant on an integer number as defined in the Shor algorithm. A security key is used. The large number of bits of size 2042 can be factorised into its prime number [6]. The total time taken by the classical computer can be factorised by 100 years. However, by the use of quantum phenomenon, principally, it can be broken down within a few minutes. Grover's algorithm's major application is used for the unstructured search problem. Grover's algorithm is fast and works in a quadratic way and also enhances the run time for a variety of other algorithms [7]. The application of Grover's algorithm is beyond the search application. In the quantum computing phenomenon, the algorithm search phenomenon can

DOI: 10.1201/9781003514312-18

be understood by the procedure of the search problem. Let us consider a set of items that have the same colour except one. The number of items is N, in a classical way $\dfrac{N}{2}$ steps are as follows and the different item can be identified or searched; in worst case, all the N steps follows but in Grover's algorithm $\sqrt{N}$ steps are followed and the item is selected. This phenomenon provides the quadratic quantum speed up for the large classical problems. This is called the amplitude amplification method.

18.2 RELATED WORK

More research has been carried out expressly on quantum computing and software safety, and the combined methods should be novel, as indicated in our review paper. During the previous decade, scientists concentrated on the development of quantum computers and the various quantity computer algorithms that increase computing phenomena. Complicated cryptographic methods may be solved in seconds with quantum computing. In the quantum computing age, the existing symmetric safety approach of classical computers and supersymmetric security procedures has collapsed. The problems and challenges of software security during the quantum era were notably highlighted in our research. The following are extensive investigations into the endurance of software and quantum computing:

According to S. Mitra et al. [8] quantum cryptography is the most promising cryptographic area as it has the potential to provide for faster, more powerful, and more secure correspondences. Cryptographic endeavour relies on quantum characteristics of light covered by quantum physics to surpass computation innovations using numerical figuring. A number of well-known encryption methods, such as Rivest, Shamir, Adleman (an asymmetric cryptography algorithm), El Gamal cryptosystem, and hash functions, are vulnerable to quantum attacks. In Shor's calculation and quantum PC compromises, a big issue may be split into several pieces.

The Internet of Things (IoT) is mentioned by M. Abomhara et al. [9]. IoT organisations will become inevitable, as will contraptions. Advanced attacks are no longer new to the IoT. Nonetheless, as the IoT is extensively affected by our lives and social systems, computerised shielding is becoming progressively more necessary. Thus, IoT must be established, and thus the dangers and attacks on IoT structures must be well understood.

X. Ma et al. [10] discuss how the quantum key movement allows for the licensing of previously inaccessible social issues to provide theoretically safe keys for information. The researchers created a security proof based on optical modes that are not similar to traditional qubit-based security affirmations. Additionally, the suggested layout is devoid of assessment devices; that is, it is immune to all recognised dangers.

It's important to keep administrative duties separate from operating roles, as Rocha et al. have noted, in order to solve any organisation's problems of access, cost, and control [11]. In their design for an organisational layer of security, the authors drew on their previous implantation of a symmetrical technique for easy connectivity and hubs called "spine." This device was meant to do a lot for overall security, handling, and navigation.

According to Song et al., survival is the primary determinant of information reclamation and assurance amid organisational disappointment. The study focused on a network association of an organisation's geography [12]. Here the diverse accessibility prerequisites and assurance of present status depend on various time settings. This design gives different degrees of administration assurance, execution, and adaptable improvement.

Alenezi et al. used the Fuzzy AHP technique on security factors to ensure the usable security factor's heaviness in programming security [13]. The paper discusses the product's realistic security factors, and the Fuzzy AHP approach determines the component's individual load. Additionally, our analysis research used a symmetrical approach based on Fuzzy AHP for assessing and examining the influence on software's security durability.

Agrawal et al. used the multi-standards dynamic methodology of fuzzy analytical hierarchy process (FAHP) for the assessment of usable security in programming. Security is an essential part of software [14]. The improvement of security and ease of use in programming are of foremost significance. To guarantee the precise evaluation of both, the creators chose the philosophy of FAHP and applied it for the assurance of the components utilised in the product to improve the security of the designer.

Kumar et al. worked on the concept that the convenience and security of the product are a trade-off between the two components of security and durability [15]. The developers selected the elements of ease of use and security according to the necessity of the product. This suggests that quantum key transmission is capable of appreciating both common sense and security.

Zheng et al. [16] stated that quantum correspondence provides full safety leeway and has evolved widely in recent years. As a major element of quantum correspondence, quantitatively secure direct correspondence provides excellent security and speed of communication over a quantum channel. Full use of a quantum convention needs the ability to effectively regulate the transmission of a message in the time area; it is thus essential that the correspondence job is carried out by combining the quantum secure correspondence with the quantum memory [17]. They reported in this letter the test demonstration of a secure quantic match with cutting-edge nuclear quantum storage without precedent for guidance. They made use of the level of polarisation of the photons as data transporters and assessed the consistency of trap degradation. Our study completes a fundamental step towards the safe direct correspondence of common sense quantum and shows a probable use for substantial distance quantum correspondence in a quantum organisation.

The aforementioned symmetric methods provide details on software safety throughout the quantum computing age. We discuss different software safety and quantity computing ages in the next section.

18.3 METHODOLOGY

Let us consider a uniform superposition $|s\rangle = \dfrac{1}{\sqrt{N}} \sum_{x=0}^{N-1} |x\rangle$; this superposition would collapse according to the fifth law of quantum [1]. The right value of colour could be 1 to 2^n. We need to try $\dfrac{N}{2} = 2^{n-1}$ times try to identify the colour. The identification procedure of different colours of the quantum computer amplification phenomenon is applied here, enhancing the probability. The procedure amplifies the different item colours and shrinks the other so that the metric of final state will be to return the right item with near certainty.

Grover's algorithm has geometrical interpretation in terms of two reflections, that can generate rotation in a two-dimensional (2D) plane [18]. The two states are different colour items (which are indicated by $|w\rangle$) and uniform superposition $|s\rangle$. These two vectors span a 2D plane in the vector space [19]. They are not perpendicular, so we consider the third vector $|s'\rangle$ perpendicular to two vector spans. The s'' is obtained from "w" and "s" by removing and scaling.

Grover's algorithm has the following steps:

> *Step 1:* The amplification procedure starts from the uniform superposition spanned by perpendicular vectors 'w' and 's'', which express the initial state as:

$$|s\rangle = \sin\theta|w\rangle + \cos\theta|s'\rangle \text{ where } \theta = arcsin\langle s|w\rangle = arcsin\dfrac{1}{\sqrt{N}}$$

> *Step 2:* In step 2 of Grover's algorithm, we apply the oracle reflection [20] to the $|s\rangle$. This geometrically corresponds to the reflection of the state $|s\rangle$ to $|s'\rangle$. This transformation makes the different colour item $|w\rangle$ negative, which means the average amplitude will change.
> *Step 3:* Now we apply the next transformation of the state with respect to the $|s\rangle$. This transformation maps up the reflection and highlights the change in colour.

Two reflections always correspond to the rotation [21]. The transformation rotates the initial state $|s\rangle$ and moves towards $|w\rangle$. This transformation boosts the negative amplitude of $|w\rangle$ three times to its original value. This procedure will be repeated until the result will arrive. This transforming rotation occurs $\sqrt{N}$ times.

18.4 OPERATION OF QUBITS

Quantum computing differs from the classical computing because of its basic operation which is based on Q algebra (Quantum algebra). The qubits (φ) are defined in the mathematical expression as:

$$|\varphi\rangle = \alpha|0\rangle + \beta|\varphi\rangle \tag{18.1}$$

The bracket notation $|0\rangle = \begin{pmatrix} 1 \\ 0 \end{pmatrix}$, $|1\rangle = \begin{pmatrix} 0 \\ 1 \end{pmatrix}$ and α, β are the complex numbers then

$$|\alpha|^2 + |\beta|^2 = 1 \tag{18.2}$$

This shows that $|\alpha|^2 = 0$ then $|\beta|^2 = 1$ or $|\alpha|^2 = 1$ and then $|\beta|^2 = 0$. Let us consider U as unitary matrix and U' its transpose:

$$UU' = U'U = I \tag{18.3}$$

I, is the identity matrix.

$$|\varphi = U|\varphi = \begin{pmatrix} U_{00} & U_{01} \\ U_{10} & U_{11} \end{pmatrix}\begin{pmatrix} \alpha \\ \beta \end{pmatrix}$$

$$= \begin{pmatrix} U_{00}\alpha + U_{01}\beta \\ U_{10}\alpha + U_{11}\beta \end{pmatrix}$$

The state of qubits is defined as a tensor product $\otimes$ (it is the multiplication of two vector spaces). This is defined as:

$$|\gamma_j\rangle = \alpha_j|0\rangle + \beta_j|1\rangle, \text{ for } j = 1, 2, 3.$$

The tensor product of the joint state is

$$| \gamma_1 \gamma_2 \gamma_3 \rangle = \gamma_1 \otimes \gamma_2 \otimes \gamma_3 \tag{18.4}$$

$$
\begin{aligned}
&= \alpha_1 \alpha_2 \alpha_3 \,|000\rangle + \alpha_1 \alpha_2 \beta_3 \,|001\rangle + \alpha_1 \beta_2 \alpha_3 \,|010\rangle \\
&\quad + \alpha_1 \beta_2 \beta_3 \,|011\rangle + \beta_1 \alpha_2 \alpha_3 \,|100\rangle + \beta_1 \alpha_2 \beta_3 \,|101\rangle \\
&\quad + \beta_1 \beta_2 \alpha_3 \,|110\rangle + \beta_1 \beta_2 \beta_3 \,|111\rangle
\end{aligned}
\tag{18.5}
$$

Equation 18.5 shows that the tensor product of three inputs will result in eight outputs, it gives the all possible outcome of the input data. Qubits can mathematically be represented by a unitary function, so the number of inputs is equal to the number of output qubits.

18.5 CONCLUSION

Software security is a major concern in the field of quantum computing. The rapid progress made in quantum technology will require inventive and highly efficacious approaches in the cyber security domain. The encryption and decryption methods can easily be matched by the qubit combination at the same time. The cryptographic algorithm used in cybersecurity will be ineffective with the advent of quantum computers; the researchers need to work on converse solutions to this problem. The lattice-based cryptographic algorithm does not have a standard library of encryption algorithms and is, yet, years away from commercial software. The production of a fully secure frame is beyond comprehension; as a result, software robustness cannot be regarded as the limit of large and safe software. The aim is therefore to reduce the support problem for functioning software over lengthy periods of time. The quantum method and longevity of the programme are mentioned in this article. The quantum-safe method guarantees a quantum attack on the current security approach to encryption. However, no assurance for perfect safety in the context of online communication can be made either by the lattice quantic algorithm or by any other means.

REFERENCES

1. McGraw, G. (2004). Software security. IEEE Security & Privacy, 2(2), 80–83.
2. Arizon-Peretz, R., Hadar, I., & Luria, G. (2021). The importance of security is in the eye of the beholder: Cultural, organizational, and personal factors affecting the implementation of security by design. IEEE Transactions on Software Engineering, 48(11), 4433–4446.

3. L. K. Grover (1996), "A fast quantum mechanical algorithm for database search", Proceedings of the 28th Annual ACM Symposium on the Theory of Computing (STOC 1996). doi:10.1145/237814.237866, arXiv:quant-ph/9605043

4. C. Figgatt, D. Maslov, K. A. Landsman, N. M. Linke, S. Debnath & C. Monroe (2017), "Complete 3-Qubit Grover search on a programmable quantum computer", Nature Communications, vol. 8, Art 1918. doi:10.1038/s41467-017-01904-7, arXiv:1703.10535

5. Arute, F., Arya, K., Babbush, R. *et al.* Quantum supremacy using a programmable superconducting processor. Nature 574, 505–510 (2019). https://doi.org/10.1038/s41586-019-1666-5

6. Shor, Peter W. "Polynomial-Time Algorithms for Prime Factorization and Discrete Logarithms on a Quantum Computer." SIAM Journal on Computing 26.5 (1997): 1484–1509. Crossref.Web.

7. Ladd, T., Jelezko, F., Laflamme, R. *et al.* Quantum computers. Nature 464, 45–53 (2010). https://doi.org/10.1038/nature08812

8. S. Mitra, B. Jana, S. Bhattacharya, P. Pal, and J. Poray, "Quantum cryptography: Overview, security issues and future challenges," 2017 4th Int. Conf. Opto-Electronics Appl. Opt. Optronix 2017, vol. 2018-Janua, pp. 1–7, Apr. 2018. doi: 10.1109/OPTRONIX.2017.8350006.

9. M. Abomhara and G. M. Køien, "Cyber Security and the Internet of Things: Vulnerabilities, Threats, Intruders and Attacks," J. Cyber Secur. Mobil., vol. 4, no. 1, pp. 65–88, Jan. 2015. doi: 10.13052/JCSM2245-1439.414.

10. X. Ma, P. Zeng, and H. Zhou, "Phase-Matching Quantum Key Distribution," Phys. Rev. X, vol. 8, no. 3, p. 031043, Aug. 2018. doi: 10.1103/PHYSREVX.8.031043/FIGURES/10/MEDIUM.

11. A. F. Rocha, E. Massad, and F. A. B. Coutinho, "Can the human brain do quantum computing?," Med. Hypotheses, vol. 63, no. 5, pp. 895–899, 2004. doi: https://doi.org/10.1016/j.mehy.2004.03.044.

12. J. L. Hsu, S. K. Chong, T. Hwang, and C. W. Tsai, "Dynamic quantum secret sharing," Quantum Inf. Process., vol. 12, no. 1, pp. 331–344, Jan. 2013. doi: 10.1007/S11128-012-0380-0.

13. M. Alenezi, M. Nadeem, A. Agrawal, R. Kumar, and R. A. Khan, "Fuzzy multi criteria decision analysis method for assessing security design tactics for web applications," Int. J. Intell. Eng. Syst., vol. 13, no. 5, 2020. doi: 10.22266/ijies2020.1031.17.

14. A. Agrawal et al., "Software Security Estimation Using the Hybrid Fuzzy ANP-TOPSIS Approach: Design Tactics Perspective," Symmetry 2020, vol. 12, no. 4, p. 598, Apr. 2020. doi: 10.3390/SYM12040598.

15. R. Kumar, M. Zarour, M. Alenezi, A. Agrawal, and R. A. Khan, "Measuring Security Durability of Software through Fuzzy-Based Decision-Making Process," Int. J. Comput. Intell. Syst., vol. 12, no. 2, pp. 627–642, May 2019. doi: 10.2991/IJCIS.D.190513.001.

16. Z. Wenhua, F. Qamar, T.-A. N. Abdali, R. Hassan, S. T. A. Jafri, and Q. N. Nguyen, "Blockchain Technology: Security Issues, Healthcare Applications, Challenges and Future Trends," Electron. 2023, vol. 12, no. 3, p. 546, Jan. 2023. doi: 10.3390/ELECTRONICS12030546.

17. H. Alyami et al., "The evaluation of software security through quantum computing techniques: A durability perspective," Appl. Sci., vol. 11, no. 24, 2021. doi: 10.3390/app112411784.
18. Preston, R. H. (2022). Applying Grover's algorithm to hash functions: a software perspective. IEEE Transactions on Quantum Engineering, 3, 1–10.
19. Gilliam, A., Pistoia, M., & Gonciulea, C. (2020). Optimizing quantum search using a generalized version of Grover's algorithm. arXiv preprint arXiv:2005.06468.
20. Gong, C., Du, J., Dong, Z., Guo, Z., Gani, A., Zhao, L., & Qi, H. (2020). Grover algorithm-based quantum homomorphic encryption ciphertext retrieval scheme in quantum cloud computing. Quantum Information Processing, 19, 1–17.
21. I. Chuang & M. Nielsen, "Quantum Computation and Quantum Information", Cambridge: Cambridge University Press, 2000.

Index

For Product Safety Concerns and Information please contact our EU
representative GPSR@taylorandfrancis.com
Taylor & Francis Verlag GmbH, Kaufingerstraße 24, 80331 München, Germany